THE UNITED NATIONS' DECLARATION ON PEASANTS' RIGHTS

This is the first book to address and review the United Nations' Declaration on the Rights of Peasants and Other People Working in Rural Areas (UNDROP), which was adopted by the United Nations General Assembly in December 2018.

Food security and sustainable agri-food systems, responsible governance of natural resources, and human rights are among the key themes of the new millennium. The Declaration is the first internationally negotiated instrument bridging these issues, calling for a radical paradigm change in the agricultural sector while giving voice to peasants and rural workers, recognised as the drivers of more equitable and resilient food systems. The book unfolds the impact of the Declaration in the wider realm of law and policy making, especially concerning the new human rights standards related to access and control of natural resources and the governance of food systems. The chapters in the book touch on a broad array of topics, including women's rights, the role of and impact on indigenous peoples, food sovereignty, climate change, land tenure, and agrobiodiversity. Voices from outstanding scholars and practitioners are gathered together to inform and trigger a further debate on the negotiation process, the innovative and potentially disruptive contents, the relations with other fields of law, and the practical scope of the Declaration. The volume concludes with a collection of case studies that provide concrete examples to help us understand the potential impacts of the Declaration at regional, national, and local levels.

This book is the first comprehensive tool to navigate the Declaration and is designed for students, researchers, and practitioners in the fields of food and agriculture law, peasant, agrarian and rural studies, human rights and environmental law, and international development and cooperation.

Mariagrazia Alabrese is Associate Professor at the Institute of Law, Politics and Development at the Sant'Anna School of Advanced Studies, Italy.

Adriana Bessa is a Brazilian lawyer, independent researcher, and guest lecturer at the Université Catholique de Lille and the Universidad de Alicante.

Margherita Brunori is Postdoctoral Research Fellow in the Department of Sociology and Social Science at the University of Trento and in the Department of Law at the University of Brescia, Italy.

Pier Filippo Giuggioli is Associate Professor in the Department of Italian and Supranational Public Law at the University of Milan, Italy.

Other books in the Earthscan Food and Agriculture Series

Deep Agroecology and the Homeric Epics
Global Cultural Reforms for a Natural-Systems Agriculture
John W. Head

Fighting for Farming Justice
Diversity, Food Access and the USDA
Terri R. Jett

Political Ecology of Industrial Crops
Edited by Abubakari Ahmed and Alexandros Gasparatos

The Sociology of Food and Agriculture
3rd Edition
Michael Carolan

The Politics of Food Provisioning in Colombia
Agrarian Movements and Negotiations with the State
Felipe Roa-Clavijo

The Governance of Agriculture in Post-Brexit UK
Edited by Irene Antonopoulos, Matt Bell, Aleksandra Čavoški and Ludivine Petetin

The United Nations' Declaration on Peasants' Rights
Edited by Mariagrazia Alabrese, Adriana Bessa, Margherita Brunori, and Pier Filippo Giuggioli

For more information about this series, please visit: www.routledge.com/books/series/ECEFA/

THE UNITED NATIONS' DECLARATION ON PEASANTS' RIGHTS

Edited by Mariagrazia Alabrese, Adriana Bessa, Margherita Brunori, and Pier Filippo Giuggioli

Routledge
Taylor & Francis Group
LONDON AND NEW YORK

earthscan
from Routledge

Cover image: © Getty Images

First published 2022
by Routledge
2 Park Square, Milton Park, Abingdon, Oxon OX14 4RN

and by Routledge
605 Third Avenue, New York, NY 10158

Routledge is an imprint of the Taylor & Francis Group, an informa business

British Library Cataloguing-in-Publication Data
A catalogue record for this book is available from the British Library

Library of Congress Cataloging-in-Publication Data
A catalog record for this book has been requested

ISBN: 978-0-367-68981-0 (hbk)
ISBN: 978-0-367-68977-3 (pbk)
ISBN: 978-1-003-13987-4 (ebk)

DOI: 10.4324/9781003139874

Typeset in Bembo
by Apex CoVantage, LLC

CONTENTS

ACKNOWLEDGEMENTS

As editors, we would like to express our deepest gratitude to the contributors. It was a privilege to work with each one of them. We are grateful to Arianna Berti and Luca Carriero, undergraduate students of 'Sciences for Peace: International Cooperation and Transformation of Conflicts' at the University of Pisa, for invaluable editing assistance in the preparation of this book. Our thanks also go to John Baddeley, editorial assistant, and others from Routledge publisher for their help in publishing this book.

CONTRIBUTORS

Priscilla Claeys is Associate Professor, Centre for Agroecology, Water and Resilience, Coventry University, UK.

Samuel Cogolati is PhD in international law (KU Leuven) and MP in the House of Representatives, Belgium.

Lorenzo Cotula is a principal researcher (law and sustainable development) at the International Institute for Environment and Development and a visiting professor at Strathclyde Law School.

Eloisa Cristiani is Associate Professor of Agricultural Law, Institute of Law, Politics and Development, Sant'Anna School of Advanced Studies, Pisa, Italy.

Marc Edelman is Professor of Anthropology, Hunter College and the Graduate Center, City University of New York, US.

Hilal Elver is Professor of International Law and Global Distinguished Fellow, Resnick Food Law and Policy Center, UCLA Law School US; former UN Special Rapporteur on the Right to Food (2014–2020).

Francesco Francioni is Professor Emeritus of International Law, European University Institute, and Professor of International Cultural Heritage Law, LUISS University, Rome.

Christine Frison is a postdoctoral research fellow (Fund for Scientific Research F.R.S.–FNRS) at the Institute for Interdisciplinary Research in Legal Sciences of the UCLouvain (Belgium), a member of the Louvain Partnership Research on Ecological and

Social Transition (LPTransition) and an associate research fellow at the Center for International Sustainable Development Law (CISDL, Montreal, Canada).

Jérémie Gilbert is Professor of Human Rights Law, Roehampton Law School, UK.

Christophe Golay is Senior Research Fellow and Strategic Adviser on Economic, Social and Cultural Rights, Geneva Academy of International Humanitarian Law and Human Rights, Switzerland.

Tadesse Kebebew is Teaching Assistant at the Geneva Academy and PhD in International Law, the Graduate Institute of International and Development Studies, Switzerland.

Robert Kibugi is Senior Lecturer in Law, Faculty of Law, University of Nairobi, Kenya.

Alessandra Di Lauro is Full Professor of Agricultural law, Department of Agriculture, Food and Environment, University of Pisa, Italy.

Joanna Bourke Martignoni is Senior Research Fellow, Geneva Academy of International Humanitarian Law and Human Rights, Switzerland.

Miguel Ángel Martín López is Professor of International Public law and International Relations, University of Sevilla, Spain.

Elisa Morgera is Director of the UKRI GCRF One Ocean Hub (OOH), University of Strathclyde Law School, and Professor of Global Environmental Law at Strathclyde Centre for Environmental Law and Governance (SCELG), University of Strathclyde Law School, UK.

Julia Nakamura is PhD candidate at SCELG, University of Strathclyde Law School.

Smita Narula is Haub Distinguished Professor of International Law, Elisabeth Haub School of Law, Pace University, US.

Naiara Posenato is Aggregate Professor in Private Comparative Law and Commercial Comparative Law, University of Milan, Italy.

Francesca Restifo is Senior Human Rights Lawyer, International Bar Association's Human Rights Institute, Switzerland.

Annalisa Savaresi is Associate Professor of International Environmental Law, Center for Climate Change, Energy and Environmental Law, School of Law, University of Eastern Finland.

Giuliana Strambi is Researcher, Institute for International Legal Studies, CNR – National Research Council of Italy, Italy.

Sandrine Le Teno holds an Advanced LLM in International Law from the Université Libre de Bruxelles, Belgium.

Mara Tignino is Reader at the Faculty of Law and the Institute for Environmental Sciences at the University of Geneva and Lead Legal Specialist of the Platform for International Water Law at the Geneva Water Hub, Switzerland.

Elsa Tsioumani is Marie Skłodowska-Curie Fellow, School of International Studies, University of Trento, Italy.

FOREWORD

During the last days of the year 2018, the long-awaited Declaration on the Rights of Peasants and other People Working in Rural Areas (UNDROP) was adopted by the member states of the United Nations. It is the Declaration that gives a voice to the voiceless majority. The UNDROP protects the rights of one-third of the world's population, which includes peasants, artisanal or small-scale farmers, either for subsistence and/or for the markets, fishing communities, indigenous peoples, rural workers, pastoralists, and landless agricultural workers. Rights set forth in the Declaration could be applied to individuals, as well as communities.

After more than two decades of struggle at every stage under the leadership of rural people's global network of La Via Campesina, FIAN, CETIM, and many small organisations, the finish line has been crossed. This is a dramatic example of direct involvement of social movements in the process of making international law from below. Therefore, it should be celebrated enthusiastically, as international law is generally considered to involve only top-down inter-governmental lawmaking. Rural communities and peasants' organisations from Indonesia to Ethiopia, from Asia to Africa to Latin America worked tirelessly and cooperated with strategic institutions, global organisations, and important decision makers, one by one, tirelessly explaining what this Declaration is about, and above all, why it is so important that rights of rural communities should be specifically protected at this historical moment.

This book is an essential reading to understand the historical struggle of the peasants, and other people are living in rural areas. The authors of the each chapter represent different constituencies, such as academics, members of the research centres and civil society, as well as international organisations. Each touched upon UNDROP's different sections. Most of the authors have been participating in the preparation and drafting of the Declaration for a long time. Therefore, this book gives a reader a complete understanding about present and future challenges of lives of peasants and long struggle of rural communities to keep their lifestyle, while protecting fragile rural ecosystem. Therefore, it seeks to nurture a mind changing in society and socio-economic dynamics from dominant free market understanding.

From its name to the content of the Declaration there have been too many complaints, criticisms, and reluctances. The UNDROP was considered too controversial by some Western governments claiming that it goes beyond the existing human rights principles. Not surprisingly, the industrial agriculture sector, big developers, and its promoters have seen the UNDROP as a dangerous precedent. Particular concerns were raised about the articulation of "collective rights" for rural workers, arguing that the Declaration "seeks to set out new rights for rural workers that others do not have."[1] Yet despite the clamour, the Declaration was accepted by majority of the members of the UN General Assembly.

Urban values and industrialisation have steadily gained power over rural communities since the inception of Modernity some centuries ago, and rural communities have been harmed in a variety of ways. The 1948 Universal Declaration of Human Rights and the two human rights covenants, especially the Covenant on Economic, Social and Cultural rights, were not directly responsive to the needs of peasants and rural communities, which became more apparent after a half century of practice.

The UNDROP is the most comprehensive human rights document to address neglected rural concerns. It skilfully combines existing human rights principles and interprets them from the perspective of peasants and rural communities, prioritising their relationship with nature, and articulating the daily struggle of peasants and related groups for surviving and standing against powerful players. It puts together many of the principles and rules of the international human rights law documents relevant to the livelihoods of rural communities and peoples, including rural women's and children's rights. It also incorporates human rights standards contained in the voluntary guidelines adopted by the UN Food and Agriculture Organization (FAO) and the World Committee of Food Security (CFS). In this sense, the Declaration sheds important light on norms and principles that are directly relevant to peasants, such as labour rights, environmental rights, rights over biological diversity, access to seeds, land and water, social security and food sovereignty.

The food sovereignty movement is becoming a highly influential global peasants movement. It strongly supported the efforts that eventuated in the Declaration. Food sovereignty gives power to peasants to make their own decisions about the way in which, what, how, and when to produce food and feed themselves. It extends the reach of food sovereignty to communal and collective rights, as well as individual rights, and strongly supports right to food.

One of the contentious rights in the UNDROP is the right to save, exchange, and sell seeds. The ownership of seeds is very important for food sovereignty and the protection of biodiversity. Without such ownership, the use, reuse, grow, store, develop, exchange, donation, or sale of seeds is simply not possible. This dimension of food sovereignty is now under serious threat due to the concerted efforts of the oligopoly of giant agricultural companies and powerful states under the banner of patent and copyrights protection. Therefore, it is vitally important to support peasant seed banks and conservation projects against market-driven efforts to substitute genetically engineered industrial seeds.

Agro-business is penetrating entire food chains not only in relation to seeds but also with respect to land, water, chemicals, as well as transportation and supermarket chains. They also exert undue influence over national and global regulatory institutions and scientific research. The courts are essential in providing adversely affected people with a potential means to resist the harm being done by such conglomerates. The UNDROP provides an additional legal tool to use in judicial systems against powerful agro-business lobbyists.

The UNDROP also recognises the role of peasants and traditional knowledge in helping to reduce climate change and protect biodiversity. The negative impact of climate change on agricultural production is enormous and likely to increase. At the same time, the El Nino years have devastating impacts on rural communities in Africa, South Asia, and Latin America. Many of the countries in Sub-Saharan Africa face renewed threats of famine. There is an irony at present. Many thought that the phenomenon of famine was behind us in the aftermath of the Second World War when confidence in the industrial wonders of modernity was at its height. We are currently witnessing the terrible impact of the hurricanes and wildfires in various parts of the world. Many of the worst victims of such weather events are always already vulnerable, poor and food insecure, mostly women and children.

In 2020, a new kind of destruction has arrived, the Covid-19 pandemic, and abruptly transformed our lives, especially those of agricultural workers who in good and normal time are barely earning enough to meet their daily needs. Covid-19 showed us that food and agricultural workers are the most essential but at the same time the most vulnerable part of the long food chain, and without them, food will not come to our table. We have seen the disastrous effects of Covid-19 on industrial meat-packing plants and farmworker communities everywhere. These workers are mostly undocumented immigrants, who left their villages for distant lands to send what they earned by working abroad to their villages. They are rarely protected by regulatory policies, and they are also left on their own and exploited by the big profit-making companies. During the pandemic they were not able to work because of quarantine restrictions. Therefore, there was no earnings to send back to their families living in their home villages. As a result, poverty and severe food insecurity among peasants and rural communities have skyrocketed in this period.

Against this background, there is nothing more timely and urgent than to acknowledge peasants' human rights and promote the UNDROP. The struggle is continuing. It is now time to urge national legislators to implement the UNDROP at regional, national, and local levels. This would give a new tool to communities in their various fights against powerful human rights violators.

The long struggle to bring the Declaration into being has always been supported by the United Nations Special Rapporteurs on the right to food, starting with Jean Ziegler. During my mandate period of six years, the Draft Declaration was a very helpful tool to initiate cordial relationships with various peasant organisations in many countries, and learn about the daily ordeals of rural communities against big money, land grabbing, eviction, losing seed and water rights, and insecure land tenure struggles. I found the courage, enthusiasm, and hopes of these peasant advocates for the future to be inspirational. Therefore, in any global platform on food systems, meaningful participation in peasants, indigenous peoples, and other rural peoples are essential.

The challenge facing all of us who seek to vindicate the struggle of peasants and rural communities is to make the UNDROP more than words of aspiration, and this will require concerted political action.

Hilal Elver

Note

1 https://news.mongabay.com/2019/02/un-passes-first-ever-declaration-for-peasant-rights/

INTRODUCTION

Mariagrazia Alabrese, Adriana Bessa, Margherita Brunori, and Pier Filippo Giuggioli

The world food crisis, the 'land grabbing' phenomenon, and the climate change emergency have ignited the debate on the lack of adequate international protection for small-scale and subsistence farmers and have multiplied claims and proposals for international human rights–based principles for food and agriculture and natural resources governance.

In December 2018, after heated debates and numerous speculations, and five years of negotiations at the Human Rights Council, the UN General Assembly (UNGA) adopted the UN Declaration on the Rights of Peasants and Other People Working in Rural Areas (UNDROP) with 121 votes in favour, 8 against, and 54 abstentions.[1]

The Declaration is the first international human rights instrument to recognise a catalogue of specific rights to small-scale and subsistence farmers, fishers, pastoralists, forest dwellers, and rural workers. This socio-economic (and, as argued in this book, cultural) group has been subject to historical and systemic discrimination in laws, public policies, and national development strategies, and disproportionately bears the burden of the expansion of industrialised large-scale agriculture and the rapid deterioration of the environment and the global climate. Despite their key contribution to global food security and biodiversity preservation, as recognised in several international documents, peasants continue to be major victims of human rights violations worldwide.

The UNDROP is, at the same time, innovative and controversial. First, it originates from a political document conceived by the leading small-scale farmers' association La Via Campesina and was developed in close collaboration with agrarian and human rights organisations. Second, it not only provides for a peasant-specific interpretation of classic human rights but also expands the catalogue by including a number of new rights, such as the right to land, seeds, and biodiversity, by affirming their individual and collective dimensions. Third, it raises sensitive questions, potentially challenging certain consolidated international norms.

DOI: 10.4324/9781003139874-1

The UNDROP is based on three main assumptions: peasants have a special relationship with the land, water, and natural resources on which they depend for their livelihoods; they provide a unique contribution to economic development, biodiversity conservation, and improvement as well as the realisation of the right to food and food security at local, national, and global levels; they suffer disproportionately from poverty, hunger, and malnutrition and from the consequences of environmental degradation.

In light of the diverse legal domains the document crosscuts, the UNDROP encourages States to take an active role in its implementation and to coordinate and harmonise its standards with other established and emerging international law instruments. This is especially the case for the accommodation of peasants' rights with the rights of indigenous peoples, international trade agreements, and existing codes of conduct for businesses and other non-state actors. The document also provides scope for a reinterpretation of natural resources governance through a human rights-based approach.

The Declaration provides international recognition and collective voice to social groups that have been largely neglected in international law and policy. At the same time, it officially introduces the concept of food sovereignty, defined as the right of peasants to choose what to produce and to determine their food systems, into the United Nations vocabulary.

The volume is divided into four parts, each of them illustrative of a specific aspect of the Declaration: the definition of rights holders, its core rights on access to, and control of natural resources, its potential impact on the international governance of agri-food systems and natural resources, and its potential influence on the ground, with thematic case studies.

An introductory chapter positions this new human rights instrument within the public international law frame and addresses fundamental questions regarding the nature of States' obligations and aspects of justiciability. It discusses the extent to which this instrument, being a declaration and therefore not formally binding, imposes legal obligations on United Nations Member States (Francesco Francioni).

Part I looks at rights holders. It delves into and critically elaborates upon the inclusion of specific social groups and categories under the scope of the Declaration and its impact on international law. It starts with the definition of peasants as proposed in the text of the UNDROP and debates on its socio-legal implications (Marc Edelman). It then acknowledges the inclusion of traditional local communities and indigenous peoples among rights holders. In this sense, it examines whether and to what extent the UNDROP differs from, competes, and concurs with the Convention on Biological Diversity and Protocols and the United Nations Declaration on the Rights of Indigenous Peoples (Adriana Bessa and Jérémie Gilbert). Part I also reflects on the specific provisions on gender equality and the rights of rural women and argues that the final version of the Declaration does not reflect specific feminist claims made during the negotiations. For the authors, reinterpretation is needed to acknowledge feminist and intersectional standpoints that transcend the confines of existing international gender equality law (Joanna Bourke Martignoni and Priscilla Claeys). The specific context of small-scale fisheries, including small-scale marine and continental capture fishing as well as small-scale aquaculture, is also addressed, with an analysis of the interplay between the UNDROP and the FAO's Voluntary Guidelines for Securing Sustainable Small-Scale Fisheries regarding the recognition, protection, respect, and full realisation of the human rights of peasants involved in small-scale fisheries (Elisa Morgera and Julia Nakamura).

Part II focuses on UNDROP innovations and discusses new standards on access to and control of natural resources from a human rights perspective. It analyses new standards on the rights to water and sanitation (Mara Tignino, Francesca Restifo, and Tadesse Kebebew) and the right to food and food sovereignty (Christophe Golay), and it engages in a critical discussion on the recognition of a new right to land (Lorenzo Cotula) and right to seeds (Sandrine Le Teno, Christine Frison, and Samuel Cogolati).

Part III focuses on several aspects of the agri-food system governance. The first chapter in this part discusses the UNDROP as a tool to rebalance power relations in food systems governance (Smita Narula). Then, it proposes reflection on how the Declaration may influence the evolution of the international climate regime (Mariagrazia Alabrese and Annalisa Savaresi). The ways in which the UNDROP engages in the discussion on sustainability and ecological and resilient agri-food production are examined (Eloisa Cristiani and Giuliana Strambi). Lastly, Part III discusses the UNDROP's demands for new models of food production and consumption by taking due account of biocultural rights (Alessandra Di Lauro).

Finally, Part IV includes a series of thematic case studies that help understand the potential impacts of the Declaration in different subject areas and in several geographical regions. The status of law and practice relating to the protection of land rights in Kenya is illustrated and contrasted with the UNDROP's new standards related to landless rural workers and women (Robert Kibugi). The Greek Peliti seed network is presented and analysed in light of the UNDROP contents and potential impact on civil society (Elsa Tsioumani). The European law and policy framework for small-scale fisheries and women fishers is also looked at in order to explore new perspectives brought by the UNDROP on the subject matter (Adriana Bessa and Miguel Ángel Martín López). The case law of the Inter-American Human Rights System is discussed with the aim of identifying insights that may illuminate the implementation of the UNDROP (Naiara Posenato). Finally, the application of the contents of the Declaration is problematised through the analysis of a case study concerning land acquisition and access to justice in Gujarat, India (Margherita Brunori and Pier Filippo Giuggioli).

The conclusion lays out the cross-cutting, underlying themes that can be drawn from these reflections on the UNDROP, such as fragmentation of international law and recognition of rights holders, the politics of law, and governance in an increasingly complex world.

Note

1 UN General Assembly, 55th plenary meeting, Monday, 17 December 2018, 10 am, New York (Official Records) UN Doc A/73/PV.55, 20–5 <https://digitallibrary.un.org/record/3795602> accessed 6 July 2021.

1

THE PEASANTS' DECLARATION

State obligations and justiciability

Francesco Francioni

Peasants: the invisible class

Travellers who visit the historic centre of the Italian city of Modena can admire images of Adam and Eve tilling the earth in the 9th century bas-reliefs by Wiligelmo on the Modena Cathedral (now a World Heritage site). These images are a moving reminder of the vicissitudes of peasant life in this world since the beginning of time. But apart from the iconography of the biblical stories of Genesis in Medieval art, peasant life has had limited visibility in modern times compared to the vast iconography of rural landscapes depicting nature's beauty and serenity. The work and life of peasants in this artistic narrative are much idealised and carry the echo of the classical world celebrated by Latin poet Virgil in his poems the *Bucolics/Eclogues*. In these poems, Virgil lauded the beauty of pastoral and rural life, probably in contrast to the ferocity and violence of the political and military struggles of urban life in the Rome of the time. But the voice and stories of peasants and rural people, of their toils and real life are hardly heard or seen in these works, which remain a lyrical transfiguration of rural life into an Arcadia that exists only in the imagination of the poet or artist.

This loss of visibility of peasants' real life in literature and visual arts corresponds with a loss of visibility and memory in historical narratives. As acutely pointed out in sociological literature, peasants have always been *'une class object'*[1] – an object rather than a subject of history – because of the prevalent attention paid by historians to cities as centres of power and protagonists of history and also because of the prevalent attention given to the socio-economic implication of the industrial revolution. This is in striking contrast with the role played by peasants in critical periods of modern history. One needs only recall the 'peasants' war' in Germany (*DeutscherBauernkrieg* 1524–1526),[2] the most important class revolt in Europe before the French Revolution (condemned by Martin Luther), and the peasant

DOI: 10.4324/9781003139874-2

movement in 19th-century revolutionary Italy that gave an important contribution to the *Cinque giornate di Milano* revolt, the five-day popular uprising that triggered the overthrowing of the Austro-Hungarian government and the beginning of the Italian *Risorgimento* (war of national liberation).

But the most powerful representation of peasant life in contemporary narratives comes from the cinema. A 1978 film by Ermanno Olmi, *The Tree of Wooden Clogs* (*L'albero degli zoccoli*), a masterpiece which won the Palme d'or at the Cannes Film Festival of 1979, is an extraordinary, compassionate story of the hard, real, day-to-day life of ordinary peasants in the valleys of Lombardy at the dawn of the 20th century. The microcosm represented on the screen, its reality and authenticity, acquires the character of a universal testimony of the plight and injustice suffered by peasants as a class of people long deprived of any right in this world. As such, it retains a profound, lasting resonance for all those concerned with the human rights of peasants and people working in rural areas.

The 2018 declaration

Against this backdrop of such loss of memory and visibility, the Declaration on the Rights of Peasants and Other People Working in Rural Areas is a welcome step by the United Nations (UN) towards giving voice and legal status to a largely dispersed and forgotten part of the human population. This step is all the more appropriate because it concerns a class of people that has become more vulnerable due to the impact of irreversible changes caused by economic and social circumstances. Capital-intensive agricultural production has profoundly modified traditional food production methods based on the work of the peasant's family and by and large on an economy of subsistence. Besides finance capital, industrial agriculture relies on advanced technology and machinery and increasingly, at least in the developed world, on a mass of migrant workers that are now taking the place of the forgotten class of peasants of yesterday. This transformation in the social structure of the peasant class makes the adoption of an expansive title of the Declaration very appropriate, which besides the term 'peasants' includes the reference to 'other people working in rural areas.' However, these 'other people,' besides those expressly indicated in Article 1 of the Declaration, i.e., indigenous peoples, local communities, and the immigrant workers whose labour brings food to the table of city dwellers, should also include a new growing class of people who reject the idea that urbanisation is the only model of social organisation capable of ensuring progress and sustainable development. These people, many of whom are young, believe that the future of humanity will depend on a gradual return to life in the countryside and work in rural areas, possibly 'in harmony with nature' as prescribed by the 1992 Rio Declaration on Environment and Development (Rio Declaration) (Principle 1). This possibility is captured by the pioneering work of Dutch architect Rem Koolhaas, who, after having spent much of his life planning urban areas and city centres, has now turned his attention to rural areas and the enormous potential that they offer to the people who live and work there to emancipate themselves from the subordinate role that they have traditionally played as 'servants' of the cities. In the digital era, they may become protagonists of a more complex, sustainable model of economic, social, and cultural development that may transcend the traditional stereotype of 'peasants.' This possible dynamic evolution of the category of peasants, implicit in the title and in Article 1 of the Declaration, is also important to the ends of counterbalancing the limited vision of other pseudo-progressive documents adopted by the UN and other organisations with regard to humanity's future in cities and mega cities. I am referring, particularly, to the UN *World Urbanization Prospect*, a report which assumes that by 2050 almost 70% of the world population will live in cities[3], and also to the so-called Eco Modernist Manifesto for the Anthropocene, which advocates intensive urbanisation as the key to a more efficient use of energy and natural resources.[4] The experience of the severe environmental degradation caused by human concentration in large metropolis and, now, the disaster caused by the Covid-19 pandemics have dramatically shown the fragility of life in densely populated metropolitan areas and the immense potentiality that vast expanses of rural area have, not only to save lives in the event of global epidemics but also for a 'healthy and productive life in harmony with nature,' in a new perspective of sustainable development in accordance with the words of Principle 1 of the Rio Declaration.

State obligations

The Declaration presents a catalogue of rights whose respect and protection depend on the legal obligations placed on certain subjects. These subjects may be the states whose jurisdiction the peasants and other peoples working in rural areas are found within or non-state actors, such as individuals, private organisations, and entities whose conduct may have an impact on the lives of peasants. The text of the Declaration makes it abundantly clear that the primary duty bearers of its provisions are the states. Each article is structured so that a provision indicating the corresponding obligation of the state follows the enunciation of the rights. Non-state actors are also bound to respect the Declaration according to Article 2, paragraph 5. However, even in this case, it is up to the state 'to take all necessary measures' to ensure that such non-state actors are effectively regulated and monitored so as to guarantee their compliance with the Declaration.

Nothing in the Declaration clarifies what the precise scope of the state obligation 'to regulate' individuals and actors other than states in view of safeguarding and strengthening the rights of peasants is. Article 2.5 contains very vague language and refers to non-state actors that states 'are in a position to regulate.' This oblique reference leaves open whether the 'position to regulate' refers to the ordinary powers that each state has under its domestic law or to a duty under international law to regulate according to a due diligence standard. Unlike the International Covenant on Civil and Political Rights (ICCPR), whose Article 2 defines the scope of the obligation by each state party referring to 'all individuals within its territory and subject to its jurisdiction,' the Peasants Declaration falls short of defining the sphere of application of the obligations thereby prescribed to states. This absence of a precise jurisdictional clause renders the Declaration similar to the Covenant on Economic, Social, and Cultural Rights (ICESCR), which, as is well known, fails to replicate the model of the jurisdictional clause of Article 2 of the ICCPR and leaves open the scope of application of state obligations to respect, protect, and fulfil the economic, social, and cultural rights proclaimed by the Covenant. This solution is understandable in light of the widely recognised non-justiciable character of many economic, social, and cultural rights, whose realisation can be only 'progressively' achieved through appropriate domestic policies and international assistance.[5] However, the Peasants Declaration goes beyond economic, social, and cultural rights to encompass fundamental civil and political rights capable of immediate implementation also by judicial means. In relation to these rights, such as the right to non-discrimination, to life, to personal liberty, to mental and physical integrity, and freedom of thought and belief,[6] a more precise definition of the jurisdictional scope of state obligations would have been appropriate.

But the reason why such a jurisdictional clause in the Declaration would have been helpful is that, in the present reality, the human rights of peasants and other peoples working in rural areas are exposed to violations not only by acts or omissions of the territorial state. Often, violations occur as a consequence of acts, decisions, or policies adopted and implemented in a foreign state by governmental authorities or non-state actors, like business corporations. There is ample evidence of this in the practice of acquisitions by foreign governmental entities of large tracts of arable land to produce export crops, which may involve the displacement of local peasant communities or the dispossession of land essential to these communities' subsistence, as recognised by Article 17 of the Declaration.[7] This is also the case for foreign investments by private actors in agricultural or forestry projects in the service of industrial or environmental interests of foreign countries, which have the effects – as in the case of bio-fuels – of reducing available productive land and access to food for peasants communities in violation of Articles 15, 16, and 17 of the Declaration. In this regard, it is worth noting that Article 17.4 provides that peasants and other people working in rural areas

> Have the right to be protected against arbitrary and unlawful displacement from their land or place of habitual residence, or from other natural resources used in their activities and necessary for the enjoyment of adequate living conditions. States shall incorporate protections against displacement into domestic legislation that are consistent with international human rights and humanitarian law. States shall prohibit arbitrary and unlawful forced eviction, the destruction of agricultural areas and the confiscation or expropriation of land and other resources, including as a punitive measure or as a means or method of war.

In addition, Article 18.5 provides that

> States shall protect peasants and other people working in rural areas against abuses by non-state actors, including by enforcing environmental laws that

contribute, directly or indirectly, to the protection of the rights of peasants or other people working in rural areas.

These are important provisions laying down an obligation for every state to adopt and implement legislation aimed at protecting peasants against abuses by non-state actors. But it would be misleading and ineffectual to limit this obligation to the territorial state where the peasants are located and where the abuses are perpetrated. Territorial states certainly have the primary responsibility to ensure respect and protection of the human rights of peasants. However, in the reality of today's inter-dependent economy and international business, abuses inevitably occur often as a consequence of decisions or policies adopted in a foreign country, which the territorial state is either unwilling, because of corruption or unequal bargaining power, or unable, because of lack of scientific knowledge, to adequately regulate and comply with applicable human rights standards. This is the reason why in international law scholarship, and at the level of international standard setting, the idea has emerged that, in the field of human rights, and especially of economic, social, and cultural rights, it is necessary to develop an extraterritorial dimension of state obligations to respect and protect such rights. This is shown by the 2012 adoption of the Maastricht Principles on Extraterritorial Obligations of States in the Area of Economic, Social, and Cultural Rights (Maastricht Principles), whose cardinal legal principle is that states are obliged to prevent third parties from violating such rights in other countries whenever they retain the legal and political means to influence the conduct of such third parties in accordance with international law.[8] This principle has been upheld by the International Court of Justice in its recent jurisprudence.[9]

The standard of conduct required in compliance with the relative obligation is 'due diligence,' that is, a duty of care towards the rights of peoples located in other states whenever such rights may risk being breached by activities taking place within the national jurisdiction. A similar approach was followed by the Human Rights Council (HRC) with the adoption in 2011 of the Guiding Principles on Business and Human Rights. These principles build on the state–business nexus to provide a hybrid legal framework of state obligations and of corporate duties to protect and respect human rights, largely based on the same concept of due diligence as that underlying the Maastricht Principles.[10]

In view of these important developments in scholarship and practice, it is submitted that, even in the absence of specific clauses to this effect, the Declaration should be interpreted and applied in such a way that the obligations to respect, protect, and fulfil the rights of peasants and of other people working in rural areas may be extended, as far as possible, to non-state actors whose activities and operating contexts are capable of posing a material risk to the rights of peasants and such other people. At the same time, a progressive interpretation of the Declaration, in accordance with the Maastricht Principles and the HRC 2011 Principles on Business and Human Rights, should lead to the conclusion that it provides a basis for the extension of state duty to respect and protect the rights of peasants in relation to activities taking place in another country whenever the state has effective influence and control over the decision-makers of such activities and over the management of the risks that such activities involve for the human rights of peasants.[11]

The legal significance of the declaration

As for all resolutions adopted by the General Assembly of the United Nations, the formal legal value of the Peasants Declaration must be assessed in light of the limits that the UN Charter places on the power of its organs to adopt binding acts. As is well known, with the exception of Article 17 of the UN Charter, concerning the approval of the budget of the Organisation, the General Assembly has the power only to undertake studies and make recommendations. Under Article 13.2, the Assembly's powers include promoting international cooperation and 'encouraging the progressive development of international law and its codification.' Paragraph 3 of the same article extends this power to

> Promoting international cooperation in the economic, social, cultural, educational and health fields, and assisting in the realization of human rights and fundamental freedoms for all without distinction as to race, sex, language or religion.

These provisions are a sound basis for the initiative taken by the General Assembly with the elaboration and adoption of the Peasants Declaration. They certainly contribute to the normative value of the Declaration. However, they are not sufficient per se to confer a legally binding value on the Declaration, which formally remains an act of a 'hortatory' character, a 'recommendation.'

Notwithstanding this, it is worth trying to go beyond this uncontested legal assessment and, for the purpose of the present analysis, ask whether in view of the formal and solemn character of this proclamation and of the important subject matter – human rights – the Declaration may be capable of producing binding obligations for states with regard to the treatment of peasants and other people working in rural areas.

The first question concerns the relevance of the title of 'Declaration' given to this General Assembly resolution. It is well known that this title is given only to a selected category of General Assembly resolutions that are meant to have a high moral, political, or legal value. Sometimes this value is further enhanced by the adjective 'universal,' as in the case of the 1948 Universal Declaration of Human Rights. So, even if General Assembly resolutions are not meant to produce binding effects, we need to ask if this special category of resolutions that are labelled 'declarations' may nevertheless play a role in the progressive development of international law or in the creation of consent-based obligations for certain states.

As far as the progressive development of international law is concerned, there is no doubt that General Assembly declarations have been on the forefront of the development of general norms and principles of international law in the field of human rights, from the 1948 Universal Declaration of Human Rights to the 2007

Declaration on the Rights of Indigenous Peoples; therefore, the Peasants Declaration may also contribute to a practice that will ultimately consolidate a catalogue of human rights for peasants and other people working in rural areas. However, as of today, we cannot consider the whole Peasants Declaration as a reflection of customary international law. Some provisions simply restate general norms of international law, such as those on non-discrimination (Article 3), the right to life, physical integrity and security of person (Article 6.1), the right not to be held in slavery or servitude (Article 6.2), and the right to be recognised as a person before the law (Article 7). Many other provisions, instead, proclaim rights that have a special significance for peasants and other people working in rural areas, such as the right to land (Article 17), the right to the conservation and protection of the environment (Article 18), the right to seeds (Article 19), or the right to the sustainable use of natural resources (Article 20). These provisions, although especially relevant for peasant communities, cannot be considered a simple reflection of customary norms of international law. At most, they may indicate standards of achievement for the progressive development of international law. This realistic assessment is confirmed by the lack of a general *opinio iuris* among the members of the General Assembly in support of the whole spectrum of rights proclaimed by the Declaration. Its adoption was not by a unanimous vote or by consensus. It was by a majority vote with 7 votes against[12] and a substantial number (47) of abstentions.[13]

The negative conclusion on the possibility of considering the whole of the Peasants Declaration as part of customary international law does not exclude that its 'bindingness' may derive from a legal basis other than customary international law. In this regard, it is to be observed that for all those states that cast a positive vote in support of the Declaration, a presumption exists that their vote amounts to an informal expression of consent to be bound by the norms and principles of the Declaration. This presumption, if founded on the logical consideration that a positive vote on such a high profile document, the result of many years of diplomatic negotiations and deliberations, cannot be expressed in the superficial conviction that it carries only a symbolic value and should, therefore, have no legal consequence. Of course, one can object that the consequences of the vote are only political and not legal, because of the political nature of the General Assembly resolution in which the Declaration is formally incorporated. But the positive vote in support of the Declaration acquires a legal relevance by virtue of the legal underpinning that the Declaration has in the UN Charter. The rights of peasants proclaimed in the Declaration have a legal basis in the commitment of the UN Charter to protect human dignity, to eliminate all forms of discrimination, including those against peasants, to promote the right to self-determination, and to progressively eliminate material obstacles that hinder economic, social, and cultural development. A breach of these rights, indirectly, constitutes a breach of the commitment undertaken with the UN Charter to respect and protect human rights and fundamental freedoms. The ample majority of states that in the General Assembly voted in favour of the Declaration must be presumed to have acted in full awareness of the nexus between the rights proclaimed in the Peasants Declaration and the principles enshrined in the UN

Charter. Thus, the positive vote given for the adoption of the Peasants Declaration may be a valid ground for an initial, limited 'bindingness' of the declaration, at least for those who have cast positive votes. This is without prejudice to the possibility that subsequent practice may eventually produce general bindingness.

This conclusion finds an important confirmation in the decision of the Supreme Court of Belize of 18 October 2007, which considered the positive vote cast by Belize in support of the General Assembly resolution adopting the Declaration on the Rights of Indigenous Peoples as a valid legal ground for the finding of a legal obligation to respect the right of the Maya people of Belize to their ancestral lands.[14]

Justiciability and implementation

The question of justiciability and the implementation of human rights and fundamental freedoms has traditionally followed the distinction between civil and political rights, on the one hand, and economic, social, and cultural rights, on the other hand. Only the first category, according to traditional thinking, would be subject to judicial implementation by courts and other enforcement bodies. The logic of this distinction is simple: civil and political rights entail negative obligations for states, which are susceptible to immediate enforcement. Economic, social, and cultural rights, instead, entail positive obligations and costs for states, and a 'progressive realization' through appropriate governmental policies, which, by their nature, are not susceptible to judicial review and judicial implementation. In the opinion of this writer, this distinction is insufficiently nuanced because all human rights are 'indivisible'[15] and all of them entail positive obligations and costs for states. All the more so in the era in which we are living, when a perfect storm is befalling us in the form of a planetary pandemic that threatens not only the right to health but also a whole range of human rights, including economic and social rights, exposes the limits of globalisation, and produces unprecedented hardship and human costs worldwide. If there was any need, the present situation dramatically reveals how inseparable economic, social, and cultural rights are from civil rights and liberties, and how illusory the latter are when the former are ignored or denied.

In coherence with this idea, the Peasants Declaration takes a holistic approach to the human rights of peasants and other people working in rural areas and provides a general framework that includes civil and political rights as well as economic, social, and cultural rights which are mutually interdependent. Consistent with its nature as a soft law instrument, the Declaration does not provide an independent mechanism of enforcement or supervision. Therefore, its justiciability depends on general methods and mechanisms that are available for all human rights either at the level of domestic law or international law.

At the level of domestic law, the Peasants Declaration may be the source of applicable law in disputes arising from alleged breaches of peasants' rights especially in those states that have voted in favour of the Declaration and have, therefore, expressed their consent to accept the relative obligations. For states that have abstained or voted against, the Declaration may not be considered the direct source of obligations which are enforceable in the forum state but may nevertheless provide an important source of interpretative criteria for the determination of obligations arising from other sources of international law, customary or treaty law. This, of course, is without prejudice to the enforceability of peasants' rights that find their basis in customary international law or on human rights treaties which are binding upon the forum state. This is, certainly, the case, *inter alia*, with Article 6 of the Declaration dealing with the right to life, to physical integrity, to personal liberty, the right not to be subject to torture or to slavery, as well with Article 7 proclaiming the right to be recognised as a person before the law. All these rights, together with the fundamental principle of non-discrimination in their exercise, as provided in Article 3,[16] have the potential for immediate judicial implementation by national courts. While these rights retain their general validity and justiciability under general norms of international law or under applicable treaties, such as the ICCPR, the added value of the Declaration consists in conferring these rights more substantive meaning and possibly new dimensions in the specific context of peasants and other people working in rural areas. This is so precisely because of the special needs of these increasingly vulnerable people and the particular socio-economic and cultural context of peasant communities in which such rights must be interpreted and implemented.[17]

As far as international implementation is concerned, we must distinguish between (a) 'direct' and (b) 'indirect' forms of implementation, and (c) implementation by 'other means,' that is, by mechanisms 'borrowed' from fields of international law other than human rights law.

(a) The direct forms of implementation can be identified, first of all, in the role of the HRC as a vehicle for promoting the human rights of peasants. A logical avenue to pursue this objective would be to set up an independent expert or special rapporteur on the rights of peasants, following the model of the special rapporteur on the rights of indigenous peoples. Another form of direct implementation is provided by the procedures offered by various Treaty Bodies relevant to the enforcement of rights contained in the Peasants Declaration. I am referring especially to the two Committees established under the ICESCR and the ICCPR. These Committees can be a valid implementation instrument especially in relation to states that have accepted the individual right to lodge complaints for alleged violations of human rights protected under the Covenants. The exercise of this right is now possible also for economic, social, and cultural rights after the entry into force in 2013 of the Protocol additional to the ICESCR.[18] In all these procedures, the Peasants Declaration can be a valuable tool to interpret the two Covenants in light of the specific provisions of the Declaration and of the needs and interests of peasants. This is true also for human rights courts operating at a regional level. This brings us to the 'indirect' form of implementation.

(b) Indirect implementation of the Peasants Declaration consists in the interpretation by human rights courts of specific provisions of applicable human rights treaties in such a way as to give recognition and protection to specific rights of peasants and other people working in rural areas even if such rights are not expressly recognised in the text of the applicable treaty. This jurisprudential approach finds its expression in the ground-breaking judgement of the Inter-American Court of Human Rights (IACHR) in *Mayagma (Sumo) AwasTingni Community v Nicaragua.*[19] In this case, the Court interpreted and applied Article 21 of the American Convention on Human Rights (the right to property) in a progressive way so as to recognise the ownership title of the Awas Tingni people to their ancestral land, based exclusively on informal customary possession. Similarly, Article 21 has been used to recognise the communal right of traditional local communities over their forest, in conjunction with Article 25 on access to justice, in the *Saramaka People v. Suriname* case. In this case, a traditional local community opposed and sought a remedy against a governmental decision to grant a logging concession to foreign investors on their ancestral land.[20] These jurisprudential precedents open the way for the indirect implementation by human rights courts and human rights treaty bodies of specific peasants rights proclaimed in the Declaration, such as the right to seeds (Article 19) and the right of access to land and to be protected against arbitrary eviction (Article 17) through an evolutive interpretation of applicable human rights treaties and customary law.

(c) Implementation of the Peasants Declaration by 'other means' may be the most important form of enforcement of the rights of peasants and of other people working in rural areas. This is so because these rights depend, to a large extent, on the attitudes that states take with respect to the activities of economic actors, especially those of business actors and their transnational operations. Today, such operations depend, to a large extent, on the freedom of movement of finance capital and on foreign investments which, under the present regime of international law, enjoy strong guarantees of protection against unfavourable acts or policies of the host state. Key to this protection is the compulsory system of investment arbitration that under bilateral investment treaties (BITs) or under regional agreements give foreign investors the right to bring complaints before arbitral tribunals to enforce their economic rights against the investment host state. While investment arbitration is essentially designed to protect the economic interests of the investors, recent practice and jurisprudential developments have shown the possibility, and, indeed, the desirability, for a human rights role in the settlement of investment disputes.[21] This has occurred when the economic rights of the investors were negatively impacted by regulatory measures or protective policies adopted by the host state to safeguard the economic, social, and cultural interests of the local population. Important examples of *due regard* paid by investment tribunals to the human rights of the local population can be found in conjunction with the rights of indigenous peoples and investors rights. In *Glamis Gold v the United States*, a claim by a Canadian company to obtain redress for economic loss allegedly caused by the adoption of regulatory measures by the United States was rejected by the arbitral tribunal *inter alia* on account of the need to protect the environmental rights of the local population and the rights of a local indigenous community to the peaceful enjoyment of their ancestral lands in the face of the risks posed by the realisation of a gold mining project.[22] The engagement of an investment tribunal with the task of reconciling the protection of investor rights with the need to respect the rights of the local population is all the more important at a time when large foreign investments in agricultural projects and the practice of 'land deals,' sometimes labelled with the pejorative expression of 'land grabbing,'[23] expose peasants and local people working in rural areas to destabilising effects on their economic, social, and cultural life.

Conclusion

Until now, peasants and people working in rural areas have been at the losing end of developing processes. All too often agricultural development, industrial development, and development in energy and mineral resources have resulted in negative economic, social, and environmental consequences for peasants, aggravating their poverty and social exclusion while the economic benefits of such development have gone to others. The Peasants Declaration is a welcome step not only towards giving international legal status to these people but also to remind states that, in the exercise of their territorial sovereignty and in their relations with both domestic and foreign business initiatives, which are carried out in their territory, they have a duty to care for the dignity and well-being of this marginalised class of people.

Notes

1 Pierre Bourdieu, 'La paysannerie une classe objet' (1977) 17–8 *Actes de la recherce in sciences sociales* 2–5.
2 Peter Blikle, *The Revolution of 1525: The German Peasants War from a New Perspective* (Johns Hopkins University Press 1981).
3 UN Department of Economic and Social Affairs, 'World Urbanization Prospects' (2018) UN Doc ST/ESA/SER.A/420.
4 The Modernist Manifesto was published by a private group of scientists, intellectuals and policy makers in April 2015 <www.ecomodernism.org>.
5 See Committee on Economic, Social and Cultural Rights (CESCR) 'General Comment No 3: The Nature of States Parties Obligations (art. 2 of the Covenant)' (14 December 1990) UN Doc E/1991/23 III. For a review of case law on the justiciability of economic, social and cultural rights, see Ben Saul, David Kinley and Jaqueline Mowbray, *The International Covenant on Economic, social and Cultural Rights, Commentary, Cases and Materials* (Oxford University Press 2014) 132; Manisuli Ssenyonjo, *Economic, Social and Cultural Rights in International Law* (2d edn, Routledge 2016) 78.
6 UN Human Rights Council, 'United Nations Declaration on the Rights of Peasants and Other People Working in Rural Areas' (8 October 2018) UN Doc A/HRC/RES/39/12, artt. 4, 6, 7 and 8.
7 For a critical evaluation of the practice of so called 'land grabbing,' see Nino, *Land grabbing e sovranità territoriale in diritto internazionale* (Editoriale Scientifica 2018).
8 The Maastricht Principles were adopted at a meeting convened by the International Commission of Jurists and Maastricht University, with the participation of legal experts from all over the world, and on the basis of a research project conducted over a period of more than ten years. ETO Consortium, Heidelberg, January 2013. For a general treatment of the subject of extraterritorial application of human rights obligations, see Marko Milanovic, *Extraterritorial Application of Human Rights Treaties* (Oxford University Press 2011).
9 *Legal Consequences of the Construction of a Wall in the Occupied Palestinian* Territory (Advisory Opinion of 9 July 2004) International Court of Justice Reports 136, 240.

10 The Commentary on Principle 4 (state–business nexus) states: 'States should encourage and, where appropriate, require human rights due diligence by agencies themselves and by those business enterprises or projects receiving their support. A requirement of human rights due diligence is most likely to be appropriate when the nature of business operations or operating context poses significant risks to human rights' UN Human Rights Council, 'Human rights and transnational corporations and other business enterprises' (17 June 2011) UN Doc A/HRC/17/L.17/Rev.1.
11 For an analogy with the duty of due diligence of the state of origin of multinational enterprises in the field of environmental protection, see Francesco Francioni, 'Exporting Environmental Hazard through *Multinational* Enterprises: Can the *State of Origin* be Held Responsible?' in Francesco Francioni and Tullio Scovazzi (eds), *International Responsibility for Environmental Harm* (Graham and Trotman/Martinus Nijihoff 1992).
12 Austria, Hungary, Israel, New Zealand, Sweden, the United Kingdom, and the United States.
13 Forty-seven abstentions including Brazil, Canada, Ethiopia, Japan, South Korea, Russia, Turkey, Ukraine, and all members of the EU except the two who voted against (Hungary and Sweden) and Portugal who voted for.
14 Decision in 'Aurelio Cal and Others v. Attorney-General of Belize and Minister of Natural Resources and Environment' (2009) 135 *International Law Reports* 78.
15 UN General Assembly, 'Vienna Declaration and Programme of Action' (12 July 1993) UN Doc A/CONF.157/23 (adopted by the UN World Conference on Human Rights) 20–46.
16 Article 3 of the Declaration prohibits discrimination 'based on any grounds such as origin, nationality, race, colour, descent, sex, language, culture, marital, status, property, disability, age, political or other opinion, religion, birth or economic, social or other status.'
17 On the matter of the adaptability of human rights standards to the socio-cultural conditions of different people and communities, Federico Lenzerini, *The Culturalization of Human Rights* (Oxford University Press 2014); Francesco Francioni, 'Human Rights and Cultural Pluralism: What Role for International Law?' (2018) 12 *Diritti Umani e Diritto Internazionale* 307–14.
18 UN General Assembly, 'Optional Protocol to the International Covenant on Economic, Social and Cultural Rights' (10 December 2008) UN Doc A/RES/63/117. At the time of writing, the Protocol has 26 state parties.
19 *Mayagma (Sumo) AwasTingni Community v Nicaragua,* Inter-American Court of Human Rights Series C No 79 (31 August 2001).
20 *Saramaka People v Suriname,* Inter-American Court of Human Rights Series C No 172 (28 November 2007).
21 Pierre-Marie Dupuy, Francesco Francioni and Ernst-Ulrich Petersmann (eds), *Human Rights in International Investment Law and Arbitration* (Oxford University Press 2009); Bruno Simma, 'Foreign Investment Arbitration: A Place for Human Rights?' (2011) 60 *International & Comparative Law Quarterly* 573–86.
22 *Glamis Gold v United States,* award of the arbitral tribunal constituted under Chapter 11 of NAFTA (8 June 2009).
23 On this subject, see the recent monograph by Nino, *Land grabbing e sovranità territoriale in diritto internazionale*, supra note 7.

Selected bibliography

Dupuy P, Francioni F and Petersmann E (eds), *Human Rights in International Investment Law and Arbitration* (Oxford University Press 2009).

Nino M, *Land Grabbing e sovranità territoriale in diritto internazionale* (Editoriale Scientifica 2018).

Ssenyonjo M, *Economic, Social and Cultural Rights in International Law* (2nd edn, Routledge 2016).

PART I

Rights holders

2

DEFINING PEASANTS IN THE UNDROP

*Marc Edelman**

Introduction

'Language is not merely a passive reflection of things as they are, but also a tool for imagining and making things as they could be', John Patrick Leary remarks in *Keywords: The New Language of Capitalism*.[1] Many international human rights instruments have a definition of the rights holders in their first article. Frequently these definitions generate considerable controversy, making it abundantly clear that even the most anodyne characterisations are social constructions – and the result of political struggles – rather than ahistorical absolute truths. The 1989 Convention on the Rights of the Child (CRC), for example, states that 'a child means every human being below the age of eighteen years unless under the law applicable to the child, majority is attained earlier.'[2] This seemingly straightforward categorisation, however, conceals considerable disagreement. In the debates that preceded the vote and in the declarations and reservations that states registered when they ratified the CRC, misgivings ranged from the incompatibility of the CRC's Article 1 with national legislation about the age of majority (e.g., Cuba and Liechtenstein) and labour law (e.g., Singapore) to concerns about provisions contrary to Islamic Shariah (e.g., Afghanistan, Iran, Iraq, Somalia, and Syria) or that might undermine parental authority (e.g., Kiribati and Poland).[3]

This fractiousness is not limited to binding conventions or treaties. It is evident in the cases of non-binding instruments as well. In the 2007 United Nations (UN) Declaration on the Rights of Indigenous People (UNDRIP), specifying the rights

* Funding for this research included grants 1024017 and 1358143 from the US National Science Foundation (Cultural Anthropology and Law and Social Sciences Programs) and grant 66785–00 44 ENHC-44–72 from the PSC-CUNY Research Awards Program. I am grateful to Adriana Bessa, Priscilla Claeys, and Christophe Golay, who provided constructive comments on an earlier draft. Any errors or shortcomings are of course my own.

DOI: 10.4324/9781003139874-4

holders was sufficiently contentious that the text does not contain any definition at all, even though jurists involved in the UNDRIP negotiations developed detailed arguments about what an article on rights holders might contain.[4] In contrast, the International Labour Organization (ILO)'s Convention 169 on Indigenous and Tribal Peoples (ILO Convention 169), adopted in 1989, contains a rigorous description of those it covers, with 'self-identification' a key criterion.[5] Nonetheless, in part because of growing scepticism from indigenous movements that viewed definitions of indigeneity as potentially excluding certain vulnerable or autochthonous groups, the ILO backtracked and later insisted that 'ILO Convention No. 169 does not define who are indigenous and tribal peoples. It only describes the peoples it aims to protect.'[6]

How UNDROP specifies the rights holders

The definition of rights holders in the UN Declaration on the Rights of Peasants and Other People Working in Rural Areas (UNDROP) evolved in a similarly disputatious process. However, unlike the controversies about the term 'indigenous' related to ILO Convention 169 and UNDRIP, the campaign for the Declaration on the Rights of Peasants and Other People Working in Rural Areas eventually took a strong stance in favour of including a specification of rights holders. Nonetheless, the initial 2002 proposal of La Via Campesina (LVC), a transnational agrarian movement, for a new international legal instrument on peasants' rights did not specify a definition at all.

This position shifted in 2008 when LVC substantially redrafted its proposed Declaration and opted to include a definition of the rights holders. Box 1 includes the definition that LVC proposed in 2009 and that the Human Rights Council Advisory Committee endorsed in its draft Declaration of 2012, alongside the definition in Article 1 of the final version of the UNDROP.

BOX 2.1 THE DEFINITION OF PEASANT

2009 – La Via Campesina	*2018 – UNDROP, Article 1*
A peasant is a man or woman of the land who has a direct and special relationship with the land and nature through the production of food and/or other agricultural products. Peasants work the land themselves, rely above all on family labour and other small-scale forms of organising labour. Peasants are traditionally embedded in their local communities and they take care of local landscapes and of agro-ecological systems.	1. For the purposes of the present Declaration, a peasant is any person who engages or who seeks to engage, alone, or in association with others or as a community, in small-scale agricultural production for subsistence and/or for the market, and who relies significantly, though not necessarily exclusively, on family or household labour and other non-monetised ways of organising labour, and who has a special dependency on and attachment to the land.

The term peasant can apply to any person engaged in agriculture, cattle-raising, pastoralism, handicrafts-related to agriculture, or a related occupation in a rural area.

This includes Indigenous people working on the land.

The term peasant also applies to landless. According to the UN Food and Agriculture Organization definition, the following categories of people are considered to be landless and are likely to face difficulties in ensuring their livelihood:

1. Agricultural labour households with little or no land;
2. Non-agricultural households in rural areas, with little or no land, whose members are engaged in various activities such as fishing, making crafts for the local market, or providing services;
3. Other rural households of pastoralists, nomads, peasants practising shifting cultivation, hunters and gatherers, and people with similar livelihoods.[7]

2. The present Declaration applies to any person engaged in artisanal or small-scale agriculture, crop planting, livestock raising, pastoralism, fishing, forestry, hunting or gathering, and handicrafts related to agriculture or a related occupation in a rural area. It also applies to dependent family members of peasants.
3. The present Declaration also applies to indigenous peoples and local communities working on the land, transhumant, nomadic and semi-nomadic communities, and the landless engaged in the above-mentioned activities.
4. The present Declaration further applies to hired workers, including all migrant workers regardless of their migration status, and seasonal workers, on plantations, agricultural farms, forests, and farms in aquaculture and in agro-industrial enterprises.[8]

What happened over the 17 years of struggle that culminated in the UNDROP and its Article 1 definition of peasants? The idea of a declaration on the rights of peasants originated in Indonesia during the exhilarating '*reformasi*' period that followed the fall of the Suharto dictatorship in 1998.[9] In 1965–66, in one of the largest and least-remarked bloodbaths of the 20th century, the Indonesian military and right-wing gangs massacred an estimated 500,000 people, including suspected members or sympathisers of the Communist Party (PKI), land reform beneficiaries, and ethnic Chinese. Peasants were the hardest hit sector and they continued to suffer under Suharto as the regime persecuted opponents and turned thousands of hectares of agrarian reform lands over to the former owners or to new agribusiness investors, who set up palm oil, rubber, and other plantations.[10]

When the Suharto regime collapsed under pressure from a mass pro-democracy movement, a wave of land invasions swept Indonesia, with peasants in many cases recuperating lands that had been stolen from them decades before.[11] Together with human rights organisations, peasant groups elaborated a national-level declaration on peasants' rights in 2001, which the following year served as a model for a proposed international instrument – initially intended as a binding convention – elaborated by the Southeast Asian region of LVC.[12] Neither of these predecessor texts to UNDROP contained a specification of the rights holders.

Only in 2009, following an LVC conference on peasants' rights in Jakarta and a meeting in Bilbao of LVC's Human Rights Commission, did the movement propose a new text for a declaration. This 2009 version's definition of the rights holders is in Box 2.1, alongside the definition in the UNDROP. The 2009 LVC definition appears verbatim in the 2012 Declaration text that the Advisory Committee of the UN Human Rights Council presented as a basis for negotiation of an UNDROP by the entire Council.[13]

A careful reader will note several differences between the 2009 LVC/2012 Advisory Committee definition and that which appears in the final version of the UNDROP of 2018. The earlier definition's reference to the peasant's 'direct and special relationship with the land and nature' becomes 'a special dependency on and attachment to the land' in UNDROP Article 1.[14] The latter phrase echoes 'accepted language' that states had already agreed to in several earlier UN normative sources.[15] 'Nature' appears only in the UNDROP preamble, very likely because its location there would be less contentious than if it were asserted as a central feature of 'peasant-ness' in the definition of the rights holders. The idea that 'peasants . . . take care of . . . local landscapes and agro-ecological systems,' present in 2009, is not part of the UNDROP definition.

The UNDROP definition alludes to two dimensions of peasant life and economy that were absent in the 2009 definition and that prefigure other sections of the 2018 Declaration. First, it indicates that peasants work '*in association with others or as a community*' (emphasis added). This clause reinforces UNDROP's frequent references to different sorts of rights that may be exercised 'individually and/or collectively,' a formulation that pushed in the direction of group rights to land and other resources and that excited intense opposition from developed-country member states, especially the European Union and the United States. Second, compared with LVC's 2009 text, the 2018 UNDROP definition is more capacious, including not just the agriculturalists, livestock producers, pastoralists, indigenous peoples, artisans, hunter-gatherers, and landless mentioned in 2009 but also fishers, foresters, and migrant, seasonal, and other rural workers, such as those who labour on plantations and agro-industrial enterprises. This greater inclusivity reflected both the Advisory Committee's decision to expand the title of the instrument before presenting it to the Human Rights Council and the growing participation, particularly in the last three years of negotiations in the Working Group sessions, of movements representing rural workers, nomads, pastoralists, small-scale fishers, and indigenous peoples.[16]

UNDROP Article 1 also contains a more explicit acknowledgement than the 2009 LVC definition that peasants often produce *both* for subsistence or self-provisioning and for the market, and that they may hire wage labour or mobilise labour not just through the family, as in the 2009 text, but via other 'non-monetized' forms of organisation. While not specified in the definition, examples of the latter might include unwaged collective work or labour exchange parties in traditional local communities, such as the Haitian *coumbite*, the Andean *minga*, the Mesoamerican *faena*, or the North American Amish barn raising. The language that encompasses these forms echoes the reference to 'non-monetized sectors of the economy' in Article 14.1 of the 1979 Convention on the Elimination of All Forms of Discrimination against Women (ICEDAW).[17] As with other sections of the UNDROP, the drafters made every effort to incorporate 'accepted language' not only that states had already agreed to but also that they would thus, at least in theory, find legible and amenable.

Fighting words

The negotiating process in the UN frequently consists of parsing the meanings of particular words and bartering ones that certain states don't like for others that they find more tolerable. It is important to emphasise at the outset, however, that the concerns states raised about the word 'peasant' typically had less to do with any abstract commitment to semantic precision than with a broader opposition to the UNDROP. Throughout the negotiations, the states most vocally opposed to the Declaration brought up not just semantic objections but also budgetary and procedural ones. In almost all cases, these interventions masked an opposition that derived more fundamentally from UNDROP's potential radical implications in areas such as intellectual property and seeds, collective rights to land, access to natural resources, and the labour and environmental rights of rural people.

English was the language of negotiation for the UNDROP, and almost immediately this introduced some significant problems when it came to the title of the document and its proposed Article 1 on the rights holders. The English word 'peasant' is not a precise equivalent to the Spanish '*campesino,*' the French '*paysan,*' or the Russian '*крестьянин*' (*krest'ianin*). First, the English term, while derived from the French, has come to imply an occupational and perhaps a class location in the agricultural sector and has largely lost the association with somebody physically or geographically located in or from the countryside, whether involved in agriculture or not. The Spanish and French terms, while broadly associated with agriculture, are etymologically more closely related to geographical location in the countryside (*campo* or *pays*). To be concrete, the owner of a small rural store might be a *campesino* or *paysan* but, in an English-speaking milieu, would be unlikely to identify with or be identified by the label 'peasant.' The Russian *krest'ianin*, on the other hand, is a cognate of *Kristian* (*Кристиан*), implying a strong identity between occupation and religion that, among other things, obscured the real religious diversity that existed among agriculturalists in the vast Russian Empire and later the Soviet Union.[18]

Second, and more importantly, the English word 'peasant' can be intensely derogatory. This negative valence is also, of course, sometimes present in Spanish, French, and Russian, though not usually to the same degree. Raymond Williams pointed out in his seminal book *Keywords* that 'peasant' is 'a loose term of abuse – in English, usually very self-conscious and exaggerated – of "uneducated" or "common" people.' He explained this 'insolent' usage as at least in part a product of the enclosures of village common lands, which meant that by the 18th century, this social group 'virtually ceased to exist in England.'[19] This 'special difficulty in the uses of the word' complicated its adoption as a form of self-ascription or self-identification among English-speaking agriculturalists in various parts of the world and thus of the global alliances that transnational agrarian movements like LVC sought to construct. It also caused difficulties in the UNDROP negotiations as different developed-country delegations called attention to the pejorative aspects of 'peasant' and adduced this as yet one more reason why the negotiations ought to be scuttled.

Perhaps not surprisingly, given that England was historically the first country in the world to eliminate its peasantry, the United Kingdom was among the states most hostile to UNDROP. This negative stance was evident on several occasions during the negotiations and also apparently manifested behind the scenes in internal discussions within the European Union (EU). An Austrian diplomat, commenting on the pejorative connotations of the word 'peasant' in English, told me that 'the British colleagues make fun of "peasant." They laugh and do impressions from a British TV show where one of the characters yells, "Out of my way, peasant!" We don't have that problem in other languages.'[20]

States' real or feigned preoccupation with the offensive connotations of 'peasant' marked a stark contrast with the way militants of LVC and other agrarian movements embraced the term as a key element of self-ascription and political identity. As I pointed out during one of the negotiating sessions:

> It's very important to acknowledge – and this should be abundantly evident to anybody who has spent the past few days in this room – that oppressed groups often appropriate the negative labels of the dominant groups, infuse them with new, positive meanings and wear them as a badge of pride. Young African Americans in my country, for example, often use the N-word and LGBT+ people celebrate the previously pejorative term 'queer.'[21]

In the third session of the Open-Ended Intergovernmental Working Group on a United Nations Declaration on the Rights of Peasants and Other People Working in Rural Areas (Working Group), Jorge Varela, a Honduran representative of the World Forum of Fisher Peoples pointed to the broad scope of the proposed definition and remarked, 'We artisanal fishers feel honoured to be included in the definition of campesinos.'[22] Perhaps the most emblematic instance of this inversion of the previously derogatory term 'peasant' occurred during the fifth session of the Working Group in 2018 when José Manuel Benítez Castaño, from the Spanish Coordination of Agriculturalists' and Livestock Producers' Organizations (Coordinadora de Organizaciones de Agricultores y Ganaderos, COAG), spoke on the proposed Article 1, in a voice brimming over with feeling:

> Mr. President, this Article 1 affects me emotionally in a special way, because I am a son, grandson and great-grandson of campesinos. Therefore I know

> well how it is to live as a peasant and I also know how this has been with the passage of time. My grandfather said with great pride that he was a campesino. He had some hilly land, a few goats and olive trees from which he got the oil for the whole year for the family. My father no longer said he was a campesino. He said he was an agriculturalist and cattleman. In his time, being a campesino was held in contempt, it was the worst thing, and that is why he didn't want me to be a campesino either. But over the years I insisted on being a campesino and working the land and with animals. I am very happy and enjoy this, because I am a lover of the land and my profession and because I also consider it a point of pride to produce quality food and in the quantity necessary to feed society. And of course, it is a very dignified profession, of which I do not have to be ashamed.[23]

This intervention constituted a powerful reassertion of both the contemporary relevance of peasant identity and the principle of self-ascription underlying the definition in Article 1.

Negotiating the definition

In preparing for the Working Group that would negotiate the Declaration text, the Mission of the Plurinational State of Bolivia, which chaired the negotiations, was acutely aware that the definition of the rights holders would be a contentious topic. In an effort to pre-empt at least some of this controversy, the Office of the High Commissioner for Human Rights invited me to be a panellist at the first Working Group session, where I was tasked with presenting a paper on definitions of 'peasant' and 'peasantries.'[24] For heuristic purposes – and to preemptively assuage the doubts or disarm the arguments of sceptical states – the paper examined four different kinds of definitions of 'peasant':

(1) Historical definitions, such as those from societies where peasants constituted an estate-like, caste-like, corporate or subordinated social group, characterised by specific restrictions on geographical or social mobility, limited rights, and obligations to provide services and perform particular deference behaviours for superordinate groups;
(2) Social scientific definitions from anthropology and sociology and from the interdisciplinary fields of peasant studies and agrarian studies;
(3) Activist definitions employed by agrarian movements, particularly LVC and its constituent organisations, that self-identify as 'peasants' (or 'campesinos,' etc.);
(4) Normative definitions, including those proposed by civil society organisations and by the Advisory Committee of the Human Rights Council.[25]

My intention in distinguishing between these kinds of definitions was to forestall any analytical muddling or slippage between types that might develop in states' perceptions or in the objections that they might raise. The paper argued that 'peasants' and 'farmers' – frequently posited as diametrically opposed categories in earlier social scientific literature – in fact existed along a continuum of technological mixes and economic aspirations, from mainly self-provisioning to exclusively entrepreneurial and commercial enterprises.[26] It also pointed out that the pejorative connotations of the word 'peasant' (and its equivalents in other languages) are indicative both of peasants' extreme subordination and of the ubiquitous elite practice of blaming peasants for a variety of economic and social ills – indeed of the very concerns that motivated supporters of the Declaration.[27]

My own and others' efforts to anticipate states' doubts about the proposed Declaration text were, at best, modestly successful. Following the first Working Group session, the Ambassador of Bolivia as Chair was asked to redraft the text and, in particular, to make it more in accordance with 'accepted language' in international law. The redrafting, in which Bolivia drew on informal consultations with states and on the advice of civil society and academic experts, was more extensive in relation to other articles of the Declaration than it was for Article 1.

Many delegations recognised early on the need for a comprehensive definition of the rights holders and some noted that civil society groups proudly identified with the term 'peasant.'[28] In 2014, for example, Brazil, then a part of the 'core group' supporting the Declaration and a key member of the GRULAC (Group of Latin American Countries), argued:

> It is important that the definition of rights holders be compatible with the range of circumstances affecting, in each and every country, the situation of peasants or other groups or persons living or working in rural areas. In Brazil, a variety of laws, policies and mechanisms are dedicated to the promotion and protection of the rights of family farmers, rural wage workers, seasonal workers, landless workers and beneficiaries of land reform, victims of slave labour, indigenous peoples, maroons and members of other traditional communities. Such diversity illustrates the importance of a comprehensive definition of rights holders, one which avoids gaps or mutually excluding categories.[29]

In the second Working Group session, a statement by a coalition of South Asian and other feminist organisations deployed the terms 'peasant' and 'farmer' largely interchangeably and called for a more explicit inclusion of women and Dalits in the definition of rights holders.[30] Both of these objectives were ultimately unsuccessful. While UNDROP Article 1 does allude to 'dependent family members of peasants,' its mentions of rural women are elsewhere, such as in Article 4. Indeed, some LVC participants in the negotiations later lamented that the final version of the UNDROP lacked the significant recognition of rural women and their rights that had been present in the 2009 LVC/2012 Advisory Committee drafts.[31] Notably, the phrase 'women and men' in the title of the 2009 LVC Declaration, inserted at the insistence of LVC's dynamic women activists, was absent in both the Advisory Committee's 2012 version and in the UNDROP, very likely a reflection of subtle and not-so-subtle patriarchal forces that periodically manoeuvred behind the scenes to excise statements related to rural women's rights and gender-based exploitation.[32]

In the third Working Group session, the US delegate noted that the word 'peasant' presented 'a linguistic challenge' and asked, 'Do all define themselves as peasants? We doubt that US farmers would consider themselves peasants.'[33] The Netherlands, in a similar vein and likely reflecting discussions within the EU, insisted that 'peasant' is generally taken as a derogatory term in English and that 'peasants' are presented as a homogeneous, single group with unitary interests.[34] South Africa, on the other hand, requested that the definition be expanded to include 'smallholder producers' and noted that the word 'peasant' is not used in its policies.

Guatemala had barely participated in the first three Working Group sessions, but in the fourth one it mounted a sustained attack on various aspects of the draft Declaration, including the definition in Article 1. 'We're putting everything in a single bag,' the Guatemalan delegate declared.

> We can't work with ambiguous and diffuse concepts. . . . In my country nearly all exports of non-traditional products are by campesinos, who aren't poor, who belong to cooperatives. . . . The only collective right that indigenous peoples have is to land. If we include indigenous people here, we're undermining their rights as peoples. Women as a collectivity don't have collective rights. . . . The definition goes beyond agriculture and now we're talking about industry, agroindustry.[35]

Switzerland took a different tack in the debate, calling for the inclusion of migrant and seasonal workers in the definition, a suggestion that ultimately became part of UNDROP Article 1 and that reflected frequent interventions from civil society organisations and rural movements arguing for greater labour rights protections and the right to social security for rural people. Ecuador, a member of the 'core group' supporting the Declaration, also sought to broaden Article 1, but its suggestion of incorporating Afro-descendants and Montubios was not adopted in the final text.[36] Brazil suggested widening the definition to include 'local communities,' a category widely used in earlier international agreements.[37] Both Switzerland and Ecuador argued for maintaining the phrase 'local communities' as a way of covering relevant groups of rights holders that were not specifically named.

Conclusion

The term 'peasants' in the title of the UNDROP and as a categorisation of rights holders proved to be controversial. At times, the contention was genuine and reflected real substantive disagreements over meanings and the scope of the definition, while at other times, the objections that states raised were thinly veiled efforts to jettison the negotiations and the entire UNDROP project. In general, states that were favourably disposed to the UNDROP favoured ample definitions and welcomed the inclusion of demands from the new sectors – nomads, pastoralists, fishers, and indigenous peoples – that were present in Geneva in increasing numbers in the last three Working Group sessions. States that opposed the UNDROP tended to push for more restrictive framings of the rights holders or even to deny entirely the validity of the 'peasant' category. Despite the inclusiveness of Article 1 and the presence of nomads, pastoralists, fishers, and indigenous peoples in the negotiations, the final UNDROP text, if not the definition itself, largely reflects the interests of small agriculturalists and rural workers. The debates around the UNDROP and the final text largely elided tensions between peasants and the agricultural workers they might employ, as well as those between peasants and pastoralists or peasants and indigenous peoples. Some vulnerable groups arguably part of the 'other people' category – such as Dalits and other caste-like groups in South Asia, North Africa, and elsewhere – did not ultimately receive explicit mention in the UNDROP's Article 1.

The rural movements – which intervened in the debates an average of 50 times during each one-week-long Working Group session[38] – were initiators and the central protagonists of the process that led to the UNDROP. In the end, they contributed mightily to authoring an international legal instrument that applied to them, an almost unprecedented occurrence in international law.[39] As the debate shifts to the implementation of the UNDROP, the ample definition of the rights holders in Article 1 takes on heightened significance. Most importantly, implementation must focus not only on peasants – however that is defined – but also on all those other categories of rural people who fought for and are named and recognised in the UNDROP, as well as for those who were not ultimately named.

A personal coda

Many years ago, in a book called *Peasants Against Globalization*, I expressed scepticism about overly operationalised definitions of 'peasants' and about how some foundational thinkers in agrarian studies – or peasant studies, as it was then called – prefaced every analysis with highly specific definitions of their object of study.[40] Rather than building an analysis on rigid definitions, it was, I suggested, more interesting to examine when, how, and for what purposes different political actors invoked and bounded the 'peasant' category.[41] The imperative of defining the rights holders in a draft human rights instrument disabused me of this 'ivory tower' academic pretension.

When I had the opportunity to present to the first Working Group session an in-depth paper on various approaches to defining 'peasants' and 'peasantry,' I took pains to ground the argument as much as possible in a substantial corpus of academic research from peasant studies and agrarian studies.[42] This was also a performative exercise inasmuch as the UN and the diplomats and functionaries who work in it have exaggerated regard for and defer to credentialed 'experts,' from within and without the UN system. As the negotiations evolved, however, the expertise of the peasants and other rural people who took part in the debates came increasingly into view. Their interventions about the occupational multiplicity and diverse economic activities and relations of rural households, for example, reinforced and gave credibility to my own analysis (and mine to theirs). While my inputs into the process derived from both extensive reading and first-hand fieldwork among peasants, mainly in the Americas, they did not come from having had to survive – as the peasants did – by farming or hunting in the morning, fishing in the afternoon, or rearing cattle in the evening, but only by criticising after dinner.

On 17 December 2018, I had the privilege of observing the UN General Assembly session that adopted the final version of the UNDROP. In some ways, the moment was anticlimactic. The General Assembly does not have space for civil society participation, as the Human Rights Council and some other UN bodies do. The peasant delegates who had lobbied in New York in the lead-up to the vote had already departed a week or earlier. I huddled in the General Assembly balcony with a handful of people: the Youth Vice-President of the National Farmers Union of Canada, who had hurriedly arranged a trip to New York, a Bolivian diplomat, a representative of the non-governmental organization Why Hunger, and a jurist colleague who had supported the campaign for a Declaration for many years. This was a small act of witnessing at the conclusion of a process that took 17 years of concerted struggle by hundreds of people, who represented millions more throughout the world. It was a minor celebration of a big victory and of a tool for the harder work that was yet to come.

Notes

1 John Patrick Leary, *Keywords – The New Language of Capitalism* (Haymarket Books 2019) 11.
2 Convention on the Rights of the Child (adopted 20 November 1989, entered into force 2 September 1990) 1577 UNTS 3 (CRC), art. 1.
3 Convention on the Rights of the Child (adopted 20 November 1989, entered into force 2 September 1990) 1577 UNTS 3 (CRC) <https://treaties.un.org/Pages/ViewDetails.aspx?src=IND&mtdsg_no=IV-11&chapter=4> accessed 7 June 2021; John Tobin (ed), *The UN Convention on the Rights of the Child: A Commentary* (Oxford University Press 2019).
4 UN Economic and Social Council, 'Study of the Problem of Discrimination against Indigenous Populations; Final Report (last part) submitted by the Special Rapporteur Mr. José R. Martínez Cobo' (30 September 1983) UN Doc E/CN.4/Sub.2/1983/21/Add.8.
5 UN General Assembly 'United Nations Declaration on the Rights of Indigenous Peoples' (2 October 2007) UN Doc A/RES/61/295; Jessie Hohmann and Marc Weller (eds), *The UN Declaration on the Rights of Indigenous Peoples: A Commentary* (Oxford University Press 2018); International Labour Organization (ILO), *Indigenous and Tribal Peoples Convention C169* (27 June 1989).
6 ILO (ed), *ILO Convention on Indigenous and Tribal Peoples, 1989 No 169: A Manual* (Revised edn, International Labour Organization 2003) 7.
7 La Via Campesina, 'Declaration of the Rights of Peasants – Women and Men' (2009).
8 UN General Assembly, 'United Nations Declaration on the Rights of Peasants and Other People Working in Rural Areas' (17 December 2018) UN Doc A/RES/73/165.
9 Marc Edelman, 'Linking the Rights of Peasants to the Right to Food in the United Nations' (2014) 10 *Law, Culture and the Humanities* 196.
10 Vincent Bevins, *The Jakarta Method: Washington's Anticommunist Crusade and the Mass Murder Program That Shaped Our World* (Public Affairs 2020); Helen-Louise Hunter, *Sukarno and the Indonesian Coup: The Untold Story* (Praeger Security International 2007); Geoffrey Robinson, *The Killing Season: A History of the Indonesian Massacres, 1965–66* (Princeton University Press 2018); Bradley Simpson, *Economists with Guns: Authoritarian Development and U.S.-Indonesian Relations, 1960–1968* (Stanford University Press 2008).

11 Anton Lucas and Carol Warren, 'The State, the People, and Their Mediators: The Struggle over Agrarian Law Reform in Post-New Order Indonesia' (2003) 76 *Indonesia* 87.
12 Konferensi Nasional Pembaruan Agraria, 'Deklarasi Hak-Hak Asasi Petani Indonesia' in Kelompok Studi Pembaruan Agraria (ed), *Meneguhkan Komitmen Mendorong Perubahan: Argumen-argumen dan Usulan Ketetapan MPR-RI tentang Pelaksanaan Pembaruan Agraria dan Pengelolaan Sumberdaya Alam yang Adil dan Berkelanjutan* (Kelompok Studi Pembaruan Agraria 2001); La Via Campesina, *Peasant rights-Droits paysans-Derechos campesinos* (2002).
13 UN Human Rights Council, 'Draft Declaration on the Rights of Peasants and other People Working in Rural Areas' (20 June 2013) UN Doc A/HRC/WG.15/1/2.
14 Neither phrase was present in the 2016 draft, which reflected revisions made by the Chair of the negotiation, the Ambassador of Bolivia, after the first two Working Group sessions. UN Human Rights Council, 'Draft declaration on the rights of peasants and other people working in rural areas' (8 March 2016) UN Doc A/HRC/WG.15/3/2, 2.
15 UN Human Rights Council, 'Draft Declaration on the Rights of Peasants and other People Working in Rural Areas' (6 March 2017) UN Doc A/HRC/WG.15/4/2, 10–1.
16 UN Human Rights Council, 'Final Study of the Human Rights Council Advisory Committee on the Advancement of the Rights of Peasants and other People Working in Rural Areas' (24 February 2012) UN Doc A/HRC/19/75; Coline Hubert, *The United Nations Declaration on the Rights of Peasants: A Tool in the Struggle for Our Common Future* (Centre Europe – Tiers Monde 2019) 38–40. Some of these movements had already elaborated their own rights declarations. For example, LIFE Network, 'Declaration on Livestock Keepers' Rights.'
17 Convention on the Elimination of All Forms of Discrimination against Women (adopted 18 December 1979, entered into force 3 September 1981) 1249 UNTS 13 (ICEDAW).
18 One of the first people to make this point was Konstantin Sergeyevich Aksakov, a prominent 19th-century Slavophile ideologist. Aleh Ivanou, Minsk, personal email communication with the author, 22 March 2018.
19 Raymond Williams, *Keywords: A Vocabulary of Culture and Society* (Oxford University Press 1985) 232.
20 Interview, 20 June 2014.
21 12 April 2018.
22 17 May 2016.
23 Benítez's intervention was on 12 April 2018. My translation from the original Spanish.
24 The official invitation came from the OHCHR, but only after the Bolivian Mission contacted me, apparently at the recommendation of FIAN and CETIM, two NGOs that were deeply involved in the negotiations.
25 Marc Edelman, 'What Is a Peasant? What Are Peasantries? A Briefing Paper on Issues of Definition' *First Session of the Intergovernmental Working Group on a United Nations Declaration on the Rights of Peasants and Other People Working in Rural Areas* (15 July 2013).
26 This argument drew significantly on Jan Douwe Van der Ploeg, *The New Peasantries: Struggles for Autonomy and Sustainability in an Era of Empire and Globalization* (Earthscan 2008).
27 On this, see Jim Handy, '"Almost Idiotic Wretchedness": A Long History of Blaming Peasants' (2009) 36 *Journal of Peasant Studies* 325.
28 For example, the intervention of the South African Mission in the second Working Group session, 3 February 2015.
29 Intervention of Carlos Eduardo da Cunha Oliveira, Counsellor, Permanent Mission of Brazil to the UN, in informal consultation of 12 November 2014. Personal email communication with the author, 24 November 2014.
30 PWESCR and SAFA, 'Statement for the Working Group on the Rights of Peasants and People Living in Rural Areas Second Session, 2nd – 6th February 2015.'
31 See Sandra Moreno Cadena, 'The Declaration and the Struggles of Peasant Women: Gender as an Unachieved Question,' in Priscilla Claeys and Marc Edelman, 'The United

Nations Declaration on the Rights of Peasants and other People Working in Rural Areas' (2020) 47 *The Journal of Peasant Studies* 1–68.

32 On women's growing participation in La Via Campesina, see Nettie Wiebe, 'Women of La Via Campesina: Creating and Occupying Our Rightful Spaces' (2013) *La Via Campesina's Open Book: Celebrating 20 Years of Struggle and Hope*. (See the chapter in this volume by Johana Bourke Martignoni and Priscilla Claeys).

33 18 May 2016.

34 17 May 2016.

35 15 May 2017.

36 Montubios – a coastal group of mixed African, European and indigenous ancestry – are one of more than two dozen 'nationalities and peoples' that the 'plurinational and intercultural' Ecuadorian state officially recognizes.

37 UN Human Rights Council, 'Draft Declaration on the Rights of Peasants and other People Working in Rural Areas,' supra note 15.

38 Hubert, *The United Nations Declaration on the Rights of Peasants: A Tool in the Struggle for Our Common Future*, supra note 16, at 38–9.

39 I deliberately say 'almost unprecedented,' because the processes that led to ICEDAW and UNDRIP also involved very substantial civil society input.

40 Teodor Shanin, *Defining Peasants: Essays Concerning Rural Societies, Expolary Economies, and Learning from Them in the Contemporary World* (Blackwell 1990); Eric R. Wolf, *Peasants* (Prentice-Hall 1966).

41 Marc Edelman, *Peasants Against Globalization: Rural Social Movements in Costa Rica* (Stanford University Press 1999) 188–93.

42 Edelman, 'What Is a Peasant? What Are Peasantries?,' supra note 25.

Selected bibliography

Claeys P and Edelman M, 'The United Nations Declaration on the Rights of Peasants and other People Working in Rural Areas' (2020) 47 *The Journal of Peasant Studies* 1–68.

Hubert C, *The United Nations Declaration on the Rights of Peasants: A Tool in the Struggle for Our Common Future* (Centre Europe – Tiers Monde 2019)

Shanin T, *Defining Peasants: Essays Concerning Rural Societies, Expolary Economies, and Learning from Them in the Contemporary World* (Blackwell 1990).

Van der Ploeg JD, *The New Peasantries: Struggles for Autonomy and Sustainability in an Era of Empire and Globalization* (Earthscan 2008).

Wolf ER, *Peasants* (Prentice-Hall 1966).

3

INDIGENOUS PEOPLES AND TRADITIONAL LOCAL COMMUNITIES IN THE UNDROP

Synergies and challenges

Adriana Bessa and Jérémie Gilbert

Introduction

The United Nations Declaration on the Rights of Indigenous Peoples (UNDRIP) was adopted in 2007 with the aim of guaranteeing that the individual and collective rights of indigenous peoples were better respected, protected, and fulfilled across the globe. A key element of the UNDRIP concerns the recognition of indigenous peoples' collective rights to land and natural resources. It represented an important rupture with the classic interpretation of human rights as exclusively individual rights. A similarly significant contribution was included in the United Nations (UN) Declaration on the Rights of Peasants and Other People Working in Rural Areas (UNDROP or Declaration). Partially building upon principles upheld in the UNDRIP, the UNDROP affirms the collective rights of other rural groups to natural resources, including land, seeds, and biodiversity. In this sense, it not only sheds important light on the collective dimension of these rights but also expands the list of rights holders.

The UNDROP also reaffirms the notion of traditional local communities as rights bearers in international human rights law. The notion of traditional local communities emerged in the documents adopted during the 1992 UN Conference on the Environment and Development (UNCED), in which they were depicted, together with indigenous peoples, as 'custodians of biodiversity.' As part of a global environmental agenda, the UNCED encouraged states to adopt measures to protect their traditional knowledge and ecological practices since these were fundamental to further strategies for nature conservation and sustainable use.

The UNDROP establishes clear connections to the rights of indigenous peoples and traditional local communities, opening up for both synergies and potential challenges to the nature of this relationship. This represents a significant opening as it integrates the rights of indigenous peoples and traditional local communities, which represents a very significant and rich body of norms, decisions, and jurisprudence relevant to these groups.

DOI: 10.4324/9781003139874-5

The chapter discusses the interactions between the indigenous peoples' and traditional local communities' rights as prescribed in the UNDRIP and environmental legal instruments and the UNDROP and reflects upon the categorisation of these groups as peasants and the legal foundation of their rights under human rights law. It also explores relevant synergies and complementarities between the UNDRIP and the UNDROP and discusses apparent dichotomies and challenges with the aim of contributing to a better understanding and protection of the rights of indigenous peoples and traditional local communities.

Indigenous peoples and the UNDROP: identity, politics, and controversies

Although there are only four mentions of indigenous peoples in the text of the UNDROP,[1] the influence of the indigenous movement has been very significant to the document.[2] During the negotiations that finally led to the adoption of the Declaration, there were some debates on the inclusion of indigenous peoples as a specific category of peasants. There were different reasons for such a debate. Although many indigenous peoples are living and working in cities,[3] there are no doubts that there are some significant numbers of rural indigenous communities that are directly engaged in agricultural practices.[4] In many countries, differentiating between some indigenous communities, peasants and small-scale farmers is often very complex.[5] For example, as noted by Toledo: 'Large numbers of indigenous peoples are, however, peasant producers and therefore can be indistinguishable from the non-indigenous peoples living nearby. In the Andean and Mesoamerican countries of Latin America, for instance, indigenous peoples farm like mestizo peasants.'[6] As such, many indigenous communities are directly affected by policies on agriculture and rural development; hence, the UNDROP is clearly an important instrument in such contexts.

The debate was not on whether indigenous can be peasants and *vice versa* but on whether there could be clashes between the rights of peasants and the rights of indigenous peoples. The issue can be particularly acute in certain countries where there have been some serious social tensions around land and agrarian reforms centred around identitarian conflicts between peasants and indigenous communities. For example, analysing the situation in Bolivia, Lorenza Belinda Fontana highlights that 'the so-called "politics of recognition" triggered processes of political ethnicization and organizational fragmentation, eventually contributing to fuelling new tensions between indigenous groups and peasant unions.'[7]

These concerns were echoed in the corridors of Geneva where the negotiations of the UNDROP took place. There was also wariness amongst the indigenous movement, notably the representatives and advocates of the movement within the UN circles, that the UNDROP could tone-down and undermine the painstaking progresses that have been achieved by indigenous peoples to support the development of the UNDRIP. Behind the debates on whether or not to include indigenous peoples was also the spectrum of the controversies on the recognition of collective rights and their impact on individual human rights, an issue that has haunted the human rights movement for a long time.[8] From this perspective, the UNDROP has benefited from decades of advocacy from indigenous peoples on collective rights, building on the UNDRIP and supportive legal interpretations from UN Treaty Bodies affirming the fundamental importance of collective rights for the enjoyment of individual human rights.[9]

Ultimately, the result is based on a progressive and pragmatic approach where indigenous peasants are specially mentioned. The key is that indigenous peoples are integrated and recognised in the UNDROP, opening avenues for synergies and complementarities between the different instruments. Importantly, the UNDROP also affirms that respect and protection of indigenous rights are important and a guiding principle in case of potential clashes between peasants and indigenous communities (see Articles 2 and 28).[10]

Traditional local communities and the UNDROP: a gateway for visibility in human rights debates

The notion of traditional local communities has been much less articulated in international fora as compared with indigenous peoples. Despite earlier references, the notion of traditional local communities or 'local communities embodying traditional lifestyles' only became popular upon the adoption of the multilateral agreements and documents during the UNCED, in 1992, especially the Rio Declaration on Environment and Development (Rio Declaration) and the Convention on Biological Diversity (CBD). These instruments made a strong case regarding communities' traditional lifestyles in tune with nature conservation strategies, and their role as traditional knowledge creators and custodians.[11] Principle 22 of the Rio Declaration called upon states to 'recognize and duly support their identity, culture and interests and enable their effective participation in the achievement of sustainable development.'

Moreover, two CBD provisions ascribed rights to indigenous and traditional local communities, which was so far unusual in environmental legal instruments. Article 8j of the CBD demanded states enact national legislation to protect and preserve communities' traditional knowledge and to regulate the equitable sharing of the benefits arising from their commercial and non-commercial utilisation. Article 10c of the CBD determined that communities' 'customary use of biological resources in accordance with traditional cultural practices' should be protected and encouraged. Following the CBD, a number of environmental legal instruments reaffirmed and expanded communities' rights. In addition to the rights to traditional knowledge and benefit sharing, reaffirmed in virtually all recent environmental agreements and UN documents,[12] the rights to participation in decision-making[13] and preferential access to natural resources subject to conservation measures have also been recognised.[14] In 2010, the Nagoya Protocol on Access to Genetic Resources and the Fair and Equitable Sharing of Benefits Arising from their Utilization to the CBD (Nagoya Protocol) affirmed communities' right to free, prior and informed consent (FPIC), or 'approval and involvement' for any access to and utilisation of their traditional knowledge and genetic resources found on their land or over which they exercise formal or customary rights.[15]

From a human rights perspective, however, traditional local communities remained invisible and in a legal limbo. This was largely due to the lack of concrete and clear elements which could help with their identification. In 2015, Adriana Bessa proposed a working definition of traditional local communities, based on elements harvested in international legal instruments and UN documents. As she proposed, the term applies to

> Communities engaged in small-scale rural activities, for economic, social and cultural purposes. They are characterized by long-lasting relations of solidarity among community members, the common use of land and other natural resources, and the [creation and] custody of traditional knowledge strongly connected with and embedded in community values and traditions, including religious beliefs, and crucial for local biodiversity conservation and sustainable management strategies. These communities do not necessarily belong to any particular ethnic, religious or linguistic minority in the countries where they reside – although in many occasions this is the case – and do not easily fit into the strict legal categories of minorities and indigenous peoples, as defined in international law and academic literature.[16]

The definition of peasants offered in Article 1 of the UNDROP has come to consolidate both the notion of traditional local communities and indigenous peoples living and working in rural settings, by underlining their common condition as peasants. It does place an important emphasis on a few elements that are fundamental to understanding the foundations of the human rights entrenched in the UNDROP's norms, and the linkages between rights to land, seeds, and biodiversity, just to name a few, and communities' survival and dignity.

First and foremost, the UNDROP underlines the fact that peasants are engaged in non-industrial and small-scale production based on the use of natural resources and rely on this production for their subsistence in short and long terms. As a result, they are extremely vulnerable to industry-based development projects, environmental disasters, or any other event that might menace the integrity of resources they depend on. Another important definitional element is the use of household labour and other non-monetised ways of organising labour that strengthens social solidarity, community identity, and socio-economic resilience.[17]

More specific elements, which are equally important to the discussion developed in this chapter, are related to the transhumant, semi-nomadic, and nomadic ways of production and lifestyles. This applies especially to indigenous and non-indigenous pastoralists and hunter-gatherer communities but may easily incorporate shifting agriculture and nomadic fishery activities.[18] Likewise, the UNDROP expands its protective scope so as to include the landless engaged in rural activities as right bearers. These groups and communities engage in informal (and all too often ecologically sustainable) agriculture and collection of forest products in private and public lands. As discussed later, the UNDROP encourages states to adopt innovative measures, including flexible forms of land tenure, possession, and ownership so as to guarantee that individual and collective human rights of these diverse rural actors and their communities are protected.

Dichotomy and debates on the distinction on the rights of indigenous and non-indigenous peasants

The rights of indigenous peoples have been construed in very firm and particular bases in international law and politics.[19] The recognition of the status as 'peoples' entails the recognition of the right to self-determination, as prescribed in the first article of the two UN human rights covenants: the ICCPR and the International Covenant on Economic Social and Cultural Rights (IESCR). This provision has been partially replicated in Article 3 of the UNDRIP and reinterpreted in Article 4 so as to better reflect the indigenous reality. While the former affirms that 'indigenous peoples have the right to self-determination' and 'by virtue of that right they [may] freely determine their political status and freely pursue their economic, social and cultural development,' the latter connects the right to self-determination to the 'right to autonomy or self-government in matters relating to their internal and local affairs.' This clearly includes the governance of land and natural resources in their territories.

This rendering of the rights of indigenous peoples as based on the right to self-determination has virtually entailed 'an ironclad barrier [that] prevent[s] other minorities from walking through [the same] doors that indigenous peoples have opened.'[20] There has been a strong push in the last decades, particularly after the adoption of the UNDRIP, to broaden the notion of indigenous peoples so as to encompass vulnerable groups that would not automatically fit the category. The case law of the Inter-American System of Human Rights, for instance, has extended the application of its jurisprudence on indigenous rights to land and natural resources to rural communities of African descent residing in the continent.[21] The same has been observed outside the Americas, where humanitarian concerns have motivated the broadening of the understanding of indigenous peoples and indigeneity to provide vulnerable groups, all too often rural communities, with the same powerful legal shelter.[22]

Having said that, the stretching of the notion of indigenous peoples has proved to be no silver bullet. The application of indigenous rights to non-indigenous peasants has had very marked limits.[23] The first one relates to the different political agendas. Indigenous rights are ultimately about self-governance. As has been rightly put, indigenous peoples do not merely want to be governed differently, rather they want to govern themselves.[24] It is the very authority of the state that is being questioned. Therefore, indigenous rights are not only about improving participation in decision-making processes and democratic governance – the main claim of non-indigenous peasants and the central motto of the UNDROP – but who exercises the power.[25] As a result, indigenous rights to land and natural resources are not simply about property rights or rights to access and use resources for subsistence or livelihoods. It is part and parcel of a much broader territorial claim.

Indeed, recent narratives involving non-indigenous peasants have been mostly based on demands for respect and protection of the diversity of socio-economic systems and communities' ecological culture within the larger society and measures to rectify the maldistribution of national wealth and distributive justice, which has been endorsed in the Rio Declaration, the CBD and its Protocols, and more recently the UNDROP. Indigenous peoples, instead, demand the restoration of their original entitlements and the redress of past injustices deriving from colonialism.[26] In this sense, their demands go beyond the recognition of socio-economic and cultural pluralism within society and agrarian reform. They demand sovereignty and non-interference in their systems of polity, law, and economy.[27]

Second, the notion of indigeneity remains linked to a particular ethnic ancestry. In this sense, the identification of who is indigenous in places and contexts where there is no marked ethnic diversity or temporal differences in land occupation is objectively difficult. The same holds true for communities of mixed heritage origins and that do not particularly identify with any ethnic or linguistic groups.

In fact, ethnic and linguistic affiliation still plays a key role in the application of indigenous rights in the practice of the human rights courts and treaty bodies. On a number of occasions, the Inter-American System of Human Rights has marked clear boundaries on the application of indigenous rights for non-indigenous peasants, especially the recognition of collective rights to property, despite the analogous degree of victims' vulnerability and the similar nature of human rights abuses.[28] Within the same lines, in a case concerning the subsistence fishing rights of artisanal fishery communities in the Chagos Archipelago in Mauritius, the European Court of Human Rights dismissed the claims of international non-governmental organisations on the applicability of indigenous rights.[29]

Likewise, the Human Rights Committee (HRC) has found that the peasant communities' exercise of small-scale and subsistence-oriented economic practices alone would not fall under the protective shelter of Article 27 of the ICCPR. This article, which provides for minority rights,[30] has been evoked in a number of cases involving indigenous peoples and competing claims over natural resources and became one of the main human rights norms to protect indigenous land rights.[31] The HRC has consolidated the understanding that 'economic activities may come within the ambit of Article 27, if they are an essential element of the culture of an ethnic community.'[32] In a case concerning commune pastoral land rights of a mixed heritage community menaced by expropriation, however, the Committee reasoned that the connection between traditional economic practices and culture remained unclear.[33]

At the same time, indigenous and non-indigenous peasants share a number of features and various challenges. The connexion of communities with the environmental agenda is the first of these. The UNDRIP has affirmed the linkage of indigenous peoples in rural settings with nature preservation in the preamble:

'Recognizing that respect for indigenous knowledge, cultures and traditional practices contributes to sustainable and equitable development and proper management of the environment.' The UNDROP has followed the same lines by recognising 'the special relationship and interaction between peasants and other people working in rural areas and the land, water and nature' and their contributions 'to conserving and improving biodiversity' (Preamble).

Second, the tensions with indigenous and non-indigenous peasants and their communities with Western concepts of development and the productivist paradigm are more and more frequent. In this sense, indigenous peoples, traditional local communities, peasants, and other rural dwellers living on subsistence and small-scale economies are usually depicted as needy and poor.[34] This idea of poverty tends to focus on communities' (poor) economic capital while overlooking their (rich) social and natural capital, which makes them resilient to economic crisis. As has been emphasised, indigenous communities alone 'live, own and occupy approximately one quarter of the world's lands and waters which represents 80% of the world's biodiversity.'[35]

This portrayal of indigenous peoples and local communities reinforces stereotypes reproducing the view of rural actors as ignorant and unable to make the best use of local natural resources. They reinforce the marginalisation of communities, their lifestyles, and modes of production and justify the usurpation of their resources and the sacrifice of their ways of life for the sake of economic development and growth.[36] The UNDROP's ultimate purpose is exactly this: to fight stereotypes, halting the systemic discrimination against indigenous and non-indigenous peasants and their communities and rebalancing power relations in and beyond rural contexts by, among other things, redefining the notion of development through a human rights-based approach.

UNDROP and UNDRIP: synergies and complementarities

There are many parallels between the UNDROP and the UNDRIP. At the core of these two UN declarations is the recognition of the fundamental importance of the protection of collective rights and the adoption of a human rights-based approach to the governance of lands and natural resources. Crucially both instruments highlight the importance of recognising customary and communal land tenure systems. From this perspective, the UNDROP puts similar emphasis on collective rights over land and natural resources as a self-standing human right[37] and an element of the rights to subsistence, an adequate standard of living and culture. One extra element added in the UNDROP is the specific focus on collective use and management systems and the natural commons (see Article 17.3), which were not specifically integrated into the UNDRIP.

Likewise, the UNDROP goes one step further as compared to the UNDRIP on the relationship between the land tenure, traditional economic and cultural (including but not restricted to religious) practices, and sustainability. Although the UNDRIP makes connections between indigenous peoples' rights and environmental sustainability, the UNDROP goes much further and adds new elements to this relationship. It recognises and proclaims the importance of concepts that so far has been practically exclusive to indigenous peoples, especially 'harmony with nature, also referred to as Mother Earth' (Preamble). Nevertheless, it adds an extra layer to sustainability by offering new approaches and synergies between peasants and indigenous traditional use of land and natural resources and traditional knowledge. This focus does not come out of nowhere in terms of international law making, as many indigenous communities have put forward the importance of traditional ecological knowledge and indigenous knowledge in the development of international biodiversity law. But the focus on ancestral knowledge of agricultural production, which is an essential focus of the UNDROP, brings a new legal approach to the rights of indigenous peasants and brings it closer to non-indigenous peasants, including traditional local communities. The UNDROP integrates a strong focus on intergenerational knowledge, practices, and beliefs passed from generation to generation by cultural transmission (see Article 20 on traditional knowledge), a clear incorporation of recent environmental law instruments which was not so explicit in previous indigenous rights specific instruments.

Moreover, the UNDROP's recognition of the human right to land and natural resources as different from and independent of the human right to property brings new light to the land rights of indigenous and non-indigenous peasants in international human rights law. As it has been rightly put, conceptually, land rights

> encompass all forms of tenure arrangement, among which there is the most complete 'bundle of rights' that is ownership (which includes the right to use, to dispose of, and to enjoy the fruits of land), together with minor rights, such as use and access rights.[38]

The ownership aspect of the right to property presupposes exclusive enjoyment of the land and resources and consequently the exclusion of others from access and use, even if for subsistence and humanitarian purposes. This sharply counters the principle of solidarity, one of the key foundations of human rights.[39] The UNDROP's Article 17 on the right to land and natural resources clearly champions the accommodation of diverse needs as well as collective arrangements, consuetudinary management regimes with the aim to guarantee all rural dwellers 'an adequate standard of living, . . . a place to live in security, peace and dignity and to develop their cultures.' This entails inclusive and participatory management systems, fluid boundaries between communities, and overlapping and compatible uses of land, i.e., seasonal farming, forest products collection, and pastoralism.

Within the same track, the UNDROP contributes to the progressive development of the notion of impact assessment studies. Not included in the UNDRIP but consolidated in the case law of international human rights bodies involving indigenous peoples, the UNDROP determined that a 'duly conducted social and environmental impact assessment' must be carried out prior to any exploitation affecting peasants' natural resources (Article 5.2(a)). In our view, it goes even further so as to suggest the need to incorporate studies on impacts of projects on peasants' culture and cultural heritage. By articulating the interlinkages between social interactions, economic activities and dynamics, ecosystems and natural cycles, and the cultural traits and heritage of communities, the UNDROP suggests that these four pillars – social, economic, environmental, and cultural – are inextricably related and any interferences in one of them necessarily affect the other.

Indeed, the UNDROP supports a much more embedded and clear focus on the right to a healthy environment as a human right with express mention of climate change (Article 18). This clear connection between human rights and environmental degradation is much more developed in the UNDROP than it was in the UNDRIP. There is a much more specific emphasis on the importance of ecological concerns when it comes to land, including the preservation of biodiversity, as well as the conservation, preservation, protection, and restoration of land to maintain its agroecological services. The same is true with respect to other natural resources. Let us take the right to seeds for example. Whereas in the UNDRIP focus is placed on the indigenous seeds as part and parcel of indigenous cultural heritage (Article 31), the UNDROP emphasises the need to preserve and enhance seed diversity (Article 19.6) and increase states' investments in research and development of orphan crops. Fundamental for food security at local and regional levels, these are highly neglected by international research institutions.

The UNDROP integrates lots of new focus on sustainable modes of agricultural productions, with agroecology being very central.[40] This is new in terms of indigenous rights as even though international human rights bodies are being pushed to integrate indigenous perspectives on agriculture, this was not done so under an explicit banner of human rights approach to agroecology. As noted by Suárez-Torres et al: 'Agroecology and ancestral knowledge are being adopted by a growing number of organisations, indigenous peoples and social groups in various parts of the world, as development alternatives that respond to local needs and worldviews.'[41] In many ways, the UNDROP makes way for more clear connections between the two movements, agroecology and indigenous, and for a new human rights-based approach to anchor this relationship.

Overall, the rights proclaimed in the UNDROP offer a solid and useful complementarity with existing indigenous peoples' rights while at the same enlarging the scope of many of these rights so as to encompass non-indigenous peasants and their communities.

UNDROP and UNDRIP: dichotomies and challenges

The major setback in terms of indigenous rights in the UNDROP is the lack of reference to the right of self-determination. Although the UNDROP supports the call for food sovereignty (see Article 15.4),[42] it does not include or refer to self-determination. Self-determination was initially integrated into the draft of the text as part of its preamble but for political reasons disappeared during the negotiations. As discussed earlier, the right to self-determination is at the core of the indigenous rights movement and its inclusion in the UNDRIP explains the long 22 years of negotiation, as the human right to self-determination is arguably the most politically charged norm in international human rights law.

The other significant setback concerns the right to FPIC. The FPIC has been highly debated and finally excluded from the final draft of the UNDROP due to the political and diplomatic tractions that took place during the negotiations. This led to a compromise, with the UNDROP's threshold as regarding focus on consultation rights much lower (see Article 5.2 referring to consultations in good faith). Although this should not affect the rights of indigenous peasants whose right to FPIC is guaranteed in the UNDRIP,[43] the exclusion of FPIC appears to question the application of this guarantee to non-indigenous peasants.

However, there has been a slow but steady development of the recognition of FPIC as a procedural guarantee in the events which might affect indigenous peoples' and traditional local communities' lives and livelihoods. Upon the adoption of the Nagoya Protocol, for instance, FPIC became mandatory for all contracting parties in cases of access to communities' genetic resources and associated traditional knowledge (Articles 6.2 and 7). The practice of regional human rights monitoring bodies has amplified the scope of FPIC so as to include land and natural resources concessions and exploitation in the lands of indigenous peoples and ethnic rural groups.[44] In this sense, the Inter-American Court of Human Rights has clarified that the rights of indigenous peoples to 'effective participation' are not limited to consultation but actually entail the right to FPIC in the event of any activities that might affect their lives and livelihoods.[45] Similarly, the African Commission on Human and Peoples' Rights in its 2012 resolution on a Human Rights-Based Approach to Natural Resources Governance stated: 'all necessary measures must be taken by the State to ensure participation, including the free, prior and informed consent of communities, in decision making related to natural resources governance.'[46]

Our last comment on the relationship between the UNDRIP and the UNDROP is not on the content of the rights, but more about the future challenges in terms of supporting the implementation of the UNDROP. The two declarations suffer from similar potential complexities when it comes to implementation for they are non-binding instruments.[47] In such a context, the UNDRIP has proven to be extremely resilient as a 'soft law' instrument. This is notably due to the fact that indigenous representatives have secured unprecedented influence in the structure and agenda of international organisations, generating institutions such as the UN Permanent Forum on Indigenous Issues (2002), the Expert Mechanisms, and many more UN structures and programmes that support the implementation of the UNDRIP. This includes specific UN human rights mechanisms, such as the Special Rapporteur on the rights of indigenous peoples. Likewise, indigenous lobbying has been quite successful in integrating the UNDRIP in intergovernmental institutions' work programmes, including those of the World Bank and regional organisations such as the Organization of American States and the European Union. It remains to be seen whether the UNDROP will benefit from similar institutional mechanisms and monitoring bodies.

Conclusion

The adoption of the UNDROP is doubtlessly a watershed in terms of the legal protection of a number of indigenous peoples' and traditional local communities' rights, especially the rights to land and natural resources. First, it expands and clarifies the contents of indigenous rights by factoring in a number of aspects related to agricultural production and sustainability. As for traditional local communities, it not only brings up valuable elements for their identification – and consequently legal protection – but also paves the way for the incorporation of new forms of land use, management, and tenure as well as the protection of collective rights in international and national law-making and public policies.

The UNDROP can definitely create positive synergies with extant indigenous rights, especially the UNDRIP, acting in a complementary manner. This new declaration is also expected to illuminate the practice of international human rights courts and monitoring bodies as regards non-indigenous peasants and their vulnerabilities. A few challenges may be envisaged, however, especially with respect to the institutional apparatus within UN agencies for the implementation of the UNDROP which, at the time being, still appears to be incipient.

Notes

1 UN General Assembly, United Nations Declaration on the Rights of Indigenous Peoples (2 October 2007) UN Doc A/RES/61/295, Preamble (reaffirming commitments to UNDRIP), art. 1 (integrating Indigenous Peoples in the definition) and artt. 2 and 28.
2 As noted by Edelman and James, the indigenous rights model was an important element for framing the campaign on peasants' rights. See Marc Edelman and Carwil James, 'Peasants' Rights and the UN System: Quixotic Struggle? Or Emancipatory Idea Whose Time Has Come?' (2011) 38(1) *The Journal of Peasant Studies* 81–108.
3 See Magdalena Ugarte, Mauro Fontana and Matthew Caulkins, 'Urbanisation and Indigenous Dispossession: Rethinking the Spatio-legal Imaginary in Chile vis-à-vis the Mapuche Nation' (2019) 9(2) *Settler Colonial Studies* 187–206; Philipp Horn, 'Emerging Urban Indigenous Spaces in Bolivia: A Combined Planetary and Postcolonial Perspective' in Philipp Horn, Paola Alfaro d'Alencon, Ana Claudia Duarte Cardoso (eds), *Emerging Urban Spaces* (Springer 2018) 43–64; Libby Porter, 'Indigenous Cities' in Dallas Rogers, Adrienne Keane, Tooran Alizadeh and Jacqueline Nelson (eds), *Understanding Urbanism* (Palgrave Macmillan 2020) 15–26.
4 For analysis, see Rodolfo Stavenhagen, *Peasants, Culture and Indigenous Peoples* (Springer 2012).
5 See definitions of 'native indigenous peasant peoples' in Bolivia for example, Andrew Canessa, 'Conflict, Claim and Contradiction in the New "Indigenous" State of Bolivia' (2014) 34(2) *Critique of Anthropology* 153–73; Jeffery R. Webber, *Red October: Left-indigenous Struggles in Modern Bolivia* (Brill 2011).
6 Victor M. Toledo, 'Indigenous Peoples and Biodiversity' (2001) 3 *Encyclopedia of biodiversity* 451–63.
7 Lorenza Belinda Fontana, 'Indigenous Peoples vs Peasant Unions: Land Conflicts and Rural Movements in Plurinational Bolivia' (2014) 41(3) *Journal of Peasant Studies* 297–319.
8 For analysis, see Analia Penchaszadeh, *The UN Declaration on the Rights of Peasants as a Tool for Promoting Collective Rights* (Friends of the Earth Europe and Friends of the Earth International 2020).

9 Several indigenous advocates made supportive statements during the negotiations on these issues, including representatives from the World Alliance of Mobile Indigenous Peoples (WAMIP) and the International Indian Treaty Council (IITC). See Saúl Vicente Vázquez, 'Collective Rights in Theory and Practice' in Claeys, Priscilla and Marc Edelman (eds), 'The United Nations Declaration on the Rights of Peasants and other People Working in Rural Areas' (2020) 47(1) *The Journal of Peasant Studies* 1–68.

10 See UN Human Rights Council, United Nations Declaration on the Rights of Peasants and Other People Working in Rural Areas (8 October 2018) UN Doc A/HRC/RES/39/12, art. 2.3: 'Without disregarding specific legislation on indigenous peoples, before adopting and implementing legislation and policies, international agreements and other decision-making processes that may affect the rights of peasants and other people working in rural areas, States shall consult and cooperate in good faith with peasants and other people working in rural areas through their own representative institutions, engaging with and seeking the support of peasants and other people working in rural areas who could be affected by decisions before those decisions are made, and responding to their contributions, taking into consideration existing power imbalances between different parties and ensuring active, free, effective, meaningful and informed participation of individuals and groups in associated decision-making processes.'

11 Convention on Biological Diversity (adopted 5 June 1992, entered into force on 29 December 1993) 1760 UNTS 69 (CBD), preamble and art. 8(j). Article 8(j) reads: 'Each contracting Party shall, as far as possible and as appropriate: . . . Subject to national legislation, respect, preserve and maintain knowledge, innovations and practices of indigenous and *local communities embodying traditional lifestyles* relevant for the conservation and sustainable use of biological diversity and promote their wider application with the approval and involvement of the holders of such knowledge, innovations and practices and encourage the equitable sharing of the benefits arising from the utilization of such knowledge innovations and practices' (our emphasis). Also see UN Human Rights Council, 'Report of the Special Rapporteur on the issue of human rights obligations relating to the enjoyment of a safe, clean, healthy and sustainable environment' (24 January 2018) UN Doc A/HRC/37/59, Framework Principle 15.

12 See, for instance, International Treaty on Plant Genetic Resources for Food and Agriculture (adopted 3 November 2001, entered into force on 29 June 2004) 2400 UNTS 303 (ITPGRFA), art. 9(a).

13 See, for example, the United Nations Convention to Combat Desertification in Those Countries Experiencing Serious Drought and/or Desertification, Particularly in Africa (adopted 17 June 1994, entered into force on December 1996) (UNCCD), art. 10.2(f) and the ITPGRFA, supra note 12, art. 9(c).

14 See International Tropical Timber Agreement (adopted 26 January 1994, entered into force on 1 January 1997 to 6 December 2011 and its 2006 version adopted by UN Conference for the Negotiation of a Successor Agreement to the International Tropical Timber Agreement on 27 January 2006 and entered into force on 7 December 2011), art. 1(j); the United Nations Agreement for the Implementation of the Provisions of the United Nations Convention on the Law of the Sea of 10 December 1982 Relating to the Conservation and Management of Straddling Fish Stocks and Highly Migratory Fish Stocks (adopted 4 December 1995, entered into force 11 December 2001) artt. 5(i) and 25.2(b).

15 Nagoya Protocol on Access to Genetic Resources and the Fair and Equitable Sharing of Benefits Arising from their Utilization to the Convention on Biological Diversity (adopted 29 October 2010, entered into force 12 October 2014) 30619 UNTS 27, artt. 7 and 5.2, respectively.

16 Adriana Bessa, 'Comunidades Locales Tradicionales y Bienes Comunes de Derecho Internacional: Debate sobre la Nueva Declaración de las Naciones Unidas sobrelos Derechos de los Campesinos y otras Personas que trabajanen Zonas Rurales' in Mercedes Ortiz García (ed), *Políticas Públicas para Fomentar una Economía y Sostenibilidad Participativa* (Tirant lo Blanch 2020) 107ff, 113 (our translation). For an appraisal of the notion

of traditional local communities in international law, see Adriana Bessa, *Traditional Local Communities in International Law* (EUI PhD thesis 2013), and Adriana Bessa, 'Traditional Local Communities: What Lessons Can Be Learnt at the International Level from the Experiences of Brazil and Scotland?' (2015) 24(3) *Review of European, Comparative & International Environmental Law* 330–40.

17 On this point, see Braden Leap and Diego Thompson, 'Social Solidarity, Collective Identity, Resilient Communities: Two Case Studies from the Rural U.S. and Uruguay' (2018) 7 *Social Sciences* 250.

18 For an appraisal of the challenges faced by nomadic fisher folks, see Eric Komlavi Hahonou, 'Pêcheurs nomades du fleuve Niger: D'une anthropologiepartagée à une anthropologie de la citoyenneté' (2017) 87(1–2) *Journal des Africanistes* 130–52; more generally for nomadic and transhumant communities, see Jérémie Gilbert, *Nomadic Peoples and Human Rights* (Routledge 2014).

19 On this point, see, for instance, Benedict Kingsbury, 'Indigenous Peoples' in International Law: A Constructivist Approach to the Asian Controversy' (1998) 92(3) *The American Journal of International Law* 414–57.

20 Will Kymlicka, 'Beyond the Indigenous/Minority Dichotomy?' in Stephen Allen and Alexandra Xanthaki (eds), *Reflections on the UN Declaration on the Rights of Indigenous Peoples* (Hart Publishing 2011) 187.

21 *Case of the Moiwana Community vs Suriname, Interpretation of the Judgment of Merits, Reparations and Costs* Inter-American Court of Human Rights Series C No 145 (15 June 2005) and *Saramaka People v Suriname Preliminary Objections, Merits, Reparations, and Costs* Inter-American Court of Human Rights Series C No 172 (28 November 2007).

22 Kymlicka, 'Beyond the Indigenous/Minority Dichotomy?' supra note 20.

23 For a detailed appraisal of the limits to the application of indigenous rights to non-indigenous rural communities in general and traditional local communities in particular, see Bessa, *Traditional Local Communities in International Law,* supra note 16.

24 Kymlicka, 'Beyond the Indigenous/Minority Dichotomy?,' supra note 20.

25 Ibid., at 189.

26 See Samson, Colin, *The Colonialism of Human Rights: Ongoing Hypocrisies of Western Liberalism* (John Wiley & Sons 2020).

27 On this point, see Jeremy Waldron, 'Indigeneity? First Peoples and Last Occupancy New Zealand' (2003) 1 *Journal of Public and International Law* 55ff. For a comparison between indigenous and minority rights, see UN Commission on Human Rights, 'Working paper on the relationship and distinction between the rights of persons belonging to minorities and those of indigenous peoples, Sub-Commission on the Promotion and Protection of Human Rights (fifty-second session), special rapporteurs Asbjorn Eide and Erica-Irene Daes' (19 July 2000) UN Doc E/CN.4/Sub.2/2000/10, para. 8.

28 See *Chengue Massacre (Colombia), Petition 1268–05, Admissibility, Report No. 45/07* Inter-American Court of Human Rights (July 23 2007); *Case of the Plan de Sánchez Massacre v. Guatemala, Reparations* Inter-American Court of Human Rights Series C No 116 (November 19 2004); *Case of the Mapiripán Massacre v. Colombia. Merits, Reparations and Costs.* Inter-American Court of Human Rights Series C No 134 (15 September 2005). More recently, clearly influenced by recent normative developments and the adoption of the UNDROP, the Court appears to be more sensitive by the case of non-indigenous subsistence peasants. As it has stated: 'The Court cannot ignore that the State has obligations towards the criollo population [non-indigenous peasants], because, given their vulnerable situation, the State must take positive steps to ensure their rights.' (*Case of the Indigenous Communities of the Lhaka Honhat Association (Our Land) v. Argentina. Merits, Reparations and Costs* Inter-American Court of Human Rights Series C No 400 (6 February 2020) para. 137). It remains to be seen, however, how the UNDROP will impact on the practice of the court regarding the collective rights of non-indigenous peasants.

29 *Chagos Islanders vs United Kingdom,* Application No. 35622/04 European Court of Human Rights 410 (20 February 2009). For an analysis of the case, see Stephen Allen,

'Self-determination, the Chagos Advisory Opinion and the Chagossians' (2020) 69(1) *International & Comparative Law Quarterly* 203–20.

30 International Covenant on Civil and Political Rights (adopted 16 December 1966, entered into force 23 March 1976) 999 UNTS 171 (ICCPR), art. 27 reads: 'In those States in which ethnic, religious or linguistic minorities exist, persons belonging to such minorities shall not be denied the right in community with the other members of their group, to enjoy their own culture, to profess and practise their own religion, or to use their own language.'

31 See, for example, UN Human Rights Committee, 'Communication No. 511/1992, Ilmari Länsman et al. v. Finland' (8 November 1994) UN Doc CCPR/C/52/D/511/1992; UN Human Rights Committee, 'Communication No. 671/1995 Jouni E. Länsman et al. v. Finland, Views of the Human Rights Committee under article 5, paragraph 4, of the Optional Protocol to the International Covenant on Civil and Political Rights (Fifty-eighth session)' (22 November 1996) UN Doc CCPR/C/58/D/671/1995. On this point, see for example, Martin Scheinin, 'Indigenous Peoples' Land Rights Under the International Covenant on Civil and Political Rights' (2004) *paper presented on the occasion of the Torkel Oppsahlsminneseminar organized by the Norwegian Centre for Human Rights, University of Oslo* 2.

32 'Communication No. 197/1985, Ivan Kitok v. Sweeden' (27 July 1988) UN Doc CCPR/C/33/D/197/1985, para. 9.2.

33 UN Human Rights Committee, 'Communication No. 760/1997, J.G.A. Diergaardt et al. v. Namibia' (6 September 2000) UN Doc CCPR/C/69/D/760/1997.

34 Gillette Hall and Ariel Gandolfo, 'Poverty and Exclusion among Indigenous Peoples: The Global Evidence' (2016) *World Bank Blogs* <https://blogs.worldbank.org/voices/poverty-and-exclusion-among-indigenous-peoples-global-evidence> accessed 26 April 2021.

35 Ibid.

36 Also see, Leap and Thompson, 'Social Solidarity, Collective Identity, Resilient Communities,' supra note 17.

37 On this point, see also Lorenzo Cotula in this volume.

38 Margherita Brunori, 'Protecting Access to Land for Indigenous and Non-indigenous Communities: A New Page for the World Bank?' (2019) 32(3) *Leiden Journal of International Law* 1–16.

39 See Carl Wellman, 'Solidarity, the Individual and Human Rights' (2000) 22 *Human Rights Quarterly* 639; Rüdiger Wolfrum, 'Solidarity' (2013) *The Oxford Handbook of International Human Rights Law* (Oxford University Press 2013) 401–19.

40 On this point, see Eloisa Cristiani and Guiliana Strambi in this volume.

41 José Suárez-Torres, José Ricardo Suárez-López, Dolores López-Paredes et al, 'Agroecology and Health: Lessons from Indigenous Populations' (2017) 4 *Current Environmental Health Reports Volume* 244–51.

42 On this point, see Christophe Golay in this volume.

43 The FPIC is referred to in various passages of the UNDRIP, supra note 10. Article 11 of the UNDRIP, for example, determines that 'No relocation [of indigenous peoples] shall take place without the free, prior and informed consent of the indigenous peoples concerned and after agreement on just and fair compensation and, where possible, with the option of return.' For analysis, see Cathal M. Doyle, *Indigenous Peoples, Title to Territory, Rights and Resources: The Transformative Role of Free Prior and Informed Consent* (Routledge 2014); Mauro Barelli, 'Free, Prior and Informed Consent in the Aftermath of the UN Declaration on the Rights of Indigenous Peoples: Developments and Challenges Ahead' (2012) 16(1) *The International Journal of Human Rights* 1–24.

44 See, for instance, *Saramaka People v Suriname Preliminary Objections, Merits, Reparations, and Costs*, supra note 21 and more recently the *Case of the Indigenous Communities of the Lhaka Honhat Association (Our Land) v. Argentina. Merits, Reparations and Costs*, supra note 28.

45 *Saramaka People v Suriname Preliminary Objections, Merits, Reparations, and Costs*, supra note 21.

46 *Resolution on a Human Rights-Based Approach to Natural Resources Governance adopted at its 51st Ordinary Session held from 18* African Commission on Human and Peoples' Rights (18 April to 2 May 2012)

47 See Francesco Francioni in this volume.

Selected bibliography

Belinda Fontana L, 'Indigenous Peoples vs Peasant Unions: Land Conflicts and Rural Movements in Plurinational Bolivia' (2014) 41(3) *Journal of Peasant Studies* 297–319.

Edelman M and James C, 'Peasants' Rights and the UN System: Quixotic Struggle? Or Emancipatory Idea Whose Time Has Come?' (2011) 38(1) *The Journal of Peasant Studies* 81–108.

Kymlicka W, 'Beyond the Indigenous/Minority Dichotomy?' in Stephen Allen and Alexandra Xanthaki (eds), *Reflections on the UN Declaration on the Rights of Indigenous Peoples* (Hart Publishing 2011) 187.

Stavenhagen R, *Peasants, Culture and Indigenous Peoples* (Springer 2012).

4

NO FOOD SOVEREIGNTY WITHOUT FEMINISM? NEGOTIATING GENDER EQUALITY IN THE UNDROP

*Joanna Bourke Martignoni and Priscilla Claeys**

Introduction

The adoption of the United Nations Declaration on the Rights of Peasants and Other People Working in Rural Areas (UNDROP) was welcomed by agrarian social movements as a landmark in the international recognition of the peasantry as political and legal subjects with the agency to define the future of agricultural, food, and environmental systems globally. The recognition of the rights to land and seeds as new human rights was hailed as a major achievement. There were a number of peasants' rights activists, however, who emphasised that the limited references to gender equality and the rights of rural women that were retained in the final document will need to be interpreted using an explicitly feminist approach in order to ensure that the Declaration fulfils its emancipatory political promise: 'Achieving recognition for a Declaration that recognises, promotes and defends peasants' rights requires the construction of a sovereign and feminist perspective so that it may really address the roots of inequalities.'[1]

Indeed, from a feminist standpoint, it can be argued that the text of the UNDROP falls short in several crucial areas. Although the provisions on the rights of rural women contained in Article 14 of the 1979 UN Convention on the Elimination of Discrimination Against Women (ICEDAW) are largely reproduced in the Declaration, it fails to move beyond these to incorporate more recent developments in international human rights law. The final version of the UNDROP does not directly engage with several key gender equality issues such as women's rights to inherit land and equal tenure rights in agrarian reform processes, including through the redistribution or allocation of communal land; women's rights to equality in marriage and in family relations; women's sexual and reproductive

* Both of the authors contributed equally to the conceptualisation, research and writing of this chapter. We are grateful to the editors and to Rita Calvario, Annette Desmarais, Marc Edelman, Sandra Moreno Cadena, Andrea Nuila, Elisabeth Prügl, Fenneke Reysoo and Sandra Schranz for their constructive feedback on earlier drafts.

DOI: 10.4324/9781003139874-6

health rights; the disproportionate burden of unpaid reproductive and agricultural labour performed by women; and gender identity as well as sexual orientation as grounds for discrimination. These were all specific claims that agrarian women's groups put forward for inclusion in the Declaration prior to and during its negotiation. More broadly, the UNDROP as adopted fails to directly mention patriarchy as a source of structural oppression and we argue here that this robs the text of much of its political power as a feminist document.

This chapter attends to the contours, achievements, and limits of feminist claim-making within the development of the UNDROP. The main question that is explored relates to the manner in which rights-based demands emanating from 'below' make their way into international instruments.[2] Discussions in the Human Rights Council (HRC) were launched in 2012 on the basis of the 'Declaration of Rights of Peasants – Women and Men' adopted by La Via Campesina (LVC) in 2008. Before being adopted by the movement as a whole, earlier drafts of what was then called the 'Peasants' Rights Charter' were discussed internally, first in Indonesia in the late 1990s and later by various peasant organisations in South East Asia in 2002. We are therefore interested in tracing the ways in which: (a) LVC and its allies formulated a shared gender equality and women's rights agenda within the UNDROP negotiation process and (b) how their demands were later met, transformed, or rejected by states within the HRC where the Declaration was drafted.

This chapter is informed by participant observation in the five sessions of the Working Group[3] tasked with the negotiation of the UNDROP between 2013 and 2018. It is also based on a textual analysis of key documents developed by agrarian social movements and women within those social movements as well as the drafting history of the Declaration as it moved through the multilateral political process in the HRC. This multi-focal study of the negotiations and their outcome allows us to draw out the conjunction of circumstances that led to the silencing of particular women's rights and gender equality demands emanating from agrarian women's groups and their allies and to highlight potential areas in which the text could be 'reclaimed' in order to achieve a peasants' rights agenda that is fully informed by feminist principles.

We organise our analysis around three central arguments that each help to explain why the provisions contained in the UNDROP that touch on gender equality and rural women's rights are not as emancipatory or politically progressive as the aspirations expressed by feminist activists within peasants' rights movements. The first looks at the period spanning 1999–2008, when discussions were taking place within LVC about peasants' rights. It shows that the three initial drafts of the Declaration, elaborated internally by LVC, failed to incorporate the strong feminist agenda that was being advanced by agrarian women's groups in other LVC spaces. The second, which looks at the early stages of the development of the UNDROP between 2013 and 2017, explores how peasant claims were translated and transformed through the inter-governmental negotiation process and by the use of agreed human rights language, to facilitate the finding of common ground. It examines the impacts of anchoring the Declaration in existing international human rights law and consequently its apolitical framing of women's rights, gender equality, and gender identity.

The third section analyses how peasant claims were received, rejected, and reshaped through interactions with states. This demonstrates how the realpolitik of achieving consensus on the text led to the last-minute deletion of a number of key provisions on women's rights, most notably the right of women to inherit land. In the conclusion, we discuss the challenges facing agrarian movements and their allies to interpret gender equality in a progressive way, building on both existing international standards and the feminist praxis of food sovereignty movements.

Feminist claims were not reflected in LVC draft Declarations

Our first contention is that the three initial drafts of the Declaration elaborated by LVC between 1999 and 2008 did not adopt a feminist approach to peasants' rights. While the provisions in the text consistently referred to both men and women as rights holders, these documents were framed within the broader context of the defence and protection of the smallholder family farm against the neoliberal/capitalist food and agricultural system. This meant that inequalities and differences *within* agrarian farming households were effectively erased. The invisibility of a distinctively feminist agenda in the text of these early instruments is something of a paradox, given that at the time of their drafting, various forms of feminism were emerging as powerful political ideas within certain LVC regions, particularly in Latin America. In this section, we contrast the articulation of feminist and gender-related claims by women in LVC, with the development of the peasants' human rights agenda in the same period in other parts of the movement.

Feminist demands by women in LVC

Women's groups within LVC have developed rapidly since the movement's inception over a quarter of a century ago. Following the failure to elect any women as coordinators at the First International Conference of LVC in 1993, at the Second Conference in Mexico three years later, the decision was made to create an International Women's Committee and to adopt specific measures to ensure gender-equitable participation and representation within the movement's governance structures. In October 2000, ahead of the Third International Conference of LVC in India, the First International Women's Assembly was organised. The Assembly agreed to focus on the objectives of achieving gender parity within all levels of decision-making and activities of LVC, strengthening the International Women's Committee, and ensuring that documents, training, events, and speeches produced by the movement were free from any sexist content or language.[4] Since that time, LVC has implemented gender parity quotas in its executive bodies, including the International Coordination Committee.

The final statement from the Second International Assembly of Rural Women (2004) focused on sexual and reproductive rights and the elimination of all forms of gender-based violence as well as steps to ensure women's 'economic autonomy, access to land, health, education and equal social status.'[5] The issue of domestic violence received great attention from 2008 onwards, when the movement launched its Global Campaign to denounce the devastating impact of various forms of violence on the lives of women and on the communities in which they live.[6] The Campaign highlighted the links between gender-based violence against women and the influence of patriarchal power relations in capitalist and neoliberal agricultural systems. One of the key messages of the Campaign was that capitalism relies on women's exploitation in order to operate, in that it renders their reproductive and productive labour invisible, making them economically dependent on men while limiting their possibilities for political participation. The Campaign identified capitalism and patriarchy as mutually reinforcing causes of violence against women and proposed an alternative vision of sovereignty that would free women from both.

In 2013, the Fourth Women's Assembly adopted the Women of Via Campesina International Manifesto.[7] At the heart of the Manifesto was the assertion that equal access to land is a key right for women with comprehensive, participatory agrarian reform being necessary to ensure that rural women can live a 'dignified and fair life.' Crucially, the document also emphasised an ecofeminist ontology of land, underlining that it 'is more than a means of production' and it is 'a space of life, culture, identity, an emotional and spiritual environment,' thereby further articulating the mutually reinforcing relationship between food sovereignty and feminism. Explicitly confronting patriarchy, the Manifesto insisted on the importance of 'fighting for the sovereignty of the land, the territory and the body and saying no to violence against women in all its forms.'[8] It thus constituted one of the first attempts to define the tenets of popular peasant feminism at the international level, although the concept had previously been discussed and developed by women within the Latin American region of LVC.

In 2017, the Fifth Women's Assembly of LVC reaffirmed its earlier Manifesto and reiterated the need for food sovereignty to be informed by feminism. This notably means ensuring that this sovereignty encompasses 'personal' sovereignty and autonomy, including bodily and sexual autonomy, and freedom from violence. The Assembly again drew attention to the continuum of violence from that experienced in the home to capitalist and corporate economic violence.[9]

Alongside this relatively well-accepted feminist perspective on peasants' rights within women's groups in LVC have been recent efforts to address systemic discrimination in agrarian communities against people with diverse gender identities and sexual orientations. This issue is one that the movement has only reluctantly begun to grapple with and much work remains to be done in order to fully infuse discussions of gender equality and women's rights with an intersectional, non-binary approach to sex- and gender-based discrimination.[10]

Developing a peasants' rights agenda in agrarian social movements

Whilst discussions of women's rights and feminism were developing within LVC, other parts of the movement were promoting the idea of an international declaration on the human rights of peasants. The following summary traces the development of successive peasants' rights declarations within agrarian social movements prior to the negotiations at the UN. The focus is on the way in which rights holders were defined and on the framing of gender equality and women's rights provisions within these internal documents.

The first enunciations of a peasants' rights agenda were developed in Indonesia in the 1990s through exchanges between various actors including farmers, students, non-governmental organizations (NGOs), rural communities, and researchers.[11] In the Peasants' Rights Charter, elaborated in 1999, the subjects or rights holders were described as 'peasants and their families.' There was no attention to women's rights or any recognition of gendered power dynamics within the peasantry or society at large. The emphasis of the document was the peasant family, which the Charter defined as 'a husband, wife, children and biological relatives, but may also include other specific social ties as long as this is accepted by the community and that they are all dependent on the same food source.' While the Charter reiterated that no discrimination should take place on the basis of sex/gender, it contained a cultural relativist clawback clause that provided that: 'Peasants' rights to these basic needs should be guaranteed by a law, not discriminated against according to sex, seniority or age, *except where such discrimination is determined by religion or culture*.'[12] Interestingly, one of the eight sections of the Charter (section G) was devoted to 'Reproductive rights (as they relate to both human reproduction and maintaining biological diversity).' This way of linking caring for other humans and for the natural world reflects the ecofeminist praxis of many agrarian movements and forms part of their contestation of neo-liberal paradigms that privilege extractivism and the exploitation of both nature and women's labour. In the area of reproductive rights, however, the emphasis in the Charter was on the rights of peasants and their families (not women) to choose family size and use birth control.

Over the next few years, Asian members of LVC continued to discuss the adoption of a peasants' rights instrument and, in 2002, regional members of LVC adopted a 'Declaration about the protection of the peasant right.' The rights holders in this Declaration were described as 'peasant men and women and their family' without defining the terms 'peasant' or 'family.' The Declaration no longer contained specific articles relating to reproduction. References to non-discrimination were found in several parts of the document, such as in Article II (3) on the right to agrarian resources which stated that 'there should be no discrimination on the grounds of gender, religion, status, race and culture.'[13]

In 2008, after consultation with other regions, LVC adopted a revised 'Declaration of the Rights of Peasants – Women and Men.'[14] The new Declaration included a definition of peasants as individual and collective rights holders and it no longer placed an emphasis on the peasant family. The text nevertheless cited reliance on family labour as a key component of peasant identity. Article 2.1 provided that 'women peasants and men peasants have equal rights,' and Article 2.3 stated that 'peasants are equal to all other people and, in the exercise of their rights, should be free from any form of discrimination, including discrimination based on . . . sex.' Article 3 contained two provisions that specifically referenced women, recognising their rights to 'be protected from domestic violence, physical, sexual, verbal and psychological' and 'control their own bodies and to reject the use of their bodies for commercial purposes.'[15] These were the only rights that were directly framed from a women's rights perspective in the document, although paragraphs 8 and 9 of Article 3 guaranteed sexual and reproductive rights as well as control over family planning for both women and men peasants. It was this draft that was presented in 2009 to the United Nations General Assembly (UNGA) during an interactive dialogue on the Global Food Crisis and which formed the basis of discussions on the initial draft UNDROP in the Human Rights Council.

Despite the assertion within many agrarian movements that feminism and radical egalitarianism form integral elements of food sovereignty, transformative claims to gender equality are often written out of or simply subsumed within movement identity-building processes that privilege an ungendered peasant subject and the family farm.[16] The LVC rights instruments developed between 1999 and 2008 focused either on the family unit or emphasised formal equality between women and men, with few articles specifically dedicated to women's rights. In the same period, demands around sexual and reproductive rights, gender-based violence, women's autonomy, and rights to land were emanating from the first LVC Women's Assemblies, but these failed to be fully integrated in the development of the peasant's rights agenda within LVC. The next section discusses the ways in which the process of inter-governmental negotiations through the use of agreed international human rights language further diluted attempts to develop an explicitly feminist approach to food sovereignty within the UNDROP.

Translating peasant claim-making into agreed human rights language

Our second argument is that the format of the negotiations in the Human Rights Council forced agrarian activists and their allies to use language drawn from 'authoritative previous documents' as a means of justifying their demands in terms that would be acceptable to states that are the traditional gatekeepers of international law.[17] In this section, we discuss how LVC claims were reframed during the negotiation process to align them with existing human rights language and what this meant for feminist perspectives on women's rights and gender equality.

The evolution of the gender provisions in the draft Declaration 2012–2017

The first draft of the UNDROP entered the Human Rights Council as an Annex to the study produced by the Council's Advisory Committee on the advancement of the rights of peasants and other people working in rural areas. The study listed gender discrimination as one of the five main causes of peasant vulnerability to human rights abuses. It also noted that this inequality is reflected in the disproportionately high numbers of women and girls suffering from hunger and malnutrition globally and in the failure to recognise women as producers or juridical equals in their control over land and productive resources. The Committee's diagnosis of the 'problem' of gender inequality in rural areas echoes mainstream United Nations (UN) discourses on rural women as victims of discriminatory norms and customs primarily located in the family and community, without examining patriarchy as a systemic form of economic and social oppression.[18] As a result, even at this early stage of the negotiating process, the subject of the peasant woman takes on a less political and agential character than the version being portrayed by women's groups within LVC that emphasise their significant contributions to the achievement of food sovereignty and their diversity as well as their autonomy.[19]

The draft Declaration proposed by the Advisory Committee and contained in the annex of its 2012 report was essentially the same as LVC's Declaration on the Rights of Peasants – Women and Men (2008). One important modification was in the title of the Declaration, which was expanded to include 'other people working in rural areas' which refers to groups such as agricultural workers, pastoralists, fisherfolk, and indigenous people working on the land.[20] This text served as the basis of discussion during the first session of the Open-Ended Inter-Governmental Working Group (OEIGWG) in 2013. In line with the LVC Declaration, the Advisory Committee's draft took a binary approach to 'women and men' that emphasised formal, rather than substantive equality, and only specifically referred to women's rights in relation to a limited number of issues. To this end, the draft by the Advisory Committee incorporated umbrella non-discrimination provisions that also applied to equal rights to land and other resources.[21]

At the second session of the OEIGWG, which took place in 2015, the Chair released a substantially revised draft Declaration, elaborated with the support of an informal team of human rights experts.[22] The objective of the new text was to reinforce its coherence with existing human rights standards and thereby respond to the objections of states concerned that they were being presented with a series of new human rights norms and obligations. This revised draft included a standalone article on gender equality (Article 4) which stated that 'gender identity is not a barrier to realizing human rights,' as well as a separate article on the rights of rural working women (Article 6). The article on rural women was based on Article 14 of the UN Convention on the Elimination of All Forms of Discrimination Against Women (ICEDAW) with some additions, including recognition of the discriminatory impact of gender-based violence against women, as well as gender identity as a ground of discrimination. This 2015 draft also contained language on non-discrimination in land rights in its Article 19, which went beyond the guarantees contained in Articles 14 and 16 of the ICEDAW to include women's equal 'right to inherit and bequeath' land tenure rights.

The revised drafts of the UNDROP issued in 2017 and 2018 were further amended to address specific objections by states and make the draft Declaration shorter. This meant that the original 2015 draft of the Declaration was significantly edited during this period, with many of the more progressive provisions on gender equality and women's rights being bargained away in the interests of textual brevity and a desire to bring more states on board.

UNDROP as a reflection of international gender equality law

The gender equality provisions in the draft versions of the UNDROP that were discussed between 2012 and 2017 were essentially drawn from the agreed language contained in the 1979 UN Convention on the Elimination of Discrimination Against Women (ICEDAW). The original text of the ICEDAW has subsequently been subjected to a series of interpretive General Recommendations by the Committee on the Elimination of Discrimination Against Women (CEDAW); however, these updated and more nuanced approaches to women's human rights were not included in the text of the UNDROP. Reflecting this, the Declaration tends to reinforce the idea that the terms sex and gender are interchangeable; these are fixed and binary categories that are grounded in innate, biological characteristics rather than malleable constructs reinforced and remade through patriarchal political, social, and economic power relations. While the apolitical approach to sex/gender that is taken for granted in most international human rights instruments is one that is acceptable to many parts (although certainly not all) of LVC's diverse membership, there were strong feminist groups within LVC pushing for a more radical, political, and intersectional approach to gender identity and women's rights. This transformative agenda proved difficult to translate into practice in the context of the UNDROP.

The impact of grounding the UNDROP in agreed international human rights language is mixed when it comes to rural women's rights and gender equality. The systematic use of the traditional women's rights framework of the ICEDAW helped move the draft Declaration away from its initial emphasis on family farming to give greater visibility to the human rights of rural women, partly reflecting the ICEDAW's framework that articulates a 1970s approach to sex discrimination focused on economic equality and political representation. As noted earlier, the provisions retained in the final version of the Declaration generally replicate Article 14 of the ICEDAW, with the addition of paragraphs on decent employment and equal remuneration, priority (not only equal) treatment of women in agrarian reform, and a direct mention of the right of women to be free from violence.

At the same time, the text of the UNDROP fails to incorporate several of the more progressive ideas in CEDAW's interpretive General Recommendation No. 34 on the rights of rural women. The General Recommendation draws attention to the 'negative and differential impact' of trade liberalisation, privatisation, and commodification of land and natural resources on the rights of rural women. It also highlights the obligations of states and other duty bearers to redress the disproportionate burden of unpaid care and agricultural work performed by women.[23] In addition, the Recommendation focuses on the importance of temporary special measures for the achievement of substantive gender equality in inheritance and user rights over land, including communal lands, and on sexual and reproductive health and rights. The adoption of this new Recommendation in 2016 provided opportunities for the incorporation of interpretative guidance on the human rights of rural women in the revised drafts of the UNDROP that were presented in 2017 and 2018.[24] However, this did not happen, and the UNDROP does not reflect the progressive ideas contained in General Recommendation 34.

Overall, the emphasis in the UNDROP negotiations on using existing language from the 1979 ICEDAW, which is not an explicitly feminist document, contributed to the erasure of many of the more far-reaching claims being made by women's rights advocates in agrarian social movements. Some of these issues include linking violence against nature to violence against women, the need to guarantee women's sovereignty and control over their own bodies, and the relationship between patriarchy and capitalist exploitation of women's productive and reproductive labour.

Negotiating gender equality in a hostile Human Rights Council

Our third argument is that due to the need for support for the Declaration from Egypt and other allies (within the non-aligned movement), not insisting upon a number of the key gender equality provisions in the initial drafts of the text was viewed as politically expedient by Bolivia and the core group charged with elaborating the UNDROP. These included women's equal rights to inherit land, temporary special measures for the achievement of gender equality, explicit acknowledgement of women's sexual and reproductive health and rights, and discrimination against peasants on the grounds of their gender identity or sexual orientation. In this section, we examine the ways in which the gender politics of the inter-governmental negotiations shaped the text of the women's rights and gender equality provisions that were retained in the final version of the Declaration. We also explore how LVC and its allies responded.

Clawing back women's rights during the inter-governmental negotiations

Several of the women's rights and gender equality provisions in the 2015 draft of the Declaration, which, as noted earlier, had been elaborated with the support of human rights experts and were based on existing human rights language, were contested by Egypt and other members of the non-aligned movement. The elements of the 2015 draft that elicited opposition were:

- The informed consent and choice of women in relation to their own bodies (Article 6.2(c));
- The reference to 'regardless of the concept of the family' in the article covering equality and justice in marriage and in family relations (Article 6.2(k));
- Provisions on 'gender' and 'gender identity' as grounds for discrimination (in Articles 4 and 6.2(m)).

At the fourth session of the OEIWG in 2017, a shortened and revised draft Declaration was released in which the previous draft Article 4 on gender equality was deleted and merged with Article 3 on equality and non-discrimination. Article 6 on rights of rural working women was reduced, becoming the new Article 4.[25] Although a number of states had come out strongly in favour of including gender identity as a prohibited ground for discrimination, all references to gender identity and sexual orientation as grounds for discrimination were removed in the 2017 draft. In addition, the new draft Article 4 no longer included the provision on the right of rural women to adequate living conditions, which was taken directly from Article 14 of the ICEDAW. Also deleted were the paragraphs on women's rights to 'autonomy, privacy, confidentiality, informed consent and choice in relation to their own bodies in all spheres of life,' as well as gender identity as a specific category within the definition of multiple forms of discrimination. Finally, the revised draft did not preserve the reference to affirmative action as a way to achieve substantive gender equality that was contained in the 2015 draft of the UNDROP and which is a key component of CEDAW's General Recommendation No. 34 on the rights of rural women.

The revised 2017 draft, however, contained a new paragraph (h) on gender equality in access to and control over land in Article 4 on rural women's rights. This reiterated similar language contained in two paragraphs of draft Article 17 on the right to land. Importantly, the addition of the equal right of women to inherit and bequeath land rights in Article 17.2 of the 2017 draft went beyond the language contained in the ICEDAW and was derived from the intergovernmentally agreed FAO Voluntary Guidelines on the Governance of Tenure (Guideline 4.6).

In advance of the fifth session of the OEIWG, which took place in April 2018, Bolivia released a revised draft containing a number of changes to facilitate dialogue with reluctant states.[26] Article 4 on the rights of peasant women and other women working in rural areas was kept intact. However, the paragraph on equality in marriage and family relations was deleted. This deletion had been requested at the fourth session by Egypt and Jordan, which had rejected paragraph 4.2(k) of the 2017 draft Declaration which guaranteed women's equality in marriage and family relations. The Arab states had argued that they had placed reservations on Article 16 of the ICEDAW, which formed the basis of that provision in the draft UNDROP. Although there were further objections from the Holy See to draft paragraph 4.2(b) on 'equal access to . . . adequate health-care facilities, information, counselling and services in family planning,' these were kept in the final draft of the UNDROP, which replicates the language in Article 14.2(b) of the ICEDAW. Meanwhile, peasant representatives from Asia insisted on the importance of land inheritance for women and of women's right to own property. As we explore later, all references to women's rights to inherit land were eventually deleted during the final round of revisions that took place in informal negotiations in mid-2018.

Drawing 'red lines' and LVC priorities in the UNDROP negotiations

The reaction of LVC to the whittling away of draft provisions in the UNDROP relating to gender equality and rural women's rights was as diverse as the movement itself. The UNDROP was widely seen by LVC and other rural participating constituencies (such as pastoralists, fishers, or indigenous peoples) as a way to assert their collective rights to land, seeds, biodiversity, and food sovereignty. Achieving recognition of these rights was the movement's priority and the LVC negotiating team clearly did not want to risk losing that battle in order to advance women's rights. LVC was acutely aware of the small window of opportunity that existed to push for the UNDROP, given that the chair of the OEIGWG and main sponsor of the Declaration, Bolivia, wished to have the text adopted prior to its presidential elections in 2019. The strategy adopted by LVC was therefore to support the 2017 draft presented by Bolivia, as long as it respected its 'red line' demands. In addition, women's rights were largely seen as covered by the inclusion of ICEDAW language in the Declaration.

That being said, various feminist groups within LVC were vocal about the need to include stronger language on women's rights and gender equality in the text throughout the negotiation process. Further, the international women's coordination of LVC provided its inputs on the evolving draft of the UNDROP on several occasions.[27] In contrast, the importance of fighting against discrimination on grounds of sexual orientation and gender identity had not achieved sufficient consensus within LVC and LVC remained silent when provisions on gender identity were deleted.

Finalising the text in mid-2018: writing out women's equal inheritance rights

In mid-2018, Bolivia held a number of informal bilateral consultations to garner support for the Declaration ahead of its finalisation. In the last revised version of the document, presented in September 2018, women's rights to inherit land were removed from both Article 4.2(h) on the rights of rural women and Article 17.2 on the right to land.[28] The former reference to 'equal tenure rights' for women and men and the related 'right to inherit and bequeath these rights' were also deleted, as well as the provision stipulating redistributive agrarian reforms 'must guarantee equal access of men and women to land, fisheries and forests' (in Article 17.6). The obligation of states to take 'appropriate measures to remove and prohibit all forms of discrimination relating to the right to land' (in Article 17.2) was left in the final version of the text and several movement actors we spoke to suggested that this could be interpreted progressively in light of the ICEDAW and General Recommendation No. 34.

At the 39th session of the Human Rights Council in September 2018, draft resolution A/HRC/39/L.16 was presented containing the final text of the UNDROP in its annex. It was adopted by the Council in a recorded vote (33 to 3, with 11 abstentions) and then by the General Assembly in December 2018. It is impossible to assess how civil society participants and diplomats would have reacted to these last-minute deletions, if they had been openly presented and discussed during the Working Group session. In conducting research for this chapter, we became aware that many LVC representatives and their allies were not alerted to the fact that numerous women's rights and gender equality provisions had been written out of the final version of the UNDROP.

Conclusion

Charting the development of women's rights within the UNDROP process provides insights into the multi-faceted gender politics of agrarian social movements and the international human rights monitoring mechanisms. Over the course of the negotiations leading to the adoption of the Declaration, proposed provisions on women's equality in land inheritance, temporary special measures, non-discrimination on grounds of gender identity and sexual orientation, and sexual and reproductive health and rights were erased or watered down. More broadly, the UNDROP does not capture many of the epistemological claims made by women peasants concerning relationships between nature and human activities and the powerful role that feminist-informed food sovereignty and agroecology might play as alternatives to patriarchal and extractivist forms of capitalism. This shows that gender transformation is a complex and constant struggle, both inside and outside agrarian movements.

The desire of peasants' movements to make use of the conservative, state-centric tools of international law to gain recognition for their identity co-exists uneasily with their food sovereignty project that would result in a radical remaking of food systems governance with increased popular democratic control.[29] This ambivalence about the use of the multilateral human rights system is something that is also common to many feminist engagements with international law, which recognise the constraints and limitations of traditional governance frameworks, while pragmatically acknowledging the need to strategically work through these in the hope of achieving incremental change.[30]

In the UNDROP negotiations, many of the equality demands emanating from feminists within LVC that sought to emphasise the linkages between neo-liberal capitalism, environmental destruction, and patriarchy were silenced. This was initially due to the emergence of the declaration in Asia and to the focus on the family farm and an ungendered peasant subject. It is also tied to an unwillingness by the movement to engage in debates about issues of gender identity and sexual orientation. Later, when non-discrimination and women's rights provisions were included in the draft text of the Declaration, these were squeezed into the confines of agreed international gender equality language which tends to reproduce a binary approach to sex/gender and seeks to insert women into existing power structures without considering the ways in which these institutions and norms might be remade.

Finally, the influence exerted by powerful blocs of states within the Human Rights Council in defining the conditions for the acceptance of gender equality demands is apparent in the negotiations for the UNDROP, as has been the case in other HRC processes. Several members of the non-aligned movement that were supportive of the Declaration in general – possibly emboldened by their success together with other states in driving forward a conservative agenda within the Human Rights Council on 'traditional values' and the 'protection of the family' – contested draft provisions on women's rights and gender equality. This form of 'counter – mobilisation' by alliances of conservative governments and supportive non-governmental actors to oppose feminist norm-making has increasingly become part of the political landscape of the Human Rights Council since its establishment in 2006.[31]

In spite of the shortcomings discussed earlier, there is potential for the UNDROP to continue to be an emancipatory project drawing on the highly participatory and bottom-up process through which it was developed. The emphasis within the UNDROP on putting human rights at the heart of food systems is a crucial step forward in the recognition of peasants and agrarian communities as political actors who must be actively involved in all stages of policy formation, implementation, and monitoring. The linkages made in the instrument between the rights to seeds, land, and food sovereignty are also significant achievements for agrarian social movements. For the UNDROP to deliver on these egalitarian promises, however, it will need to be reclaimed and defined in practice to include the more transformative feminist demands for gender equality and rural women's rights that have been developed within agrarian social movements and which are not mentioned in the text of the Declaration. Several peasants' rights and feminist groups are promoting the idea that the implementation of the UNDROP should be more firmly anchored in a progressive interpretation of international human rights law. These approaches could build on CEDAW General Recommendation 34 and the equal land inheritance rights guarantees in the Voluntary Guidelines on Land Tenure (VGGT). In addition, many agrarian women's groups are pushing for stronger connections between gender justice, radical food politics, and agroecology that would create more effective synergies with their feminist agendas and methodologies.[32] Such an intersectional and political approach would enable a stronger integration of key issues at the heart of the feminist and food sovereignty paradigms, including patriarchy as a form of oppression that intersects with race, social class, and sexual orientation; the need to address the continuum of violence from intimate to corporate and economic; women's rights to sexual and reproductive autonomy; and valuing relations of care for humans and nature.

Notes

1 Sandra Moreno Cadena, 'The Declaration and the Struggles of Peasant Women: Gender as an Unachieved Question,' in Priscilla Claeys and Marc Edelman, 'The United Nations

Declaration on the Rights of Peasants and other People Working in Rural Areas' (2020) 47 *The Journal of Peasant Studies* 1–68.

2 Priscilla Claeys, 'The Rise of New Rights for Peasants. From Reliance on NGO Intermediaries to Direct Representation' (2018) 9(3–4) *Transnational Legal Theory* 386–99.

3 The UNDROP negotiation process took place within the UN Human Rights Council in an Open-Ended Intergovernmental Working Group, chaired by Bolivia, open to all UN member states and to participation by civil society organizations with consultative status.

4 Annette A. Desmarais, *La Via Campesina. Globalization and the Power of the Peasantry* (Pluto Press 2007).

5 International Assembly of Women Farmers, 'Declaration of the Second International Assembly of Rural Women' (2004).

6 LVC leaflet on the Global Campaign (2012).

7 Women of Via Campesina International Manifesto (2013) <https://viacampesina.org/en/women-of-via-campesina-international-manifesto-2/>.

8 La Via Campesina, *Report of the IV Women's Assembly of la Via Campesina, Jakarta* (La Via Campesina 2014).

9 La Via Campesina, 'VII International Conference: Women's Assembly Declaration' <www.cadtm.org/VII-International-Conference-Women> (2017).

10 Paula Gioia, 'Coming Out! Gender Diversity in the Food System' (2019) *The Right to Food and Nutrition Watch* 6–15.

11 Mansour Fakih, Toto Rahardjo and Michel Pimbert, *Community Integrated Pest Management in Indonesia. Institutionalising Participation and People Centred Approaches* (IIED 2003), A. 14, our emphasis.

12 Ibid., A. 14, our emphasis.

13 La Via Campesina, 'Declaration of Via Campesina South East Asia and East Asia about the Protection of the Peasants' Rights' (2002).

14 La Via Campesina, 'Declaration of the Rights of Peasants – Women and Men' (2009).

15 Ibid., art. 3.2 and 3.3.

16 Clara Mi Young Park, Ben White and Julia, 'We Are Not All the Same: Taking Gender Seriously in Food Sovereignty Discourse' (2015) 36(3) *Third World Quarterly* 584–99.

17 Rashila Ramli and Zawiah Yahya, 'Language of Negotiation for Agreed Conclusions at the UN 57th session of the Commission on the Status of Women: A Case Study' (2014) 11 *Procedia, Social and Behavioral Sciences* 389–403.

18 UN Human Rights Council, 'Final study of the Human Rights Council Advisory Committee on the Advancement of the Rights of Peasants and other People Working in Rural Areas' (24 February 2012) UN Doc A/HRC/19/75 para 23; Joanna Bourke Martignoni, 'Engendering the Right to Food? International Human Rights Law, Food Security and the Rural Woman' (2018) 9(3–4) *Transnational Legal Theory* 400–14.

19 Esther Vivas, 'La Via Campesina, Food Sovereignty and the Global Feminist Struggle' *Food First* (Oakland, 12 August 2012) <https://foodfirst.org/la-via-campesina-food-sovereignty-and-the-global-feminist-struggle/> accessed 7 June 2021.

20 It should be noted here that several requests were made, unsuccessfully, by civil society actors to implement inclusive language in the UNDROP, in particular in the title. For example, the title in French and Spanish explicitly excludes women peasants.

21 UN Human Rights Council 'Draft Declaration on the rights of peasants and other people working in rural areas' (20 June 2013) UN Doc A/HRC/WG.15/1/2.

22 UN Human Rights Council 'Advanced Version of the Draft Declaration on the rights of peasants and other people working in rural areas' (27 January 2015). This text was only made available in English and does not have an official UN document number. See also Claeys, 'The Rise of New Rights for Peasants,' supra note 2, 386–99.

23 UN Committee on the Elimination of Discrimination Against Women, 'General Recommendation No 34 on the rights of rural women' (4 March 2016) UN Doc CEDAW/C/GC/34, para. 11.

24 References to General Recommendation No 34 were made in the Study that was released in 2017 by the UN Office of the High Commissioner for Human Rights. For

more see: UN Human Rights Council, 'Study on the Normative Sources and Rationale Underlying the Draft Declaration on the Rights of Peasants and Other People Working in Rural Areas' (4 May 2017) UN Doc A/HRC/WG.15/4/3.

25 UN Human Rights Council, 'Draft Declaration on the rights of peasants and other people working in rural areas' (6 March 2017) UN Doc A/HRC/WG.15/4/2.

26 UN Human Rights Council, 'Draft Declaration on the rights of peasants and other people working in rural areas' (12 February 2018) UN Doc A/HRC/WG.15/5/2.

27 For example, FIAN, LVC and CETIM sent Bolivia their consolidated comments on the 2017 draft before it was made publicly available. In this document, they insist on the importance of keeping strong language around women's choice in relation to their bodies, harassment and violence against women in their working spaces especially for migrants and seasonal and rural women workers, and on equal access of men and women to land in redistributive agrarian reforms.

28 UN Human Rights Council, 'Draft United Nations Declaration on the rights of peasants and other people working in rural areas' (10 September 2018) UN Doc. A/HRC/WG.15/5/3.

29 Emma Larking, 'Mobilising for food Sovereignty: The Pitfalls of International Human Rights Strategies and an Exploration of Alternatives' (2017) 23(5) *The International Journal of Human Rights* 758; Elisabeth Philipose, 'Decolonizing the Racial Grammar of International Law,' in Robin L. Riley, Chandra Talpade Mohantyand Minnie Bruce Pratt (eds), *Feminism and War: Confronting U.S. Imperialism* (Zed Books 2008) 103–16.

30 Hilary Charlesworth and Christine Chinkin, *The Boundaries of International Law* (Manchester University Press 2000); Brooke Ackerly, *Universal Human Rights in a World of Difference* (Cambridge University Press 2008); Sari Kouvo and Zoe Pearson (eds), *Feminist Perspectives on Contemporary International Law. Between Resistance and Compliance?* (Hart 2011).

31 Conny Roggeband, 'International Women's Rights: Progress under Attack?' (2019) *KFG Working Paper Series No 26, Berlin Potsdam Research Group.*

32 LVC, FIAN International et al, *Cooking Up Political Agendas: A Feminist Guide on the Right to Food and Nutrition for Women in Rural Areas* (FIAN July 2020).

Selected Bibliography

Bourke Martignoni J, 'Engendering the Right to Food? International Human Rights Law, Food Security and the Rural Woman' (2018) 9(3–4) *Transnational Legal Theory*.

Claeys P, *Human Rights and the Food Sovereignty Movement. Reclaiming Control* (Routledge 2015)

Claeys P, 'The Rise of New Rights for Peasants. From Reliance on NGO Intermediaries to Direct Representation' (2018) 9(3–4) *Transnational Legal Theory* 386–99.

Larking E, 'Mobilising for Food Sovereignty: The Pitfalls of International Human Rights Strategies and an Exploration of Alternatives' (2017) 23(5) *The International Journal of Human Rights* 758–77.

Moreno Cadena S, 'The Declaration and the Struggles of Peasant Women: Gender as an Unachieved Question,' in Priscilla Claeys and Marc Edelman, 'The United Nations Declaration on the Rights of Peasants and other People Working in Rural Areas' (2020) 47 *The Journal of Peasant Studies* 1–68.

5

SHEDDING A LIGHT ON THE HUMAN RIGHTS OF SMALL-SCALE FISHERS

Complementarities and contrasts between the UNDROP and the Small-Scale Fisheries Guidelines

Elisa Morgera and Julia Nakamura*

Introduction

The UN Declaration on the Rights of Peasants and Other People Working in Rural Areas (UNDROP)[1] expressly covers any person, including indigenous peoples and local communities, engaged in artisanal or small-scale fishing and related handicrafts in rural areas.[2] There is no doubt, therefore, that the Declaration applies to small-scale fisheries (SSF)[3] and affirms their human rights. It embodies the recognition[4] of the need for a coherent interpretation and application of existing international human rights standards to a broad range of actors involved in SSF, to ensure their full enjoyment of all human rights.[5] It also represents the commitment of States to promptly take steps to enhance the protection of all rights spelled out therein.[6]

Before the adoption of the UNDROP, the UN Food and Agriculture Organization (FAO)'s Voluntary Guidelines for Securing Sustainable Small-Scale Fisheries in the context of food security and poverty eradication (SSF Guidelines)[7] was the only international instrument specially dedicated to small-scale fisheries and making express references to fishers' rights. The SSF Guidelines were developed as part of FAO's long-standing efforts to contribute to the coherent interpretation and application of international instruments on sustainable fisheries, at the crossroads of international fisheries law and international environmental law.[8] International human rights law also informed and was incorporated in the SSF Guidelines,

* The authors are thankful to Margherita Brunori, Jackie Sunde, Bernadette Snow, Lorenzo Cotula, Patrick Vrencken, and Ana Suarez Dussan for their valuable comments and inputs to an earlier version of this chapter. This chapter has been prepared under the UKRI GCRF One Ocean Hub (The One Ocean Hub is a collaborative research for sustainable development project funded by UK Research and Innovation (UKRI) through the Global Challenges Research Fund (GCRF) (Grant Ref: NE/S008950/1). GCRF is a key component in delivering the UK AID strategy and puts UK-led research at the heart of efforts to tackle the United Nations Sustainable Development Goals.

DOI: 10.4324/9781003139874-7

which offer specific guidance on States' obligation with respect to the rights of women,[9] indigenous peoples,[10] migrants,[11] and children.[12]

Notwithstanding the importance of the SSF Guidelines, very little has been said about this instrument and its relationship with the UNDROP. Originating from forums that frame issues and respective responses differently, and which involve individuals and specialised international bodies with distinct interests and areas of expertise[13] (fisheries and human rights, respectively),[14] these two instruments offer their own particular perspectives on how international law currently addresses the challenges and contributions of SSF to sustainable fisheries overall. This chapter explores the extent to which they can and should be read together, with a view to illuminating a mutually supportive[15] interpretation of multiple – and traditionally disconnected – sources of international law, all of which apply to SSF. To that end, this chapter will analyse and compare the two instruments with respect to the processes that led to their development and adoption, their notion of SSF and their scope of application, the recognition and protection of customary tenure of fishery resources, the protection of small-scale fishers' traditional knowledge, and the relevance of procedural rights. These findings are then brought together to demonstrate how the mutually supportive interpretation of the two instruments can support States in the realisation of multiple Sustainable Development Goals (SDGs), which include targets on providing access of small-scale fishers to land, marine resources, productive resources and inputs, knowledge, financial services, markets, and opportunities for value added.[16]

The development of the UNDROP and the SSF Guidelines

The making and adoption of the UNDROP and the SSF Guidelines provide concrete examples of participatory international law-making processes led by distinct international UN bodies with different mandates,[17] memberships, and processes. The UNDROP was the result of long-standing efforts driven by international agrarian movements and human rights organisations, which eventually led to political and intellectual engagement with the Human Rights Council (HRC), a subsidiary body of the UN General Assembly (UNGA).[18] The SSF Guidelines, in turn, were the outcome of a multi-stakeholder process, pledged with the support of small-scale fishers and fishing communities, and ultimately endorsed by the Committee on Fisheries (COFI),[19] a subsidiary body of FAO Council.

The formal negotiation of the UNDROP commenced in 2012 with the establishment by the HRC of an open-ended intergovernmental working group (HRC Working Group) dedicated to drafting and negotiating the Declaration, which was finalised and adopted by the UNGA in 2018.[20] The process that led to the negotiation and adoption of the UNDROP started decades earlier, outside of the UN: La Via Campesina (LVC) took the lead in the organisation of a global, bottom-up agrarian movement 'to articulate the vision of a locally controlled food system' that takes stock of 'wide-ranging livelihood strategies' embedded in close relationships with rural areas.[21] As a result, the Declaration of LVC on the Rights of Peasants – Women and Men was adopted. The document was used as a starting point for the UNDROP drafting process. LVC and other international organisations representing rural constituencies from across the world continued to provide 'extensive inputs' to the HRC Working Group.[22]

Stakeholder engagement in the development of the SSF Guidelines, in turn, was managed by FAO. In 2006, the UNGA mandated FAO to develop an instrument on SSF,[23] which was supported in 2008 by the Global Conference on SSF.[24] The development of the text, however, was only formally initiated by a decision of COFI in 2011,[25] from which a series of global and regional consultative meetings and similar events engaged over 4,000 stakeholders, including small-scale fishers and their communities.[26] Then, the draft text was subjected to a two-session high-level technical consultation, and eventually FAO Member States and the European Union (EU) negotiated and agreed on the final text, considering the inputs of intergovernmental organisations, non-governmental organisations, and civil society organisations that followed the negotiations as observers.[27] Finally, in 2014, the SSF Guidelines were endorsed by consensus by representatives of 110 FAO Member States, from all regions.[28]

The level of intergovernmental support was therefore initially much broader for the SSF Guidelines than for the UNDROP, and, considering the EU support for the SSF Guidelines, broader overall. Following a majority vote at the HRC,[29] the Declaration gained 121 countries votes in its favour at the UNGA; Australia, Guatemala, Hungary, Israel, New Zealand, Sweden, the UK, and the US voted against the Declaration; EU Member States (Portugal excepted), Canada, Japan, Norway, the Republic of Korea, and the Russian Federation abstained.[30] The lack of support for the UNDROP from high-income countries may be due to the fact that the UNDROP's text is articulated in prescriptive terms (consistent use of the verb *shall*) and explicit references to *having the right to*, to reflect the explicit intent of clarifying how existing international legally-binding instruments on human rights apply in the specific case of peasants.[31] Those countries characterised by industrialised food production may have thus been reluctant towards the UNDROP's recognition of peasants' 'vision' embedded in traditional local food production.[32] In contrast, the SSF Guidelines are framed as recommendations (characterised by the predominant use of *should* and *recognise*), which may have facilitated its adoption by consensus.[33]

Another consequence of the differing levels of intergovernmental support for these instruments is that the UNDROP and the SSF Guidelines are subject to different follow-up processes, which has relevance to the ends of understanding their respective legal strength as *soft law*.[34] FAO's COFI has supported the implementation of the SSF Guidelines through planning instruments, including the FAO's Global Assistance Programme in 2014, its Umbrella Programme for the promotion and application of the SSF Guidelines, established in 2015, and the development of an SSF Guidelines Global Strategic Framework from 2016 onwards.[35] The SSF Umbrella Programme, for instance, has four interrelated components aimed at awareness-raising, strengthening the science-policy interface, empowering stakeholders, and supporting implementation.[36] These activities by FAO, governments, and stakeholders have advanced the implementation of the SSF

Guidelines,[37] including through providing legislative guidance.[38] With respect to a follow-up mechanism for the UNDROP, scholars have put forward recommendations ranging from establishing a new specific UN Special Procedure (e.g., UN Special Rapporteur, Independent Expert, or Working Group), engaging and reinforcing existing human rights monitoring mechanisms to assess the UNDROP's implementation, and drafting implementing legislation.[39]

It would be desirable to see the mutual supportiveness of these two instruments in this respect as well. FAO's activities on SSF could support the UNDROP's implementation by deepening the understanding of human rights and building capacity to respect and monitor them in the fisheries sector at different scales. Monitoring activities of international human rights bodies could also rely on the SSF Guidelines to better understand how human rights might be supported or hindered in the fisheries sector, particularly focusing on peasants involved in fishing and fishing-related activities. To that end, FAO, UNHRC, as well as the diverse UN Human Rights treaty bodies and special procedures should explore forms of collaboration.

The nuances underpinning the concept of 'small-scale fisheries'

There is no universally agreed definition of 'small-scale fisheries.' This term is multi-fold as academic scholarship has developed various methodologies to measure 'small-scale' or 'artisanal' fisheries,[40] which are often associated with numerous other terminologies like coastal, nearshore, customary, and traditional fisheries as well. Experts in the field underscore that a single, universal, and rigid definition of SSF would be unpractical and ineffective in capturing the diverse and dynamic forms of SSF around the world.[41] The flexibility of the notion, instead, allows for conveying the 'considerable diversity of contexts, from the social differentiation that typically exists in rural areas, of the blurred lines between rural and urban worlds, and from overlapping and often shifting registers of social identity.'[42]

Accordingly, the SSF Guidelines, the UNDROP, and other relevant international instruments[43] do not define SSF, which creates uncertainty as to who is entitled to special protection and preferential rights. Each country and, pursuant to the respective political and legal systems, different levels of government within the same country (e.g., state, province, and municipality) can thus develop a definition of SSF that corresponds more closely to the realities of SSF locally. In practice, however, public institutions do not take SSF into account appropriately,[44] let alone provide clear and consistent statements of what SSF are in their spheres of operation. It follows that policy-, law-, and decision-makers, as well as fisheries' managers, are likely to refer and treat SSF based on subjective, possibly ideological, and often undesirable interpretations.[45]

The SSF Guidelines provide guidance to governments and other actors by illustrating what small-scale fishing, small-scale fisher, or small-scale fish worker[46] relate to. The SSF Guidelines refer to SSF as a fisheries *subsector*,[47] which 'serves as an economic and social engine' (for local economies and communities' livelihoods)[48] 'along the value-chain – pre-harvest, harvest and post-harvest.'[49] In addition, the

SSF Guidelines avoid differentiating[50] between small-scale commercial artisanal, semi-industrial, subsistence, and arguably recreational, fishing activities.[51] They also note that small-scale fishers and fish workers are often 'self-employed and engaged in directly providing food for their household and communities as well as working in commercial fishing, processing and marketing.'[52] Finally, the SSF Guidelines do not limit the understanding of SSF actors to those solely engaged in fishing operations and related activities, but rather include 'fishers, fish workers, their communities, traditional and customary authorities, and related professional organisations and [CSOs].'[53]

Under the UNDROP, in turn, the definition of a 'peasant' includes a person who 'engages or who seeks to engage alone, or in association with others or as a community' in artisanal and small-scale fishing or a 'related occupation in a rural area' 'for subsistence and/or for the market.'[54] Similar to the SSF Guidelines, this definition is broad enough to encompass people involved in multiple activities associated with SSF, including related preparatory works and cultural practices (e.g., net making and mending, boat maintenance) conducted by fishers and fishing communities on land.[55] The Declaration also applies to hired workers, and broadly encompasses professional small-scale fishers along the supply and value-chain, and includes all migrant and seasonal workers on aquaculture farms,[56] who may have an alternative occupation as small-scale fishers.[57]

In addition, the UNDROP sets out two parameters for understanding the concept of 'peasants': (i) reliance on family labour or other non-monetised way of organising labour and (ii) special dependency on and attachment to the land.[58] So one initial difficulty arises as to whether references to *land* could be interpreted, literally, as referring only to *inland* small-scale fisher peasant – who conducts SSF or small-scale aquaculture in inland waters, e.g., rivers, lakes, lagoons, etc.[59] Other articles of the UNDROP, however, support a broader interpretation that encompasses *marine* SSF as well,[60] including in marine areas, over which coastal States and archipelagic States exercise sovereignty and jurisdiction in similar ways to those exercised on their land mass, as these actually represent an extension of the *land* seawards. This argument is supported by key concepts of the UN Convention on the Law of the Sea, which represent customary international law:[61] internal waters, the territorial sea, and the archipelagic waters are areas over which coastal States and archipelagic States exercise sovereignty and jurisdiction,[62] and where the majority of SSF and small-scale aquaculture operate due to their proximity to land and the technical inability of small-scale operations to occur offshore.[63] These legal concepts should be related to peasants' traditional or customary relationship with land utilised for their SSF activities, including *inland* and *marine* waters falling under the sovereignty of coastal States and archipelagic States, all waters of which correspond to their *fishing grounds*.[64] This interpretation of the UNDROP as broadly applying to inland and marine SSF resonates with the SSF Guidelines, which, while focusing on marine capture fisheries,[65] apply to SSF 'in all contexts,'[66] 'in both marine and inland waters,'[67] as well as to small-scale aquaculture.[68] By setting out these two parameters, the UNDROP also emphasises the 'intimate connection' of peasants with fishery resources and fishing grounds, either in inland or at sea as 'a source of social identity.'[69]

The linkages between fisheries and land become clearer in the UNDROP provisions on the governance of tenure, discussed later.[70] It is worth noting that the SSF Guidelines recognise the importance of land tenure rights in the coastal/waterfront area 'for ensuring and facilitating access to the fishery, for accessory activities (including processing and marketing), and for housing and other livelihood support.'[71] It also underscores that '[m]any small-scale fishers, fish workers and their communities – including vulnerable and marginalized groups – are directly dependent on access to fishery resources and land.'[72] The references, in both the UNDROP and the SSF Guidelines, to dependency on land[73] need to be understood, in as far as certain small-scale fishers also self-identify as indigenous peoples.[74] As such, they also enjoy the human right of indigenous peoples to 'maintain and strengthen their distinctive spiritual relationship with their traditionally occupied and otherwise used lands, territories, waters and coastal seas and other resources,' as provided by the UN Declaration on the Rights of Indigenous Peoples (UNDRIP).[75] Similar considerations apply in as far as SSF communities self-identify as 'traditional communities' under the UN Framework Principles on Human Rights and the Environment[76] and/or 'local communities' under the Convention on Biological Diversity (CBD).[77]

On the whole, the UNDROP and the SSF Guidelines are aligned in providing a broad and flexible notion of SSF, while calling attention to key features that provide the rationale for international human rights protection. This reflects observations in the literature that, '[w]hat is interesting and important about small-scale fisheries, therefore, is not their scale per se, but all that they are associated with . . . inherent and related features.'[78] The choice of the term 'peasant' under the UNDROP serves to capture the multiple grounds of discrimination experienced by individuals and communities in SSF such as expropriation of land, forced evictions and displacement, gender discrimination, the absence of agrarian reform and rural development policies, the lack of minimum wage and social protection, and the repression and criminalisation of movements protecting their rights.[79] These persisting discriminatory treatments have hindered the ability of peasants, including small-scale fishers and fish workers, to have their voice heard, defend their human rights and tenure rights, and secure sustainable use of natural resources that they depend on.[80] They have also precluded the appreciation and recognition of peasants' present and future contributions to development, conserving, and improving biodiversity, which constitutes the basis of food production throughout the world.[81] The SSF Guidelines also shed light on *vulnerable and marginalised groups* in SSF, pointing to their dependency on access to fishery resources and land, unequal power relations, isolation, limited opportunities, poverty, environmental threats and impacts, as well as the high interdependence and competition of SSF with large-scale fisheries, tourism, large-scale aquaculture, agriculture, energy, mining, industry, and infrastructure development.[82]

Mutual supportiveness between the UNDROP and the SSF Guidelines

Both the UNDROP[83] and the SSF Guidelines[84] support the application of international human rights standards to SSF, but they do so from different (albeit, arguably, compatible) perspectives. The UNDROP articulates directly specific international human rights standards from the perspectives of peasants as rightsholders, thereby clarifying States' obligations. The SSF Guidelines, in turn, focus on advancing food security and environmental sustainability, taking a 'human rights-based approach.'[85] On that basis, the SSF Guidelines identify good practices in the use and management of natural resources, emphasising the role of small-scale fishers as agents in environmental conservation and management[86] and as benefit holders of the social development, security, and safety, which is the State's responsibility to deliver.

The UNDROP makes the link more explicit between small-scale fishing communities' right to an adequate standard of living and 'concrete, productive dimensions' and 'real-life factors'[87] such as having the right to facilitated access to means of production, production, and processing, as well as technical assistance, credit, and insurance, which are necessary to gain access to local, national, and regional markets at prices that guarantee them a decent income and livelihood.[88] State discretion is clearly limited by human rights standards under the UNDROP, clarifying States' specific duties vis-à-vis peasants in SSF and the inter-linkages between concurrent human rights obligations (such as non-discrimination, access to effective remedies, and the protection of human rights defenders). Consequently, the UNDROP points towards the need to address systemic and engrained sources of discrimination, the multiple dimensions of poverty, and the underlying need to support the voice and vision of small-scale fishers, including their control over fishery resources, as part of the protection, respect, and full realisation of their human rights. In comparison, the SSF Guidelines arguably support a more accessible translation of international human rights obligations into action points that specialised managers and decision-makers are expected to implement, focusing on sector-specific means to address the real-life factors upon which the respect and enjoyment of small-scale fishing communities' human rights depend.

Against this background, the following sub-sections will identify similarities and differences, as well as opportunities for mutually supportive interpretation of the two instruments, in relation specifically to land tenure and control over natural resources, traditional knowledge, and procedural rights.

Land tenure and control over natural resources

One fundamental feature of both instruments is their recognition and protection of legitimate tenure rights to land (which, as argued earlier, extends to fishing grounds in both inland and marine waters). Recalling the FAO Voluntary Guidelines on the Responsible Governance of Tenure of land, fisheries, and forests in the context of national food security (Tenure Guidelines),[89] both the SSF Guidelines and the UNDROP consider legitimate tenure rights as including customary tenure rights as in non-conventional and traditionally regulated tenure systems, of lands[90] and natural resources,[91] as well as the restoration of the relevant peoples' access to their land in cases of natural disasters and/or armed conflict.[92] The two instruments safeguard SSF peasants from arbitrary displacement from their land or from other natural resources.[93] The instruments, however, diverge in the terms in which they address access to resources and environmental sustainability. They also differ in how they refer to the need for impact assessments, consultation/consent, and benefit sharing to ensure control over the use of resources.

Both the UNDROP and the SSF Guidelines provide for the facilitation of equitable access through redistributive reform.[94] UNDROP contains an express obligation to give *priority* to small-scale fishers, together with other peasants, in the allocation of public lands, fisheries, and forests, in line with the immediate relationship it establishes between land (including fishing grounds, as discussed earlier) and the attainment of an adequate standard of living.[95] In addition, under the UNDROP, the 'right to land' establishes a more immediate relationship between the *control* of fishing grounds and the attainment of an adequate standard of living for small-scale fishers.[96] This arguably provides 'a more explicit normative foundation for redistributive agrarian reforms.'[97] In comparison, the SSF Guidelines call for the appropriate granting of *preferential access* of small-scale fishers to land, fishery resources and to fish in waters under national jurisdiction, as well as 'exclusive zones for small-scale fisheries,'[98] and 'co-management.'[99]

The UNDROP's provision on the right to have access to, sustainably use and manage 'land and the water bodies, fisheries, pastures and forests therein'[100] serves to bring attention to the various habitats in which SSF communities may conduct activities, limited to not only fishing but also alternative or complementary livelihoods during closed seasons, closed areas, or in the event of disasters and crisis where fishing is not an option. The UNDROP also recalls the CBD's duty to conserve and sustainably use biodiversity, connecting it with the need 'to promote and protect the full enjoyment of the rights of peasants and other people working in rural areas.'[101] Such environmental standards are reflected in the SSF Guidelines through the ecosystem approach to fisheries. The latter entails involving small-scale fishers in sustainably managing fishery resources for the benefit of habitat, biodiversity, and ecosystem well-being, and ensuring that small-scale fishers can benefit from fishery resources maintained by healthy ecosystems.[102]

Safeguards

Both instruments refer to three key safeguards for the respect and protection of the rights to land tenure (including fishing grounds) and natural resources: impact assessments, consultation, and benefit sharing. These standards have emerged at the crossroads of international human rights law jurisprudence on indigenous peoples and international biodiversity law[103] and are also recognised in the UN Framework Principles on Human Rights and the Environment.[104]

The first safeguard is the undertaking of prior impact assessments, which cover the analysis of potential environmental impacts, as well as the *social* and *cultural* impacts that peasants and people in SSF may suffer. The UNDROP more broadly requires such studies prior to the granting of permission regarding *any* exploitation affecting natural resources that peasants may hold or use,[105] which is in line with the UN Framework Principles, whereas the SSF Guidelines limit this requirement to the implementation of large-scale projects.[106] According to the UN Framework Principles and other international human rights bodies, these assessments should be in accordance with the CBD Akwé: Kon Voluntary Guidelines, which provide concrete and systematic step-by-step indications on how impact assessments should include consideration of the exercise of customary laws regarding land tenure, traditional systems of natural resource use, maintenance of genetic diversity through customary management, as well as gender, inter-generational considerations, health, safety, food, livelihoods, social cohesion, and mobilisation.[107] Notably, the CBD Akwé: Kon Voluntary Guidelines recommend considering not only negative but also positive impacts, including from the viewpoint of indigenous peoples and local communities.[108]

The second safeguard concerns consultation and participation in decision-making processes. According to both instruments, SSF communities, indigenous peoples, peasants, and other people working in rural areas have the right to 'active, free, effective, meaningful and informed participation' prior to the adoption of decisions that may affect them, also taking into account relevant instruments on indigenous peoples.[109] The UNDROP requires good-faith consultation broadly, before carrying out natural resources exploitation, and participation in the preparation and implementation of food safety, labour, and environmental standards.[110] The SSF Guidelines, in turn, require consultation specifically prior to the implementation of large-scale projects; the adoption of policies and management measures related to migration of fishers and fish workers, international trade, climate change and disasters, inland and marine spatial planning, and the setting of research priorities.[111] Neither provides for the standard of free, prior and informed consent (FPIC) which is instead emphasised in the UNDRIP and the UN Framework Principles.[112]

Indeed, international human rights law instruments on indigenous peoples' rights and international biodiversity law have called more specifically for consulting with indigenous peoples to obtain their FPIC before taking or approving any measures that may affect their lands, territories, or resources, on the basis of access to all relevant information in understandable and accessible forms.[113] The Inter-American Court of Human Rights (IACHR) underscored the need for 'special and differentiated' consultation processes when the interests of indigenous and tribal peoples may be affected,[114] with the notion of public interest set at a higher threshold because their physical and cultural survival is at stake.[115] Accordingly, FPIC should arguably guarantee a 'distinguishable voice' for indigenous and tribal peoples within a pluralistic and democratic society in light of their right to decide their own development priorities.[116] Other international legal materials have recognised FPIC as part of the human rights of communities other than indigenous peoples.[117] Consensus guidance under the CBD, which benefitted from significant inputs from indigenous peoples' representatives,[118] provided further insights on the 'prior' and 'free' components of FPIC as a *continual* process building mutually beneficial and *ongoing* arrangements[119] that should be free from 'expectations or timelines that are externally imposed.'[120] This formulation fleshes out the characterisation of FPIC as a 'constant process of dialogue' advanced in the Inter-American context.[121]

With regard to the 'informed' dimension of FPIC, international human rights bodies have also clarified that FPIC should be based on an understanding of the full range of issues and implications entailed by the activity or decision in question. As discussed earlier, the relationship between FPIC and impact assessment has been explored by human rights bodies with a view to providing indigenous peoples with 'full and objective information about all aspects of the project that will affect them, including the impact of the project on their lives and environment.'[122] That said, certain States continue to object to the FPIC standard.[123]

The third safeguard is fair and equitable benefit sharing. The UNDROP calls upon States to 'take measures to ensure that any exploitation affecting the natural resources that peasants traditionally hold or use is permitted based on . . . modalities for the fair and equitable sharing of benefits of such exploitation.'[124] This is in line with international human rights jurisprudence seeking to ensure this safeguard for the maintenance of the cultural and physical survival of indigenous peoples.[125] The SSF Guidelines, in turn, support the 'equitable distribution of the benefits yielded from responsible management of fisheries and ecosystems, rewarding small-scale fishers and fish workers, both men and women.'[126] In doing so, it emphasises the role of benefit sharing towards ecosystem stewards as part of an ecosystem approach.[127] It also conceptualises benefit sharing as an *incentive* for ecosystem stewards' positive contribution to human well-being deriving from the ecosystem services they provide, maintain, or restore.[128] Benefit sharing can also be understood as *recognition* for past and present contributions of SSF communities to global environmental objectives and food security, with a view to ensuring that their traditional practices continue in the future.[129] The SSF Guidelines explicitly link benefit sharing with non-discrimination (based on gender, discussed later, and against indigenous peoples),[130] which is also a dimension recognised under the UN Framework Principles.[131] Furthermore, the SSF Guidelines note the cross-scale dimensions of benefit sharing, by reference to the need to ensure that SSF communities benefit from wider economic developments at the local level (such as tourism)[132] and international trade.[133] According to the CBD Akwé: Kon Voluntary Guidelines, benefit sharing should be explored as early as during environmental impact assessments and be considered intertwined with the partnership-building process of FPIC.[134]

These three safeguards (impact assessments, consultation, and benefit sharing) are also relevant to prevent and address gender discrimination[135] in the context of reforms and resettlement schemes, according to the UNDROP,[136] and in the designing, planning, and implementation of management measures, according to the SSF Guidelines,[137] which specifically note the role of benefit sharing from the management of fisheries and ecosystems for *women* in SSF.[138] These safeguards can also address other grounds for non-discrimination and ensure consideration of the human rights of children.[139]

Those three key safeguards (impact assessments, consultation, and benefit sharing), however, have all been criticised because they may arguably be 'consistent with the penetration of commercial forms of production' and 'operate in ways that are co-extensive with extractivist models.'[140] In particular, benefit sharing is often associated with an offer of money or other economic advantages (for instance, employment) in exchange for obtaining consent,[141] which 'encourages a climate of disrespect towards indigenous peoples.'[142] Benefit sharing has resulted in 'attempts to undermine social cohesion of affected communities' through bribes to community leaders or selective negotiations tactics.[143] In addition, monetary benefit sharing is known to 'destruct the social network' of indigenous groups,[144] putting in place self-defeating or paternalistic mechanisms that are not responsive to communities' specific needs.[145] Regional human rights bodies have thus had occasion to point out situations in which promised benefit sharing was not delivered,[146] or benefit-sharing arrangements were originally in place but broke down, and/or were weakened by ineffective State monitoring of outsiders' activities.[147]

An application of these guarantees, however, that genuinely builds upon international human rights and biodiversity law[148] can arguably make space for different worldviews of nature and development[149] embodied in SSF communities' distinctive ways of life.[150] In that way, the safeguards can serve to challenge mainstream conceptions of economic development and tackle attempts to bottle SSF communities' worldviews into neo-liberal or neo-colonial agendas.[151] To that end, impact assessments, consultation, and benefit sharing need to support communities' *agency* as tools for a concerned and iterative dialogue aimed at understanding and accommodating different worldviews, enhancing communities' choice and capabilities.[152] That transformative potential can be supported by a better understanding of SSF communities' needs, values, and priorities, as reflected under the UNDROP, so as to inform case-by-case assessments to ensure culturally appropriate implementation of international human rights law.

In turn, the SSF Guidelines provide a finer-grained understanding of the barriers and opportunities within fisheries governance. They do not indicate how to avoid imposing certain views of development upon SSF communities that could endanger their cultural or physical survival. The SSF Guidelines nevertheless can contribute to the transformative application of the three safeguards by emphasising the need for providing *support* for the exercise of SSF communities' rights, in addition to protecting or enhancing communities' *control* over natural resources (as emphasised by the UNDROP). Together, the two instruments can shift the practice of impact assessments, consultation, and benefit sharing away from a technocratic, damage-control approach, towards collaboratively identifying opportunities for creating positive impacts in addition to avoiding potential negative impacts, taking into account SSF communities' views.[153]

Traditional knowledge

The importance of traditional knowledge is recognised by both instruments.[154] These considerations about knowledge are essential to understand the tensions around environmental sustainability in SSF: whose knowledge (scientific, modern, or traditional) determines environmental sustainability approaches in the fisheries sector more broadly? How are the knowledge contributions of small-scale fishers to sustainable fisheries assessed and taken into account in decision-making processes? And whose knowledge determines the limitations imposed upon SSF to ensure sustainable practices?

The UNDROP, which recalls the rights to culture, practices, and knowledge, including ways of life, methods of production, or technology,[155] specifically emphasises traditional ways of fishing and community-based commercialisation systems;[156] and it emphasises the *rights* to use and protect peasants' traditional medicines, and to maintain their health practices using plants, animals, and minerals for medicinal use.[157] Complementing the UNDROP with a focus on means of implementation, the SSF Guidelines include as one of their objectives 'to enhance public awareness and promote the advancement of knowledge on the culture, role, contribution and potential of small-scale fisheries, considering ancestral and traditional knowledge, and their related constraints and opportunities.'[158] The SSF Guidelines then point to the need for technical and financial assistance to maintain, organise, exchange, and improve traditional knowledge of aquatic living resources and fishing techniques.[159]

The UNDROP provides a clear emphasis on the use of traditional knowledge of peasants, including small-scale fishers, in the design and implementation of climate change adaptation and mitigation policies,[160] which is less explicit in the SSF Guidelines. The UNDROP affirms their rights in contributing through the 'use of practices and traditional knowledge' in such endeavours whilst the SSF Guidelines do not mention 'traditional knowledge' in its section devoted to disaster risks and climate change. The UNDROP rather calls for full effective consultation with fishing communities, including indigenous peoples, men and women, in the development of policies and plans to address climate change in fisheries.[161] The SSF Guidelines underline the need for support in addressing climate change, through measures that secure disaster preparedness, emergency response, relief, and rehabilitation, and the role of small-scale fishers in supporting energy efficiency in the subsector.[162] They further call for transparent access to adaptation funds, facilities, and culturally appropriate technologies for climate change adaptation.[163]

All these provisions on traditional knowledge are essential to effectively contributing to the recognition of small-scale fishers and their communities' contribution to sustainable development and environmental protection, and equally to give them voice in decision-making processes that may affect their ways of life and livelihoods.[164] International guidance underscores that respecting traditional knowledge requires valuing it equally with, and complementary to, scientific knowledge, in order to promote the full respect for the cultural and intellectual heritage of indigenous and local communities relevant to the conservation and sustainable use of biodiversity.[165] The actual impact on decision-making, however, depends on whether traditional knowledge holders have sufficient procedural access to the relevant decision-making processes,[166] as discussed later. In addition, even if traditional knowledge is recognised as a form of science, its integration in various decision-making processes may impose unfair burdens on traditional knowledge holders, constraining the further development of knowledge systems in light of changed circumstances, including changes to traditional lifestyles within which traditional knowledge is rooted.[167] These considerations point to the need for caution in engaging with traditional knowledge holders and the underlying requirements for impact assessment, consultation, and benefit sharing with traditional knowledge holders, which is clearly recognised in both international human rights jurisprudence on indigenous peoples and international biodiversity law[168] but is not explicitly recalled in the UNDROP nor the SSF Guidelines.[169] Nevertheless, the relevance of these safeguards can be read into the UNDROP based on the combined effect of its provisions on traditional knowledge and that on the right of access to natural resources,[170] as the development and transmission of traditional knowledge are intrinsic to the land tenure (including fishing grounds) and customary governance of natural resources.[171]

Arguably in line with the requirements of consultation and benefit sharing,[172] the UNDROP points to the role of the State in encouraging 'equitable and participatory partnerships' with scientists.[173] This is an important point with regard to the interface between SSF communities' traditional knowledge and technology transfer in the fisheries sector.[174] While there may be growing political awareness of the benefits that could arise from marine technology transfer to SSF communities, unequal attention has been paid to actual and potential risks, particularly with regard to technologies that seek to enhance the cost efficiency of fishing activities. SSF communities' own local technologies are more likely to meet local needs for food,[175] and be sensitive to the location in which they are applied, the relative abundance of fishery resources, and the complex, traditional resource use rights.[176] SSF community–scientists partnerships could thus focus on examining and evaluating 'local-specific, small-scale technologies, coupled with community-oriented, participatory measures to protect the ecological integrity of the living coastal resources,' with a view to facilitating 'technology blending' to take advantage, where appropriate, of the positive aspects of other technologies.[177]

Returning to the matter of whether the protection of traditional knowledge helps address tensions around the environmental sustainability of SSF, the UNDROP frames these issues in terms of the environmental *rights* of peasants,[178] their entitlements to the conservation and sustainable use of rural areas and biodiversity as part of the efforts to protect their traditional knowledge and their right to livelihoods.[179] The SSF Guidelines, in turn, spell out the environmental *duties* of people in SSF (e.g., utilise non-destructive fishing practices and energy-efficiency operations),[180] once again taking the viewpoint of States as implementers of international fisheries law and applying the human rights-based approach to sector-specific contexts. What the UNDROP adds, in that connection, is underscoring the need to 'challenge ingrained prejudices about the "backwardness," or "modernity," of different systems of livelihoods and beliefs, and of different forms of natural resource use.'[181] These prejudices indeed 'underpin the structural discrimination that peasants and indigenous peoples experience in many legal systems,'[182] and that may be reflected in the mainstream approaches to environmental sustainability in the fisheries sector.

Procedural rights

Access to information, justice, and effective remedies are essential procedural rights that bolster the guarantees on tenure and control over natural resources, as well as the protection of traditional knowledge just discussed.[183] The SSF Guidelines and the UNDROP are both attentive to the need of small-scale fishing communities and peasants to have access to timely and adequate information, especially concerning production, processing, marketing, and distribution of their products.[184] And that information should be provided in a language, form, and through means adequate to their culture, so that all the interested people can be offered the opportunity to decide to appropriately participate in decision-making processes.[185] Once again focusing on sector-specific means of implementation, the SSF Guidelines emphasise the need for information to ensure the sustainability of SSF, including data on illegal, unregulated and unreported fishing, disaster risks, climate change, livelihoods, and food security,[186] as well as the need to develop information systems for data-poor conditions.[187] This is in line with the broader recognition, under the SSF Guidelines, of the importance of access to information for helping small-scale fishing communities to 'cope with existing problems and empower them to improve their livelihoods.'[188]

Provisions on access to justice are, in turn, more elaborated in the UNDROP than in the SSF Guidelines. Both require access to impartial and competent judicial and administrative bodies, to timely, affordable, and effective means of resolving disputes, and provision for prompt remedies such as right of appeal, restitution, indemnity, compensation, and reparation.[189] The UNDROP, however, also highlights the right to legal assistance, providing for legal aid as an additional measure for peasants who would otherwise not have access to administrative and judicial services.[190] Whereas the SSF Guidelines are limited in only providing for the protection of SSF communities from arbitrary eviction and protection of their legitimate tenure rights from extinction or infringement, the UNDROP is broader in requiring effective mechanisms for the prevention of and redress for *any* action aimed at or resulting in the violation of peasants' human rights.[191]

All these procedural dimensions serve to support the voice of small-scale fishers and their communities in the making of decisions that could affect them and to recognise their contributions to sustainable fisheries and sustainable development more broadly.[192] They are all essential underpinnings of the human right to natural resources, including its interface with gender discrimination and the protection of traditional knowledge discussed earlier. In other words, while the UNDROP does not recognise the right to food sovereignty as such,[193] it provides substantive and procedural dimensions underpinning SSF communities' right to determine their own fisheries systems[194] and with that underscores the opportunity to also address poverty as a 'multi-dimensional phenomenon that can be underpinned by marginalisation and lack of voice as well as low incomes.'[195] Attention to practical means of implementation as provided under the SSF Guidelines, however, remains essential, particularly as many small-scale fishing communities are located in remote areas and may not be able to take advantage of their procedural rights for logistical reasons.[196]

Linking the UNDROP and the SSF Guidelines to the Sustainable Development Goals

Both the UNDROP and the SSF Guidelines seek to support the application of international human rights instruments and standards to SSF, and in doing so, they are each remarkable in their own field. The UNDROP expands, to SSF, the understanding of the relevance of international human rights law for natural resources, which has mainly been concerned with land-based activities. The SSF Guidelines, in turn, stand out as an innovative international fisheries law instrument, as in that area of international law, there is still insufficient integration with human rights.[197]

The UNDROP and the SSF Guidelines engage with similar issues, although from different perspectives. The Declaration directly articulates specific international human rights standards from the perspectives of the rightsholders, including the need to address systemic sources of discrimination, the multiple dimensions of poverty, and the underlying need to support the voice, vision, and control of resources of small-scale fishers as part and parcel of the protection, respect, and full realisation of their human rights. The SSF Guidelines, in turn, mainly take the perspective of duty bearers.[198] By applying the human rights-based approach for identifying good practices and sector-specific means of implementation, this instrument thereby seeks to balance the progressive realisation of small-scale fishers' rights with the need to ensure the sustainable management of resources. While the SSF Guidelines may be seen as glossing over the need for more structural approaches, they usefully complement the UNDROP by providing a fine-grained picture of the technicalities that need to be tackled in the fisheries sector and a more accessible translation of human rights for the specialised managers and decision-makers that are in practice implementing the human rights obligations of States.[199]

Their combined interpretation and application are therefore called for, even when governments may not have supported the UNDROP, given the universal, interconnected, indivisible nature of the underlying international commitments to ensure inclusive and fair sustainable development.[200] To some extent, the SDGs provide the ground for doing so, by including specific targets on access of small-scale fishers to land, marine resources, productive resources and inputs, knowledge, financial services, markets, and opportunities for value added.[201] But the SDGs address these issues from a technical perspective and the lack of explicit reference to relevant human rights calls into question the suitability of these targets to focus attention on the systemic causes of discrimination that prevent small-scale fishing communities from gaining access to these resources, and the underlying need to control resources and production as part of the protection and realisation of these communities' human rights.

The complexities underlying these targets, however, can be brought to light by reflecting on the widely acknowledged need to pursue the SDGs in a coherent manner and exploring the nexus between SSF-related targets and other SDGs. The respective contributions of the UNDROP and the SSF Guidelines can thus guide governments, development operators, and businesses in respecting human rights, as recommended by the HRC,[202] supporting efforts to realise the SDGs through policy coherence, for instance, with reference to SDGs 1 (poverty), 2 (food security), 5 (gender equality), 13 (combating climate change), and 16 (access to justice). Both the UNDROP and the SSF Guidelines offer specific guidance on how to address all these issues, thereby supporting the understanding of the inter-dependencies of the SDGs in the SSF context.

In conclusion, the UNDROP supports an understanding of the daily realities of marginalisation of small-scale fishers that connects the good practices identified in the SSF Guidelines to the need to pursue the SDGs in a coherent manner. We thus argue that this understanding should help overcome the challenge of the limited support for the UNDROP by developed states, due to their resistance to embrace the rights holders' demands to recognise the 'relation between social justice claims and human rights norms' as well as on 'collective rights, on control over the means of production and agricultural value chains.'[203] Could the opportunities of a mutually supportive interpretation of international fisheries law, international human rights law, and international environmental law to the benefit of realising multiple SDGs rather be the basis for a more constructive engagement of duty bearers with the rightsholders involved in SSF? This chapter has partially explored this question by analysing two key instruments for SSF, with a focus on tenure and control of natural resources, as well as the protection of traditional knowledge and procedural rights. Concrete implementation in coherent and complementary ways can be further supported by FAO and international human rights bodies, in their operational, advisory, and monitoring activities, respectively.

Notes

1 UN Human Rights Council, 'United Nations Declaration on the Rights of Peasants and Other People Working in Rural Areas' (8 October 2018) UN Doc A/HRC/RES/39/12 (adopted by 33 votes to 3 – Australia, Hungary and the UK; 11 abstentions).
2 Ibid., art. 1.2 and 1.3.
3 The acronym SSF is used in this book chapter for both 'small-scale fisheries' and 'small-scale fishing' when referred to small-scale fishing communities.
4 The UNDROP was adopted by majority voting at the UN General Assembly (UNGA) and at the Human Rights Council (UNHRC), as discussed in the following section on the development of the UNDROP and the SSF Guidelines.
5 UNDROP, supra note 1, art. 3.1.
6 Ibid., art. 2.1.
7 FAO, *Voluntary Guidelines for Securing Sustainable Small-scale Fisheries in the Context of Food Security and Poverty Eradication* (FAO 2015) 18, (adopted at the 31st Session of the Committee on Fisheries, 9–13 June 2014).
8 FAO, 'Code of Conduct for Responsible Fisheries' (31 October 1995) Resolution 4/95.
9 In reference to women throughout many parts of the SSF Guidelines, as will be detailed later, and specifically to the Convention on the Elimination of All Forms of Discrimination against Women (adopted 18 December 1979, entered into force 3 September 1981) 1249 UNTS 13 (CEDAW).
10 In reference to indigenous peoples in many parts of the SSF Guidelines, as will also be detailed later and specifically to the UNGA, 'United Nations Declaration on the Rights of Indigenous Peoples' (2 October 2007) UN Doc A/RES/61/295 (UNDRIP).

11 In reference to migrants in SSF. No specific international instrument dedicated to migrant workers is mentioned in the SSF Guidelines, which generally refers to relevant ILO instruments. These may include the Work in Fishing Convention (adopted 14 June 2007, entered into force 16 November 2017) (C188); the Migration for Employment Convention (Revised) (adopted 01 July 1949, entered into force 22 January 1952) (C097).

12 In reference to children engaged in SSF and, in specific, to the Convention on the Rights of the Child (adopted 20 November 1989, entered into force 2 September 1990) 1577 UNTS 3 (CRC).

13 That is, expertise from a range of actors, from government level to community level. Both the UNDROP and the SSF Guidelines, as detailed further later, were the outcomes of participatory processes, proving participation by non-State actors, notably small-scale fishers' representatives, in processes that have been traditionally technocratic and States-driven, possible to obtain.

14 While fisheries' management has traditionally been the focus of international fisheries instruments, modern international fisheries law (to which FAO is inherently associated with as the main UN agency driving the legal developments of this domain) has been changing over the past decades to encompass environmental and social concerns in fisheries. For a comprehensive collection that brought about various aspects of this evolving international law field, which draws attention to environmental considerations and evolution in international fisheries law, see Erik J. Molenaar and Richard Caddell, *Strengthening International Fisheries Law in an Era of Changing Oceans* (Hart Publishing 2019). For linkages between international fisheries law and the protection of people at sea, including fishers, see Irini Papanicolopulu, *International Law and the Protection of People at Sea* (Oxford University Press 2018).

15 That is, 'mutual supportiveness' among different international regimes to be 'understood and applied as reinforcing each other with a view to fostering harmonization and complementarity, as opposed to conflictual relationships,' as argued in Riccardo Pavoni, 'Mutual Supportiveness as a Principle of Interpretation and Law-Making: A Watershed for the "WTO-and-Competing-Regimes" Debate?' (2010) 21 *European Journal of International Law* 649.

16 UNGA, 'Resolution 70/1 of 25 September 2015, Transforming our world: the 2030 Agenda for Sustainable Development' (21 October 2015) UN Doc A/RES/70/1, SDGs 14(b) and 2.3.

17 See, respectively, UNGA, 'Resolution adopted by the General Assembly on 15 March 2006, Human Rights Council' (3 April 2006) UN Doc A/RES/60/251; FAO, 'Rules of Procedure of the Committee on Fisheries' in *Basic Texts of the Food and Agriculture Organization of the United Nations* (Volumes I and II, FAO 2017) 109–16.

18 UNHRC, 'Report of the Human Rights Council on its Thirty-Ninth Session' (23 November 2018) UN Doc A/HRC/39/2, 191.

19 FAO, 'Report of the Thirty-first Session of the FAO Committee on Fisheries' (9–13 June 2014) FIPI/R1101/2015.

20 The resolution was adopted by a recorded vote of 23 to 9, with 15 abstentions. Those countries voting against were Austria, Belgium, Czech Republic, Hungary, Italy, Poland, Romania, Spain, and the United States of America. See UN Human Rights Council, 'Promotion and Protection of the Human Rights of Peasants and other People Working in Rural Areas' (11 October 2012) UN Doc A/HRC/RES/21/19.

21 Lorenzo Cotula, 'Between Hope and Critique: Human Rights, Social Justice and Re-imagining International Law from the Bottom Up' (2020) 48(473) *Georgia Journal of International and Comparative Law* 505–8.

22 Ibid., 506; Eibe Riedel, Gilles Giacca and Christophe Golay, 'The Development of Economic, Social, and Cultural Rights in International Law' in Eibe Riedel, Gilles Giacca and Christophe Golay (eds), *Economic, Social, and Cultural Rights in International Law: Contemporary Issues and Challenges* (Oxford University Press 2014) 5–6.

23 See UNGA, 'Sustainable fisheries, including through the 1995 Agreement for the Implementation of the Provisions of the United Nations Convention on the Law of the Sea of 10 December 1982 relating to the Conservation and Management of Straddling Fish Stocks and Highly Migratory Fish Stocks, and related instruments' (8 December 2006) UN Doc A/RES/61/105, paras. 12 and 97.
24 See FAO, 'Report of the Global Conference on Small-Scale Fisheries Securing Sustainable Small-scale Fisheries: Bringing Together Responsible Fisheries and Social Development' (13–17 October 2008) FIEP/R911 (Tri).
25 COFI members 'approved the development of a new international instrument on small-scale fisheries that would draw on relevant existing instruments, complementing the Code [the CCRF].' See FAO, 'Report of the Twenty-Ninth Session of the FAO Committee on Fisheries' (31 January – 4 February 2011) FIPI/R973, VII-VIII, para. 56.
26 Similarly, other FAO international fisheries instruments have been considered by technical consultations first, followed by consensus-based negotiation at COFI and final adoption by FAO Council. See Alan Boyle and Christine Chinkin, *The Making of International Law* (Oxford University Press 2007) 126–8.
27 FAO, *Securing Sustainable Small-Scale Fisheries: Update on the Development of the Voluntary Guidelines for Securing Sustainable Small-Scale Fisheries in the Context of Food Security and Poverty Eradication (SSF Guidelines)* (FAO 2014) 4–5.
28 FAO, 'Report of the Thirty-first Session of the FAO Committee on Fisheries,' supra note 19.
29 UNHRC, 'Promotion and Protection of the Human Rights of Peasants and other People Working in Rural Areas,' supra note 20. Note that the HRC is composed by 47 Member States: UNHRC, 'Current Membership of the Human Rights Council for the 14th cycle, 1 January – 31 December 2020' (2020) <www.ohchr.org/en/hrbodies/hrc/pages/currentmembers.aspx> accessed 30 April 2021.
30 UNGA, 'Report of the Third Committee, 55th Plenary Meeting' (17 December 2018) UN Doc A/73/PV55.
31 UNDROP, supra note 1, preamble.
32 Cotula, 'Between Hope and Critique: Human Rights, Social Justice and Re-imagining International Law from the Bottom Up,' supra note 21, at 505.
33 FAO, 'Chairperson's Report of the Technical Consultation on International Guidelines for Securing Sustainable Small-Scale Fisheries – Thirty-first Session of the FAO Committee on Fisheries' (9–13 June 2014) COFI/2014/Inf10, Para 32.
34 It has been argued that this feature has blurred the boundaries between *hard* law and *soft* law, where the latter mimics the former's compliance mechanisms. See Catherine Redgwell, 'International Soft Law and Globalization' in Barry Barton et al (eds), *Regulating Energy and Natural Resources* (Oxford Scholarship Online 2006).
35 In the subsequent meetings of COFI, both the 32nd and 33rd sessions held, respectively, in 2016 and 2018, the attending Member States have discussed the implementation of the SSF Guidelines. See FAO, 'Securing Sustainable Small-Scale Fisheries: Towards Implementation of the Voluntary Guidelines for Securing Sustainable Small-Scale Fisheries in the Context of Food Security and Poverty Eradication (SSF Guidelines)' COFI 32nd Session (April 2016) Doc COFI 2016/7; FAO, *Small-Scale and Artisanal Fisheries Governance* (COFI 33rd Session, Doc COFI/2018/7, March 2018). See also FAO, *Global Strategic Framework in Support of the Implementation of the Voluntary Guidelines for Securing Sustainable Small-Scale Fisheries in the Context of Food Security and Poverty Eradication* (FAO 2020).
36 FAO, 'Securing Sustainable Small-Scale Fisheries: Towards Implementation of the Voluntary Guidelines for Securing Sustainable Small-Scale Fisheries in the Context of Food Security and Poverty Eradication (SSF Guidelines),' supra note 35, at para. 18; FAO, *Enhancing the contribution of small-scale fisheries to food security and sustainable livelihoods* (FAO 2019).
37 Notably, some countries (Cambodia, Costa Rica, Ecuador, Guinea, Peru, Sri Lanka, and Tanzania) have been identified in taking specific steps to implement the SSF Guidelines

at national level, including the review and development of legal frameworks as in the case of Cambodia and Costa Rica. See FAO, 'Update on Progress to Develop the Global Strategic Framework in Support of the Implementation of the Voluntary Guidelines for Securing Small-Scale Fisheries in the Context of Food Security and Poverty Eradication (SSF-GSF)' COFI 33rd Session (July 2018) COFI/2018/SBD23, 6. See also Julia Nakamura, Ratana Chuenpagdee and Mustafa El Halimi, 'Unpacking Legal and Policy Frameworks: A Step Ahead for Implementing the Small-Scale Fisheries Guidelines' (2021) 129 *Marine Policy* 1.

38 See FAO, *Legislating for Sustainable Small-Scale Fisheries – A guide and considerations for implementing aspects of the Voluntary Guidelines for Securing Sustainable Small-Scale Fisheries in the Context of Food Security and Poverty Eradication in national legislation* (FAO 2020).

39 Priscilla Claeys and Marc Edelman, 'The United Nations Declaration on the Rights of Peasants and other People Working in Rural Areas' (2020) 47(1) *The Journal of Peasants Studies* 10–11 and 43.

40 See Darah Gibson and Rashid Sumaila, 'Determining the Degree of "Small-Scaleness" Using Fisheries in British Columbia as an Example' (2017) 86 *Marine Policy* 121; Yannick Rosseau et al, 'Defining Global Artisanal Fisheries' (2019) 108(103634) *Marine Policy*; Hillary Smith and Xavier Basurto, 'Defining Small-Scale Fisheries and Examining the Role of Science in Shaping Perceptions of Who and What Counts: A Systematic Review' (2019) 6 *Frontiers in Marine Science* 1.

41 They stem from the context-specific feature of SSF, which varies largely by and within communities at any level of governance; the numerous SSF definitions that can be found in scientific literature and legislation, which, in turn, are based on different criteria and mostly on technological parameters (e.g., boat size, type of fishing gear, motor power, etc.); the diversity and dynamism in SSF fishing activities. See Smith and Basurto, 'Defining Small-Scale Fisheries and Examining the Role of Science in Shaping Perceptions of Who and What Counts: A Systematic Review,' supra note 42; Ratana Chuenpagdee and Svein Jentoft, 'Transforming the Governance of Small-Scale Fisheries' (2018) 17 *Maritime Studies* 101, 105–6; David J. Mills et al, 'Small-scale Fisheries in the Developing World' in Robert Pomeroy and Nail L. Andrew (eds), *Small-Scale Fisheries Management: Frameworks and Approaches for the Developing World* (CAB International 2011) 1; Derek S. Johnson, 'The Values of Small-Scale Fisheries' in Derek S. Johnson and others (eds), *Social Wellbeing and the Values of Small-scale Fisheries* (Vol 7 Springer International Publishing AG 2018) 3–7; Simon Funge-Smith, *Towards Statistical Definition of Small-Scale Fisheries: A Matrix Scoring Approach to Characterization of the Scale of Fishing Units* (Working paper CWP-IS/2019/11 Twentyseventh Meeting of the Fisheries Subject Group, Coordinating Working Party on Fishery Statistics, Rome, 15–18 May 2019, 2019).

42 Cotula, 'Between Hope and Critique: Human Rights, Social Justice and Re-imagining International Law from the Bottom Up,' supra note 21, at 514.

43 For instance, the Agreement on Port State Measures Agreement to Prevent, Deter and Eliminate Illegal, Unreported and Unregulated Fishing (adopted on 22 November 2009, in Rome, Italy, entered into force on 5 June 2016) exempts 'artisanal subsistence fishing' from the scope of its application without defining it (art. 3(1)(a)(b)). This also points to the need for a definition of 'subsistence': D. Owen pers. comm. on 05 June 2019.

44 Steven W. Purcell and Robert S. Pomeroy, 'Driving Small-scale Fisheries in Developing Countries' (2015) 2 *Frontiers in Marine Science* 1; David J. Mills et al, 'Under-reported and Undervalued: Small-scale Fisheries in the Developing World' in Robert Pomeroy and Nail L. Andrew (eds), *Small-scale Fisheries Management: Frameworks and Approaches for the Developing World* (CAB International 2011).

45 SSF is commonly perceived under a negative narrative that places them as part of the problem in fisheries' sustainability, instead of elevating their contributions to the solution (in social, economic, and environmental terms). The importance of changing this

narrative to be a more positive and inclusive one was highlighted in a recent symposium held at FAO. See FAO, *Proceedings of the International Symposium on Fisheries Sustainability: Strengthening the Science-policy Nexus* (FAO Headquarters, 18–21 November 2019 FAO Fisheries and Aquaculture Proceedings No. 65) (Rome 2020) 27–8.
46 FAO, SSF Guidelines, supra note 7, sec. 2.4.
47 Ibid., preface, para. 5.
48 Ibid., foreword, para. 1–2.
49 Ibid., preface, para. 3.
50 Though such distinction remains important for the production of disaggregated fisheries catch data. See Daniel Pauly and Dirk Zeller, 'Catch Reconstructions Reveal that Global Marine Fisheries Catches are Higher than Reported and Declining' (2016) *Nature Communications* 7.
51 While recreational fishing is often seen as a different sector to SSF, recreation may be part of the 'multi-functionality' practices in SSF. See FISHINMED, *Small-Scale Fisheries Multifunctionality Best Practices* (Report prepared under the ENPI CBC Mediterranean Sea Basin Programme 2007–2013, 2013).
52 FAO, SFF Guidelines, supra note 7, preface, para. 4.
53 Ibid., preface, para. 2.3.
54 UNDROP, supra note 1, arts. 1.1 and 1.2.
55 See Svein Jentoft and Arne Eide (eds), *Poverty Mosaics: Realities and Prospects in Small-Scale Fisheries* (Springer 2011).
56 UNDROP, supra note 1, art. 1.3.
57 The migratory move of small-scale fishers occurs for different purposes other than fishing, often constituting a way of life. See Innocent Ngao Wanyonyi and others, 'Artisanal Fisher Migration Patterns in Coastal East Africa' (2016) 119 *Ocean & Coastal Management* 93.
58 UNDROP, supra note 1, art. 1.1 reads: 'For the purposes of the present Declaration, a peasant is any person who engages or who seeks to engage alone, or in association with others or as a community, in small-scale agricultural production for subsistence and/or for the market, and who relies significantly, though not necessarily exclusively, on family or household labour and other non-monetized ways of organizing labour, and who has a special dependency on and attachment to the land.'
59 In reality, the UNDROP applies to all workers on 'farms in *aquaculture*,' thereby dispelling the misconception that small-scale fisher peasants use their land only for capture fisheries. Aquaculture is often categorised as a fishing-related activity, and it does occur at artisanal and small-scale in certain parts of the world. For example, see Aya Suzuki and Vu Hoang Nam, 'Better Management Practices and their Outcomes in Shrimp Farming: Evidence from Small-Scale Shrimp Farmers in Southern Vietnam' (2018) 26 *Aquaculture International* 469; Luis A. Henriquez-Antipa and F. Carcamo, 'Stakeholder's Multidimensional Perceptions on Policy Implementation Gaps regarding the Current Status of Chilean Small-scale Seaweed Aquaculture' (2019) 103 *Marine Policy* 138. See Edelman's Chapter in this book.
60 UNDROP, supra note 1, arts. 17–1 and 21.4.
61 Robin Rolf Churchill and Vaughan Lowe, *The Law of the Sea* (Dominic McGoldrick ed, 3rd edn, Manchester University Press 1999) 60–1, 77–80 and 129–30.
62 United Nations Convention on the Law of the Sea (Adopted 10 December 1982, entered into force 16 November 1994) 1883 UNTS 397 (LOSC), arts. 2, 3, 48, and 49.
63 In this respect, the UN Convention on the Law of the Sea is limited to provisions that aim to protect fisheries in zones under sovereignty from unpermitted foreign fishing vessels or to require regulation on the conservation of living resources. See Martin Tsamenyi and Quentin Hanich, 'Fisheries Jurisdiction under the Law of the Sea Convention: Rights and Obligations in Maritime Zones under the Sovereignty of Coastal States' (2012) 27 *The International Journal of Marine and Coastal Law* 783, 785–6.
64 The term 'fishing grounds' is referred to in the FAO, Code of Conduct for Responsible Fisheries, supra note 8. According to art. 6.18 of the Code, 'States should appropriately

protect the rights of fishers and fishworkers, particularly those engaged in subsistence, small-scale and artisanal fisheries, to a secure and just livelihood, as well as preferential access, where appropriate, to traditional fishing grounds and resources in the waters under their national jurisdiction.'
65 FAO, SSF Guidelines, supra note 7, para. 2.2.
66 Ibid., para. 2.1.
67 Ibid., para. 2.2.
68 For more on small-scale aquaculture, see supra note 60, and Ben Belton and David C. Little, 'Contemporary Visions for Small-Scale Aquaculture' in Ratana Chuenpagdee (ed), *World Small-Scale Fisheries: Contemporary Visions* (Eburon Delft 2011).
69 Cotula, 'Between Hope and Critique: Human Rights, Social Justice and Re-imagining International Law from the Bottom Up,' supra note 21, at 513.
70 See the following subsection on 'land tenure and rights to natural resources.'
71 FAO, SSF Guidelines, supra note 7, preface.
72 Ibid., preface.
73 Ibid., preface; UNDROP, supra note 1, art. 1.1.
74 'Self-identification' and the 'special relationship with ancestral lands' are the '[o]nly two of the [six] listed criteria essential to be considered essential for a community to be considered as an indigenous people': ILA, *Final Report on the Rights of Indigenous Peoples (International Law Association, SOFIA Conference* 2012) 2–3.
75 UNDRIP, supra note 10, art. 25.
76 UNHRC, 'Report of the Special Rapporteur on the Issue of Human Rights Obligations Relating to the Enjoyment of a Safe, Clean, Healthy and Sustainable Environment' (24 January 2018) UN Doc A/HRC/37/59, Framework Principle 15.
77 Convention on Biological Diversity (adopted 5 June 1992, entered into force 29 December 1993) 1760 UNTS 79 (CBD), art. 8(j) and 10(c).
78 See Svein Jentoft, 'Walking the Talk: Implementing the International Voluntary Guidelines for Securing Sustainable Smalls-scale Fisheries' (2014) 13(1) *Maritime Studies* 3.
79 Riedel, Giacca and Golay, 'The Development of Economic, Social, and Cultural Rights in International Law,' supra note 22, at 8.
80 UNDROP, supra note 1, preamble.
81 Ibid., preamble.
82 FAO, SSF Guidelines, supra note 7, preface.
83 UNDROP, supra note 1, artt. 2.4 and 3.1.
84 FAO, SSF Guidelines, supra note 7, para. 3.1(1).
85 Ibid., supra note 7, para. 1.2.
86 Ibid., supra note 7, paras. 5.5, 5.13, 5.14, and 11.6.
87 Cotula, 'Between Hope and Critique: Human Rights, Social Justice and Re-imagining International Law from the Bottom Up,' supra note 21, at 510.
88 UNDROP, supra note 1, art. 16.
89 FAO, *Voluntary Guidelines on the Responsible Governance of Tenure of land, fisheries and forests in the Context of National Food Security (Adopted at the 38th (Special) Session of the Committee on World Food Security, in Rome, on 22 May 2012)* (2012).
90 FAO, SSF Guidelines, supra note 7, paras. 5.1 and 5.4; UNDROP, supra note 1, art. 17.3.
91 FAO, SSF Guidelines, supra note 7, paras. 5.1, 5.3 and 5.4; UNDROP, supra note 1, arts. 5, 12.5 and 17.7.
92 FAO, SSF Guidelines, supra note 7, para. 5.12; UNDROP, supra note 1, art. 17.5.
93 FAO, SSF Guidelines, supra note 7, para. 5.9; UNDROP, supra note 1, art. 17.4.
94 FAO, SSF Guidelines, supra note 7, para. 5.4; UNDROP, supra note 1, art. 17.6.
95 See Cotula, 'Between Hope and Critique: Human Rights, Social Justice and Re-imagining International Law from the Bottom Up,' supra note 21, at 508–13.
96 Ibid., at 508–13, which refers more generally to the notion of terrestrial land.
97 Ibid., at 509.

98 For example, Angola's fisheries act restricts the four nautical miles and the inland waters to fisheries of subsistence, scientific research, and recreational and/or artisanal natures. In Ghana, an Inshore Exclusive Economic Zone limits the six nautical miles to small semi-industrial, canoes and recreational fishing vessels. See, respectively, Aquatic and Biological Resources Law (New Fishing Act) No. 6-A/2004 (Angola, Promulgated on 3 September 2004, published in the Republic Gazette No. 81, Series I, Supplement on 8 October 2004), arts. 33, 235(2), and 237; Ghana. Fisheries Act No. 625/2002; para. 81 and schedule.

99 FAO, SSF Guidelines, supra note 7, paras. 5.1, 5.3–5.5, 5.7, and 5.15.

100 UNDROP, supra note 1, art. 17.1.

101 Ibid., art. 20.1.

102 See FAO, *The Ecosystem Approach to Fisheries* (FAO Technical Guidelines for Responsible Fisheries No. 4, Suppl 2, FAO 2003) 112. See also the definition of 'ecosystem' under the CBD, supra note 77, art 2: 'a dynamic complex of plant, animal and micro-organism communities and their non-living environment interacting as a functional unit' and of an ecosystem approach in Decisions V/6 and VII/11: 'a strategy for the integrated management of land, water and living resources that promotes conservation and sustainable use in an equitable way.' See also Elisa Morgera, 'The Ecosystem Approach and the Precautionary Principle' in Elisa Morgera and Jona Razzaque (eds), *Encyclopedia of Environmental Law: Biodiversity and Nature Protection Law* (Edward Elgar 2017) 70–80.

103 Elisa Morgera, 'Under the Radar: Fair and Equitable Benefit-sharing and the Human Rights of Indigenous Peoples and Local Communities Connected to Natural Resources' (2019) 23 *The International Journal of Human Rights* 1098.

104 UNHRC, 'Report of the Special Rapporteur on the Issue of Human Rights Obligations Relating to the Enjoyment of a Safe, Clean, Healthy and Sustainable Environment,' supra note 76, Framework Principle 15.

105 UNDROP, supra note 1, art. 5.2.

106 FAO, SSF Guidelines, supra note 7, para. 5.10.

107 CBD CoP7, Decision VII/16 'Article 8(j) and related provisions,' UNEP/CBD/COP/DEC/VII/16 (13 April 2014), annex, F, para. 19.

108 Elisa Morgera, 'Justice, Equity and Benefit-Sharing Under the Nagoya Protocol to the Convention on Biological Diversity' (2015) 24 *The Italian Yearbook of International Law Online* 113.

109 FAO, SSF Guidelines, supra note 7, para. 3.1(6); UNDROP, supra note 1, art. 2.3.

110 UNDROP, supra note 1, arts. 5.2(b) and 10.2.

111 FAO, SSF Guidelines, supra note 7, paras. 5.10,6.10, 7.7, 7.9, 9.2, 9.6 and 11.9.

112 UNHRC, 'Report of the Special Rapporteur on the issue of human rights obligations relating to the enjoyment of a safe, clean, healthy and sustainable environment,' supra note 76, Framework Principle 15.

113 UNHRC, 'Report of the Special Rapporteur on the issue of human rights obligations relating to the enjoyment of a safe, clean, healthy and sustainable environment,' supra note 76, as summarised in Framework Principle 15.

114 *Case of Kichwa Indigenous Communitiy of Sarayaku v Ecuador*, Inter-American Court of Human Rights Series C No 245 (27 June 2012), paras. 165–6.

115 *Centre for Minority Rights Development (Kenya) and Minority Rights Group International on behalf of Endorois Welfare Council v. Kenya* (4 February 2010) African Commission on Human and Peoples' Rights 276/2003, para. 212. Compare with K. Gover, 'Settler – State Political Theory, 'CANZUS' and the UN Declaration on the Rights of Indigenous Peoples' (2015) 26 *European Journal of International Law* 345, 372.

116 In light International Labour Organization (ILO), Indigenous and Tribal Peoples Convention (27 June 1989) C169, art. 7.1; Alejandro Fuentes, 'Judicial Interpretation and Indigenous Peoples' Rights to Lands, Participation and Consultation. The

Inter-American Court of Human Rights' Approach' (2015) 23(1) *International Journal on Minority and Group Rights* 39, 74–6 and 79.

117 For example, the UNHRC, 'Report of the Special Rapporteur on the issue of human rights obligations relating to the enjoyment of a safe, clean, healthy and sustainable environment,' supra note 76, Framework Principle 15; UNGA, 'Interim Report of the Special Rapporteur on the right to food' (8 August 2012) UN Doc A/67/268, para. 39; ECOWAS, Directive on the Harmonization of Guiding Principles and Policies in the mining Sector (2009); UN-REDD Programme, 'Guidelines on Free, Prior and Informed Consent' (2013) 11–2. See discussion in Morgera, 'Justice, Equity and Benefit-Sharing Under the Nagoya Protocol to the Convention on Biological Diversity,' supra note 108, at 1111–3; Lorenzo Cotula and Kyla Tienhaara, 'Reconfiguring Investment Contracts to Promote Sustainable Development' in Karl P. Sauvant (ed), *Reconfiguring Investment Contracts to Promote Sustainable Development* (Oxford University Press 2013) 301 and 303.

118 See generally Morgera, 'Justice, Equity and Benefit-Sharing Under the Nagoya Protocol to the Convention on Biological Diversity,' supra note 108.

119 CBD CoP13, Decision XIII/18 'Article 8(j) and related provisions,' CBD/COP/DEC/XIII/18 (17 December 2016), Mo'otz Kuxtal Voluntary Guidelines, para. 6.

120 Ibid., para. 14.

121 *Case of Kaliña and Lokono Peoples v Suriname*, Inter-American Court of Human Rights Series C No 309 (25 November 2015), Joint Concurring Opinion of Judges Sierra Porto and Ferrer Mac-Gregor Poisot.

122 See *Saramaka People v Suriname* Inter-American Court of Human Rights Series C No 172 (28 November 2007), para. 134; Alessandro Fodella, 'Indigenous Peoples, the Environment, and International Jurisprudence: Essays in Honour of Tullio Treves' in Nerina Boschiero et al (eds), *International Courts and the Development of International Law* (Asser Press 2013) 356 and 360.

123 See the convoluted title of the CBD, Mo'otz Kuxtal Guidelines, supra note 119 (n 122).

124 UNDROP, supra note 1, art. 5.2(c).

125 For example, *Saramaka People v. Suriname*, supra note 123, para. 91; *Mayagma (Sumo) Awas Tingni Community v Nicaragua* Inter-American Court of Human Rights Series C No 79 (31 August 2001), para. 149.

126 FAO, SSF Guidelines, supra note 7, para. 5.1.

127 Elisa Morgera, 'The Need for an International Legal Concept of Fair and Equitable Benefit Sharing' (2016) 27 *The European Journal of International Law* 353; CBD, supra note 77.

128 CBD CoP5, Decision V/6 'Ecosystem Approach,' UNEP/CBD/COP/5/23 (22 June 2000), para 9.

129 Morgera, 'Justice, Equity and Benefit-Sharing Under the Nagoya Protocol to the Convention on Biological Diversity,' supra note 108.

130 FAO, SSF Guidelines, supra note 7, para. 6.2.

131 UNHRC, 'Report of the Special Rapporteur on the issue of human rights obligations relating to the enjoyment of a safe, clean, healthy and sustainable environment,' supra note 76, Framework Principle 15. para. 9; Elisa Morgera, 'A Reflection on Benefit-sharing as a Framework Principle on Human Rights and the Environment proposed by UN Special Rapporteur John Knox (Part I)' (2018) *BeneLex Blog* <https://benelexblog.wordpress.com/2018/04/08/a-reflection-on-benefit-sharing-as-a-framework-principle-on-human-rights-and-the-environment-proposed-by-un-special-rapporteur-john-knox-part-i/> accessed 6 October 2020.

132 FAO, SSF Guidelines, supra note 7, para. 6.8.

133 Ibid., supra note 7, paras. 7.8 and 7.10.

134 Morgera, 'Justice, Equity and Benefit-Sharing Under the Nagoya Protocol to the Convention on Biological Diversity,' supra note 108, based on combined reading of the CBD Akwé: Kon Voluntary Guidelines and Mo'otz Kuxtal Guidelines.

135 UN Committee on the Elimination of Discrimination against Women, 'General Recommendation No 34 on the Rights of Rural Women' (7 March 2016) UN Doc CEDAW/C/GC/34.
136 UNDROP, supra note 1, art. 4.2(h).
137 FAO, SSF Guidelines, supra note 7, para. 5.15.
138 Ibid., para. 5.1, emphasis added.
139 UNHRC, 'Report of the Special Rapporteur on the Issue of Human Rights Obligations Relating to the Enjoyment of a Safe, Clean, Healthy and Sustainable Environment' (24 January 2018) UN Doc A/HRC/37/58.
140 Cotula, 'Between Hope and Critique: Human Rights, Social Justice and Re-imagining International Law from the Bottom Up,' supra note 21, at 514 and 520.
141 *Case of Kichwa Indigenous Communitiy of Sarayaku v Ecuador*, supra note 114, para. 194.
142 Ibid., paras. 193–4.
143 Ibid., para. 186; Jérémie Gilbert and Cathal Doyle, 'A New Dawn over the Land: Shedding Light on Collective Ownership and Consent' in S. Allen and A. Xanthaki (eds), *Reflections on the UN Declaration on the Rights of Indigenous Peoples* (1st edn, Hart 2011) 289.
144 Nieves Gomez, 'Indigenous Peoples and Psychosocial Reparations: The Experience with Latin American Indigenous Communities' in Federico Lenzerini (ed), *Reparations for Indigenous Peoples: International and Comparative Perspectives* (Oxford University Press 2008) 143 and 158.
145 Gabriella Citrioni and Karla I Quintana Osuna, 'Reparations for Indigenous Peoples in the Case of the Inter-American Court of Human Rights' in Federico Lenzerini (ed), *Reparations for Indigenous Peoples: International and Comparative Perspectives* (Oxford University Press 2008) 317–24 and 340.
146 *Centre for Minority Rights Development (Kenya) and Minority Rights Group International on behalf of Endorois Welfare Council v. Kenya*, supra note 115, para. 274.
147 *Saramaka People v Suriname*, supra note 122, paras. 77–84 and 183.
148 Morgera, 'Justice, Equity and Benefit-Sharing Under the Nagoya Protocol to the Convention on Biological Diversity,' supra note 108.
149 Alonso Barros, 'The Fetish Mechanism: A Post-Dogmatic Case Study of the Atacama Desert Peoples and the Extractive Industries' in Corinne Lennox and Damien Short (eds), *Handbook of Indigenous Peoples' Rights* (Routledge 2016) 223 and 231–2.
150 Gaetano Pentassuglia, 'Towards a Jurisprudential Articulation of Indigenous Land Rights' (2011) 22 *European Journal of International Law* 165-76; Deborah McGregor, 'Living Well with the Earth: Indigenous Rights and the Environment' in Corinne Lennox and Damien Short (eds), *Handbook of Indigenous Peoples' Rights* (Routledge 2016) 167–75; Ellen Desmet, *Indigenous Rights Entwined with Nature Conservation* (vol 8, Intersentia 2011) 58–175; Francesco Francioni, *Reparation for Indigenous Peoples: Is International Law Ready to Ensure Redress for Historical Injustices?* (Oxford University Press 2008).
151 Elsa Reimerson, 'Between Nature and Culture: Exploring Space for Indigenous Agency in the Convention on Biological Diversity' (2013) *Environmental Politics* 22; *Ylva Uggla, 'What is This Thing Called 'Natural'? The Nature-culture Divide in Climate Change and Biodiversity Policy' (2009) Journal of Politics Ecology* 17–79.
152 Morgera, 'Justice, Equity and Benefit-Sharing Under the Nagoya Protocol to the Convention on Biological Diversity,' supra note 108. See also a discussion on consultation and free prior informed content at ibid., 14–6.
153 Ibid.
154 FAO, SSF Guidelines, supra note 7, paras. 3.1(2), 5.18, 11.4, and 11.7; UNDROP, supra note 1, art. 18.3, 19.1(2), 20.2, and 26.
155 UNDROP, supra note 1, art. 26.
156 Ibid., art. 16.1.
157 Ibid., art. 23.2.

158 FAO, SSF Guidelines, supra note 7, para. 1.1(f).
159 Ibid., paras. 11.4 and 11.7.
160 UNDROP, supra note 1 art 18.3. See Alabrese-Savaresi's chapter in this book.
161 FAO, SSF Guidelines, supra note 7, para. 9.2.
162 Ibid., paras. 9.2–9.8
163 Ibid., paras. 9.7 and 9.9.
164 Elisa Morgera, 'Fair and Equitable Benefit-sharing at the Crossroads of the Human Right to Science and International Biodiversity Law' (2015) 4 *Laws* 803.
165 Tkarihwaié:ri Code of Ethical Conduct on Respect for the Cultural and Intellectual Heritage of Indigenous and Local Communities, CBD Decision X/42, annex, preamble.
166 Suman Seth, 'Putting Knowledge in its Place: Science, Colonialism and the Postcolonial' (2009) 12(4) *Postcolonial Studies* 373–88.
167 Saskia Vermeylen, George Martin and Roland Clift, 'Intellectual Property, Rights Systems and the Assemblage of Local Knowledge Systems' (2008) 15 *International Journal of Cultural Property* 201–7.
168 UNHRC, 'Report of the Special Rapporteur on the Issue of Human Rights Obligations Relating to the Enjoyment of a Safe, Clean, Healthy and Sustainable Environment,' supra note 76, Framework Principle 15. Elisa Morgera, 'The Need for an International Legal Concept of Fair and Equitable Benefit Sharing', supra note 127.
169 Though it has been subject to FAO's technical scrutiny in later studies. See Johanne Fischer and others (eds), 'Fishers' Knowledge and the Ecosystem Approach to Fisheries: Applications, Experiences and Lessons in Latin America' (2015) 591 *FAO Fisheries and Aquaculture Technical Paper* 278.
170 UNDROP, supra note 1, art 5. See section on social rights and related provisions.
171 This has been recognised internationally, for instance, in the Preamble of the Nagoya Protocol the CBD: 'the interrelationship between genetic resources and traditional knowledge, their inseparable nature for indigenous and local communities.' See Nagoya Protocol on Access to Genetic Resources and the Fair and Equitable Sharing of Benefits arising from their Utilization to the Convention on Biological Diversity (adopted 29 October 2010, entered into force 12 October 2014) 3008 UNTS 3.
172 Morgera, 'Fair and Equitable Benefit-sharing at the Crossroads of the Human Right to Science and International Biodiversity Law,' supra note 164.
173 UNDROP, supra note 1, art. 25.3.
174 Elisa Morgera and Maria Ntona, 'Linking Small-scale Fisheries to International Obligations on Marine Technology Transfer' (2018) 93 *Marine Policy* 295.
175 Derek Stephen Johnson, 'Category, Narrative, and Value in the Governance of Small-scale Fisheries' (2006) 30 *Marine Policy* 747, 27.
176 Conner Bailey, Dean Cycon and Michael Morris, 'Fisheries Development in the Third World: The Role of International Agencies' (1986) 14 *World Development* 1269–71.
177 John Kurien, 'Small-scale Fisheries in the Context of Globalisation' (1998) *CDS Working Papers Series* 29.
178 That is, the 'right to conservation and protection of the environment and the productive capacity of their lands, and of the resources that they use.' UNDROP, supra note 1, art. 18.1. The UNDROP is limited in providing environmental duties, with the exception of the duty not to store and dispose of hazardous material, substance or waste on peasants' land. See art. 18.4.
179 Ibid., art. 20
180 FAO, SSF Guidelines, supra note 7, paras 5.14 and 9.8.
181 Cotula, 'Between Hope and Critique: Human Rights, Social Justice and Re-imagining International Law from the Bottom Up,' supra note 21, at 520.
182 Ibid., at 520.

183 UN Human Rights Council, 'Report of the Special Rapporteur on the Issue of Human Rights Obligations Relating to the Enjoyment of a Safe, Clean, Healthy and Sustainable Environment,' supra note 76, Framework Principles 7 and 10.
184 FAO, SSF Guidelines, supra note 7, para. 7.10; UNDROP, supra note 1, art. 11.2 and 25.3.
185 FAO, SSF Guidelines, supra note 7, paras. 3.1(7)(8) and 11.8; UNDROP, art. 11(2).
186 FAO, SSF Guidelines, supra note 7, para. 11.5.
187 Ibid., para. 11.6.
188 Ibid., para. 11.4.
189 Ibid., para. 5.11; UNDROP, supra note 1, art. 12.2.
190 UNDROP, supra note 1, art. 12.3.
191 Ibid., art. 12.5.
192 Cotula, 'Between Hope and Critique: Human Rights, Social Justice and Re-imagining International Law from the Bottom Up,' supra note 21, at 490–7.
193 UNDROP, supra note 1, art. 15.4 reads 'Peasants and other people working in rural areas have the right to determine their own food and agriculture systems, recognized by many States and regions as the right to food sovereignty.'
194 Cotula, 'Between Hope and Critique: Human Rights, Social Justice and Re-imagining International Law from the Bottom Up,' supra note 21, at 505.
195 Ibid., at 55.
196 Justice may be brought to SSF communities in remote areas, for instance, as opposed to expecting remote communities to seek justice in central locations. For instance, in Brazil, the so-called itinerant justice enabled communities in remote areas to have access to legal advice and proceedings initiated by civil servants. See Julia Nakamura and Fábio Hazin, 'Assessing the Brazilian Federal Fisheries Law and Policy in light of the Voluntary Guidelines for Securing Sustainable Small-scale Fisheries' (2020) 113 *Marine Policy* 1–8.
197 Ibid., 2–3.
198 The elaboration on the human rights-based approach and the roles played by duty-bearers and right-holders was further supported by FAO in subsequent activities related to the SSF Guidelines implementation. See Sisay Yeshanew, Nicole Franz, and Lena Westlund, *FAO Fisheries and Aquaculture Proceedings. No. 53. Exploring the Human Rights-based Approach in the Context of the Implementation and Monitoring of the SSF Guidelines* (FAO 2016).
199 The need to introduce human rights concepts and standards to fisheries professionals is highlighted by Song and Soliman, who also note the hurdle of overcoming mainstream understandings of 'fishing rights' as individual tradable commodities, which are geared towards wealth accumulation as opposed to protecting access to fisheries as an 'indivisible part of their culture, survival and wellbeing' (20). On the other hand, the authors seem to ignore the international human rights approach to balancing of rights and the implications in that context of the international recognition of the inter-dependence of human rights and a healthy environment (22–3). Andrew M. Song and Adam Soliman, 'Situating Human Rights in the Context of Fishing Rights – Contributions and Contradictions' (2019) 103 *Marine Policy* 19.
200 For a dedicated analysis of these elements of the SDG, their meaning in terms of normativity and in *inter alia* elevating human rights, justice, and political and moral considerations, see Graham Long, 'Underpinning Commitments of the Sustainable Development Goals: Indivisibility, Universality, Leaving No One Behind' in Duncan French and Loiuse J. Kotzé (eds), *Sustainable Development Goals: Law, Theory and Implementation* (Edward Elgar Publishing 2018).
201 UNGA, 'Resolution 70/1 of 25 September 2015, Transforming our world: the 2030 Agenda for Sustainable Development,' supra note 16.
202 UNHRC, 'Report of the Special Rapporteur on the Issue of Human Rights Obligations Relating to the Enjoyment of a Safe, Clean, Healthy and Sustainable Environment,' supra note 76, Framework Principles 2, 8, 12, and 14.

203 Cotula, 'Between Hope and Critique: Human Rights, Social Justice and Re-imagining International Law from the Bottom Up,' supra note 21, at 513.

Selected bibliography

Cotula L, 'Between Hope and Critique: Human Rights, Social Justice and Re-imagining International Law from the Bottom up' (2020) 48 *Georgia Journal of International and Comparative Law* 49.

Morgera E, 'Under the Radar: Fair and Equitable Benefit-sharing and the Human Rights of Indigenous Peoples and Local Communities Connected to Natural Resources' (2019) 23 *International Journal of Human Rights*.

Nakamura J and Hazin F, 'Assessing the Brazilian Federal Fisheries Law and Policy in Light of the Voluntary Guidelines for Securing Sustainable Small-scale Fisheries' (2020) 113 *Marine Policy* 1.

PART II

Natural resources access and control

6

THE RIGHT TO LAND

Lorenzo Cotula

A turning point in human rights law

When the United Nations General Assembly adopted, by a large majority, the United Nations Declaration on the Rights of Peasants and Other People Working in Rural Areas (UNDROP),[1] it established a new milestone in the historical development of human rights law. The UNDROP is the first global instrument to specifically affirm the rights of peasants – the millions of small-scale agricultural producers who, in the words of the declaration, rely significantly on family labour and have a special attachment to the land.[2] And while non-governmental organisations often front human rights advocacy, the UNDROP is the culmination of a long-term struggle that peasant movements directly led from grassroots to global level, developing a novel approach to agency and representation in international policymaking.[3]

Compared to other human rights instruments, the UNDROP more explicitly connects rights to economic relations and control over the means of production. The affirmation of peasants' human right to land exemplifies this shift,[4] as does the prominence given to land issues throughout the Declaration. With land playing a foundational role in rural livelihoods and ways of life, the right to land crystallises peasants' deep sociocultural connection to it and its central place in peasants' struggles. Although contexts vary widely, land struggles often revolve around demanding agrarian reform, challenging privatisation, and protecting small-scale producers from large-scale plantations or increased agribusiness control over value chains.

In claiming the right to land, peasant movements have sought to overcome the limitations that conventional human rights constructs have long displayed in sustaining those struggles. For example, the internationally recognised right to property can protect peasants' land rights, including those based on customary systems,[5] but landed elites can – and do – invoke the same right to prevent redistributive reforms.

DOI: 10.4324/9781003139874-9

And while indigenous peoples have reconfigured the right to collective property as part of efforts to resist extractive industry projects, they often struggle to translate judicial wins into real change, while the very concept of property can contrast with indigenous conceptions of the relationship between humans and nature.[6]

Because land can provide the basis for livelihoods, shelter, ways of life and social identity, the internationally recognised rights to food, housing and culture may require states to address land issues.[7] For example, where people source their food directly from the land, protecting their land rights may be a key part of a state's obligation to progressively realise their right to adequate food.[8] These intersections between land and human rights are compelling, but they are also indirect, mediated by the role of land in providing food, housing or the foundations of culture.

International instruments recognise indigenous peoples' rights over their ancestral territories and link control over land to the collective right to self-determination.[9] As recognised in the UNDROP,[10] peasants can be of indigenous descent and thus entitled to these rights. But leaving aside the limitations of these rights,[11] and of the ways they are 'domesticated' in national legal systems,[12] the special nature of such arrangements excludes many people with strong connections to the land but no claim to indigeneity.

Meanwhile, the Voluntary Guidelines on the Responsible Governance of Tenure (VGGT) call on states to recognise, respect, and protect all legitimate tenure rights, including those not protected by law.[13] In 2012, the (United Nations) Committee on World Food Security (CFS) endorsed the VGGT as the most comprehensive global soft-law instrument on land and resource governance. As well as protecting peasants' land rights where national legal systems marginalise them, implementing the VGGT would also guide land restitution and redistribution.[14] The guidelines establish a strong bond between tenure and human rights, and human rights bodies have referenced them when interpreting human rights treaties.[15] But the VGGT are non-binding, primarily address tenure – rather than human rights – and are located outside the international human rights system.

Building on this tapestry of human rights configurations, the UNDROP's affirmation of peasants' right to land represents a turning point in longstanding efforts to connect land to human rights. It raises questions about the legal contours of the right to land and how this newly affirmed right complements, and intersects with, other human rights. Although such questions are theoretical, they also have practical implications. Restrictive interpretations can hollow out the emancipatory potential of a new right from within, while clarifying the full reverberations of that right can facilitate its use in processes of change. It is therefore important to explore the social, economic, and political circumstances that led to the affirmation of the right to land, what the right to land entails in practice and its potential for bringing about the change peasant movements are fighting for around the world.

The land question and peasants' rights

The right to land is steeped in decades of agrarian struggles. While recognising the limitations of legal processes, peasant movements have long mobilised rights not only to challenge the building blocks of economic ordering,[16] but also to articulate, in more proactive terms, their vision of agrarian reform and locally controlled food systems.[17] Transnational agrarian movement La Via Campesina has been at the forefront of these efforts, integrating rights language in its advocacy and calling for a United Nations instrument that would solemnly affirm the rights of peasants.[18]

Part of wider political mobilisation that gained momentum from the early 1990s, this global advocacy responded to policy and market forces that increased pressure on the rural world, particularly in the global South, including: 'structural adjustment' programmes imposed by international financial institutions, which reduced state support for farmers in poorer countries; the decline of public intervention in agricultural markets, which increased farmers' exposure to commodity price fluctuations; growing corporate concentration in, and control of, agricultural value chains; and the reconfiguration of international trade in agriculture under terms that favour agribusiness interests.[19]

The resulting socio-economic transformations squeezed small-scale rural producers, threatened their connection to the land, and undermined their demands for agrarian reform. An emerging body of evidence points to growing inequality in control over land worldwide.[20] Starting in the mid-2000s, the decade-long global surge in transnational deals for monoculture plantations came to epitomise the mounting pressures on peasants' lands.[21] In many contexts, mining, petroleum, and infrastructure projects have compounded these pressures; in others, industrialisation programmes have converted agricultural land into special economic zones for manufacturing activities.[22] Although legislation varies in different jurisdictions, recurring features of the law facilitate land transfers to these large-scale projects. From the structural marginalisation of peasants' land rights, to national reforms that encourage industrial, mining and agribusiness activities, and international treaties that protect foreign investment, these trends make millions of peasants vulnerable to dispossession from 'legalised land grabs.'[23] Many activists working to defend land rights under pressure from large-scale projects have faced repression and intimidation.[24]

Meanwhile, demand for food from sprawling urban centres and the integration of more competitive small-scale producers into global value chains have fostered agricultural intensification and commercialisation in many rural areas. In the presence of deep-seated social differentiation – shaped by unequal access to income and means of production, patriarchal social reproduction arrangements, and the weight of age in land decision-making, to mention but a few examples – these processes have underpinned what Ghanaian scholar Kojo Amanor called 'dispossession from below.'[25] Those farmers with better resources, information and connections have been able to seize market opportunities, albeit often under terms imposed by buyers. But vast segments of the rural population face barriers in accessing markets, credit, inputs and services, fostering land concentration and dispossession within the smallholder sector.[26]

National land policies in the global South – including those supported or imposed by international financial institutions – have often compounded these problems, advancing approaches that contrast sharply with the social foundations of rural realities. For example, policies to individualise and commercialise landholdings have created tensions with customary systems traditionally premised on collective landholding and 'nested' rights, on the deep connection between land, family organisation and social identity, or on the imperative of livestock mobility in the face of variable rainfall patterns.[27] While social, economic and cultural changes have eroded or transformed many traditional systems,[28] this process is often accelerated by official policies that enable urban elites to register rural and peri-urban lands in their names.[29] As a result, the land rights of marginalised rural people have become increasingly precarious, while youths and the landless often struggle to access the land they need to start a family.

In these processes of economic reorganisation and social dislocation, reappropriating the notion of 'peasants' and advocating for international recognition of their rights became an important strategy in agrarian movements' struggle for an alternative model of agricultural development.[30] The right to land featured prominently from the start. When, in the late 1990s, grassroots-level deliberations in West Java, Indonesia, initiated the process of drafting a 'Peasants' Rights Charter,'[31] the resulting text affirmed that '[p]easants and their families have the right to enough land to support them,'[32] and 'the right to legal protection of their farmland, dwellings, natural resources and biological diversity.'[33] Amendments proposed during early consultations with peasants and supportive organisations indicated the need to more explicitly recognise '[t]he right for farming laborers to gain ownership of adequate farming land,'[34] highlighting that the land question mattered not only to small-scale producers but to farm workers as well.

And when national peasant organisation and Via Campesina member Serikat Petani Indonesia escalated its advocacy, first to regional level and then to global level,[35] and LVC adopted the 'Declaration of Rights of Peasants – Women and Men' to federate the movement around a clear set of global demands,[36] the right to land was a key policy ask. The declaration frames the connection to land as a fundamental feature of peasant identity, defining peasants as 'a man or woman of the land, who has a direct and special relationship with the land.'[37] Under the heading 'Right to land and territory,' the Via Campesina Declaration affirms peasants' right to own land, collectively or individually, for housing and farming; farm their own land and any state land they depend on for their livelihoods; have security of tenure and protection from forcible eviction; reject the acquisition and conversion of land for economic purpose; and benefit from land reform.[38]

Grounded in the life experiences of peasants around the world, the right to land made its way into United Nations talks, featuring in discussions at the OEIGWG on a United Nations Declaration on the Rights of Peasants and Other People Working in Rural Areas. The United Nations Human Rights Council established this working group in 2012 as a result of LVC's advocacy and on the initiative of sympathetic states.[39] The working group drafted the declaration that would ultimately be adopted by the Human Rights Council,[40] and the General Assembly,[41] enshrining the right to land in an international human rights instrument for the first time.

The right to land in the UNDROP

The land question features prominently in the UNDROP, starting in the preamble, which recognises the special relationship between peasants and land,[42] and which includes land problems in the list of circumstances that justify the affirmation of peasants' rights – conveying alarm at the growing number of peasants who are forcibly displaced every year;[43] stressing that several factors make it difficult for peasants to defend their land rights;[44] stressing that peasant women are often 'denied tenure and ownership of land';[45] recognising that access to land 'is an increasing challenge for rural people';[46] and expressing concern about the growing repression of land rights activists.[47]

Following in the footsteps of the Via Campesina Declaration, the UNDROP affirms the deep connection between people and land in its definition of 'peasants.' As well as listing characteristics related to livelihood systems and labour organisation, the UNDROP identifies peasants as people who have a 'special dependency on and attachment to the land.'[48] It also extends rights to 'other people working in rural areas,' clarifying that it applies to any person engaged in land-based activities, such as small-scale agriculture, pastoralism, and hunting or gathering. This includes landless people, agricultural workers, and 'indigenous peoples and local communities *working on the land*.'[49]

The UNDROP's substantive provisions place considerable emphasis on the relation between human rights and peasants' land and resource rights. And while all rights are indivisible on the juridical plane and the Declaration explicitly rules out hierarchies of rights,[50] it seems politically significant that peasants' right to access the natural resources 'that are required to enjoy adequate living conditions'[51] precedes the affirmation of other rights – including the rights to life, physical and mental integrity, and liberty and security of person – in the structure of the Declaration.[52]

Article 17 affirms that peasants and other people living in rural areas have an individual and a collective right to land.[53] This encompasses the right 'to achieve an adequate standard of living, to have a place to live in security, peace and dignity and to develop their cultures.'[54] Other international human rights instruments partly frame land issues in the context of strategies to implement other human rights, such as the right to an adequate standard of living and the right to be free from hunger.[55] But the UNDROP establishes a more direct connection between people and land and inverts the conventional approach by inscribing the attainment of an adequate standard of living within the framework of an encompassing human right to land – though the Declaration also recognises a separate, free-standing right to an adequate standard of living for peasants and their families.[56] By affirming the value of collective rights and clarifying that the right to land can be exercised collectively as well as individually, the UNDROP aims to reverse the juridical marginalisation of the collective landholding arrangements that often sustain peasants' production systems.[57]

Article 17 also outlines the practical implications of the right to land. Echoing the language of the VGGT, it calls on states to 'take appropriate measures to provide legal recognition for land tenure rights, including customary land tenure rights not currently protected by law.'[58] It also calls on states to 'ensure that peasants and other people working in rural areas are not arbitrarily or unlawfully evicted and that their rights are not otherwise extinguished or infringed.'[59] Where violations have occurred, peasants and other people working in rural areas have the right 'to return to their land . . . whenever possible, or to receive just, fair and lawful compensation.'[60] A separate provision on peasants' access to justice calls on states to provide effective redress for violations, including actions that have the 'aim or effect' of arbitrarily dispossessing peasants of their land and natural resources.[61]

These protective aspects of the right to land are comparable to 'progressive' interpretations of the right to property, when applied to protecting peasants' land rights. They also converge with, and reinforce, the many provisions of the VGGT that call for recognising, respecting and protecting all legitimate tenure rights.[62] But the UNDROP goes further than the VGGT, affirming an overarching right to land that pertains to peasants by virtue of their distinctive connection to the land. In addition, the Declaration articulates the imperative of protecting peasants' land rights in direct, tangible terms (referring to 'land' rather than 'property'), without the commodifying baggage that the notion of property inherently entails.[63]

As the UNDROP applies to agricultural workers and the landless as well as small-scale producers, the right to land provides the normative foundations for agrarian reform, including land redistribution and restitution.[64] Article 17 calls on states, 'where appropriate,' to implement agrarian reforms aimed at facilitating 'broad and equitable access to land' and limiting 'excessive concentration and control of land,'[65] provided the measures comply with the parameters established in the UNDROP, such as 'securing due recognition and respect for the rights and freedoms of others.'[66] Further, the UNDROP calls on states to prioritise landless peasants, young people, and other rural workers when allocating public lands.[67]

Other international human rights instruments refer to agrarian reform in connection with initiatives to promote 'the most efficient development and utilization of natural resources.'[68] But the UNDROP explicitly connects agrarian reform to land concentration and justice. Meanwhile, the VGGT provide guidance on redistributive reform but frame it as an option for states to consider on economic, social or environmental grounds, rather than as a human rights issue.[69] Unlike other international instruments, the UNDROP also calls on states to consider the 'social function' of land as a basis for agrarian reform.[70]

In the early 1900s, French jurist Léon Duguit developed the notion of social function to highlight that property serves public as well as private interests, at once legitimising and limiting its legal protection and configuration.[71] Since then, several constitutional clauses – particularly those emerging in the context of major political change – have linked property to social function. Over the years, public authorities in different countries have relied on social function provisions to justify land redistribution, with varying degrees of success.[72] The UNDROP's reference to the social function of land taps into this juristic tradition to affirm the inherent limitations of property, not explicitly found in other international texts,[73] and to strengthen the case for agrarian reform.

Recognising social differentiation in land relations and its partial grounding in structural discrimination under national laws, Article 17 also calls on states to 'remove and prohibit' discrimination relating to the right to land. This includes discrimination resulting from a 'change of marital status, lack of legal capacity or lack of access to economic resources.'[74] Article 3 enshrines a more general prohibition of discrimination and includes property among the prohibited grounds of discrimination.[75]

Other more general provisions of the UNDROP also have a bearing on the right to land. Carefully worded clauses call on states to ensure that peasants and other people working in rural areas have a say in decisions that could affect 'their lives, land and livelihoods,' and to recognise the role of 'strong and independent' organisations that represent peasants and people working in rural areas.[76] When affirming the right to land, Article 17 explicitly references the general limitations clause in Article 28,[77] which states that peasants' rights can only be subject to limitations that are determined by law, comply with international human rights obligations, and are necessary to respect the rights of others or for 'meeting the just and most compelling requirements of a democratic society.'[78]

From words to practice

Faced with profound changes that threaten peasants' lands, livelihoods and ways of life, agrarian movements have wielded the language of rights to *affirm* peasants' connection to land as a defining feature of their identity; *defend* this bond in the face of encroachment and dispossession; and *fulfil* it through land redistribution or restitution. Successive drafting efforts have evolved from the aspirational ambitions of grassroots deliberations to the politics of inter-governmental negotiations.[79] The affirmation of the right to land is the story of social movements shaping human rights from the ground up. And by providing a universal construct that abstracts from the great diversity of land problems and arrangements, the right to land establishes a new channel for translating local struggles into global action. Compared to other international instruments, the UNDROP is more explicit in articulating the protection of land rights in human rights terms, recognising diverse tenure models – including collective forms of landholding – and establishing the normative foundations for agrarian reform inspired by principles of social justice.

As a result, the right to land cuts across recognition, representation and material distribution. In emphasising peasants' 'special attachment' to land, the UNDROP highlights the strong connection between place, history and culture and the fundamental role of land in shaping social identity and a shared sense of belonging.[80] At the same time, the right to land reflects a concern about control over the means of production. This is part of an overarching 'right to produce' that runs through the Declaration, starting with its definition of peasants as people who engage in agricultural production,[81] through its emphasis on the value of work itself (including 'people working in rural areas' in its title) and the place of work as a source of social identity (defining peasants by their reliance on family or household labour[82]), to its affirmation of peasants' right to work.[83]

More than establishing safety nets for secure access to basic goods such as food and housing, the UNDROP affirms the agency of peasants to advance their own vision of rural development, in the face of inequitable relations affecting production and trade. This framing creates strong connections between the right to land and other rights related to the spheres of work and production – because land and labour struggles often intersect,[84] because farm workers may claim land redistribution or restitution,[85] and because effective agrarian reform requires not only transferring land but also improving access to credit, inputs and extension services.[86]

There are difficult challenges to realising the right to land. Pressures on peasants' lands originate from structural transformations that affect global commodity markets, food systems, the organisation of government, and relations between rural and urban worlds. These processes are hard to counter, let alone reverse. Although the public policies that underpin them are potentially reversible, or at least adjustable, international soft-law instruments do not, in themselves, alter entrenched patterns of land ownership or the legal arrangements that protect the landholdings of transnational businesses in the face of demands for consultation, restitution or redistribution.[87] The majority vote at the United Nations General Assembly, with several higher-income countries voting against the adoption of the UNDROP or abstaining, echoes the challenges that states in the global South have long faced when challenging aspects of economic governance, such as through historical advocacy on the New International Economic Order and the right to development.[88]

The real challenge, then, lies in moving from the right to land to concrete changes in the policies and processes that shape control over land. Legal strategies may have a role to play – for example, by mobilising the right to land to influence the interpretation of binding law, including other human rights, in legal proceedings. International human rights bodies such as the HRC and the IACHR have already started considering the UNDROP in their interpretation of binding human rights treaties.[89] National courts could follow this lead. The right to land could also inspire novel approaches for reforming national laws and institutions.

But land is eminently political, and vested interests at local to global levels are likely to resist real change. The repression many land rights defenders have suffered for their advocacy in the face of agribusiness and extractive industry projects worldwide is a reminder of the often brutal opposition to land rights claims.[90] In most cases, addressing complex land questions will require political strategies, with rights approaches helping reframe the issues, catalyse collective action, and shift the prevailing 'common sense.'[91]

How activists will mobilise the right to land, in discursive practices or legal proceedings, and with what results, will determine the bearing the UNDROP will have on land relations in the longer term. In the meantime, there is little doubt that the right to land constitutes an important milestone in efforts to connect land to human rights. Its legal contours show that rights constructs can significantly depart from the traditional canon, while its affirmation from the ground up highlights the power of agency in shaping human rights.[92] The struggle for affirming the right to land must now transition to the struggle for implementation.

Notes

1 UN General Assembly, 'United Nations Declaration on the Rights of Peasants and Other People Working in Rural Areas' (17 December 2018) UN Doc A/RES/73/165. Voting patterns largely reflected dividing lines between global North and South: Australia, Israel, New Zealand, Sweden, the United Kingdom, and the United States were among the eight states that voted against, while Austria, Belgium, Canada, France, Germany, Italy, Japan, the Netherlands, the Republic of Korea, the Russian Federation, Singapore, and Spain were among the 54 states that abstained. See UN General Assembly Official Records,'73rd Sess., 55th plenary meeting at 25' (17 December 2018) UN Doc A/73/PV.55. This chapter builds on ideas presented in Lorenzo Cotula, 'Between Hope and Critique: Human Rights, Social Justice and Re-Imagining International Law from the Bottom Up' (2020) 48(2) *Georgia Journal of International and Comparative Law* 473–521.

2 'For the purposes of the present Declaration, a peasant is any person who engages or who seeks to engage, alone, or in association with others or as a community, in small-scale agricultural production for subsistence and/or for the market, and who relies significantly, though not necessarily exclusively, on family or household labour and other non-monetized ways of organizing labour, and who has a special dependency on and attachment to the land' (UNDROP, art. 1.1).

3 Priscilla Claeys, 'The Rise of New Rights for Peasants: From Reliance to NGO Intermediaries to Direct Representation' (2018) 9(3–4) *Transnational Legal Theory* 394–7.

4 UN Human Rights Council, 'United Nations Declaration on the Rights of Peasants and Other People Working in Rural Areas' (8 October 2018) UN Doc A/HRC/RES/39/12, art. 17.1. On the right to land in the UNDROP, see also Christophe Golay, 'Research Brief: The Right to Land and Other Natural Resources' (2020) *Geneva Academy of International Humanitarian Law and Human Rights*. For reflections preceding the adoption of the declaration, see Melik Özden, *The Right to Land* (Europe-Third World Centre – CETIM 2014); Sofia Monsalve Suárez, *The Right to Land and Other Natural Resources in the United Nations Declaration on the Rights of Peasants and Other People Working in Rural Areas* (FIAN International 2015); Kaitlin Cordes, 'Is there a Human Right to Land?' (8 November 2017) Columbia Centre on Sustainable Investment blog <https://blogs.ei.columbia.edu/2017/11/08/is-owning-land-a-human-right/> accessed 28 April 2021; Priscilla Claeys, 'The Right to Land and Territory: New Human Right and Collective Action Frame' in Olivier De Schutter and Balakrishnan Rajagopal (eds), *Property Rights from Below: Commodification of Land and the Counter-Movement* (Routledge 2020) 131–48. For a broader discussion of land and natural resources in international law, see Elisa Morgera and Kati Kulovesi (eds), *International Law and Natural Resources* (Edward Elgar 2016).

5 See also Christophe Golay and Ioana Cismas, 'Legal opinion: The Right to Property from a Human Rights Perspective' (2010) *SSRN* <https://papers.ssrn.com/sol3/papers.cfm?abstract_id=1635359> accessed 25 April 2021.

6 For a fuller discussion, see Lorenzo Cotula, 'Between Hope and Critique: Human Rights, Social Justice and Re-Imagining International Law from the Bottom Up' (2020) 48(2) *Georgia Journal of International and Comparative Law* 473–521. See also Thomas M. Antkowiak, 'Rights, Resources, and Rhetoric: Indigenous Peoples and the Inter-American Court' (2013) 35 *University of Pennsylvania Journal of International Law* 113–87; Maximiliano Mendieta Miranda and Julia Cabello Alonso, *Advancing Indigenous Peoples' Rights through Regional Human Rights Systems: The Case of Paraguay* (International

Institute for Environment and Development 2017); Joel E. Correia, 'Indigenous Rights at a Crossroads: Territorial Struggles, the InterAmerican Court of Human Rights, and Legal Geographies of Liminality' (2018) 97 *Geoforum* 73–83.

7 See also Jérémie Gilbert, 'Land Rights as Human Rights: The Case for a Specific Right to Land' (2014) 10(18) *Sur – International Journal on Human Rights* 113–35; Cordes, 'Is there a Human Right to Land?,' supra note 4.

8 UN General Assembly, 'Report of the Special Rapporteur on the Right to Food' (11 August 2010) UN Doc A/65/281; Oliver De Schutter, 'The Emerging Human Right to Land' (2010) 12 *International Community Law Review* 303–34. See also Lorenzo Cotula (ed), *The Right to Food and Access to Natural Resources. Using Human Rights Arguments and Mechanisms to Improve Resource Access for the Rural Poor* (FAO 2008).

9 International Labour Organization (ILO), Indigenous and Tribal Peoples Convention (27 June 1989) C169, artt. 13–9; UN General Assembly, 'United Nations Declaration on the Rights of Indigenous Peoples' (2 October 2007) UN Doc A/RES/61/295. See also Megan Davis, 'Indigenous Struggles in Standard-Setting: The United Nations Declaration on the Rights of Indigenous Peoples' (2008) 9(2) *Melbourne Journal of international Law* 439–71; Jesse Hohmann and Marc Weller (eds), *The U.N. Declaration on the Rights of Indigenous Peoples: A Commentary* (Oxford University Press 2018); Jérémie Gilbert and Corinne Lennox, 'Towards New Development Paradigms: The United Nations Declaration on the Rights of Indigenous Peoples as a Tool to Support Self-Determined Development' (2019) 23 *International Journal of Human Rights* 104–24.

10 UNDROP, supra note 4, art. 1.3.

11 Karen Engle, 'On Fragile Architecture: The UN Declaration on the Rights of Indigenous Peoples in the Context of Human Rights' (2011) 22(1) *European Journal of International Law* 141–63.

12 See, for example, Armando Guevara Gil and Carla Cabanillas Linares, 'Mineralizing the Right to Prior Consultation: From Recognition to Disregard of Indigenous and Peasant Rights in Peru' (2020) 20(1) *Global Jurist*.

13 FAO CFS, 'Voluntary Guidelines on the Responsible Governance of Tenure of Land, Fisheries and Forests in the Context of National Food Security' (VGGT) (11 May 2012) CL 144/9 Appendix D. See, e.g., paras. 1.1, 2.2.3.2, 4.4, and 5.3.

14 Ibid., ss. 14 and 15.

15 UN Committee on the Elimination of Discrimination against Women, 'General Recommendation No 34 on the Rights of Rural Women' (7 March 2016) CEDAW/C/GC/34, para. 36(a).

16 See, for example, Yudha Fathoni, *Challenging the Constitutionality of Indonesia's Investment Law* (International Institute for Environment and Development 2014).

17 See, for example, Eric T. Hoddya and Jonathan E. Ensorb, 'Brazil's Landless Movement and Rights "from Below"' (2018) 63 *Journal of Rural Studies* 74–82; George Mészáros, *Social Movements, Law and the Politics of Land Reform: Lessons from Brazil* (Routledge 2013).

18 Marc Edelman and Carwil James, 'Peasants' Rights and the UN System: Quixotic Struggle? Or Emancipatory Idea Whose Time Has Come?' (2011) 38(1) *Journal of Peasant Studies* 81–108. See also Priscilla Claeys, 'Food Sovereignty and the Recognition of New Rights for Peasants at the UN: A Critical Overview of La Via Campesina's Rights Claims over the Last 20 Years' (2014) 12(4) *Globalizations* 452–65.

19 For a discussion of some of these trends, see Saturnino M. Borras Jr., 'The Politics of Transnational Agrarian Movements' (2010) 41(5) *Development and Change* 771–803; Peter Rosset, 'Re-thinking Agrarian Reform, Land and Territory in La Via Campesina' (2013) 40(4) *Journal of Peasant Studies* 721–75; Michael Fakhri, *Sugar and the Making of International Trade Law* (Cambridge University Press 2014); Jennifer Clapp, 'The Problem with Growing Corporate Concentration and Power in the Global Food System' (2021) *Nature Food* 1–5. See also Edelman and James, 'Peasants' Rights and the UN System,' supra note 18, at 88–9.

20 Ward Anseeuw and Giulia M. Baldinelli, *Uneven Ground: Land Inequality at the Heart of Unequal Societies* (International Land Coalition 2020).

21 See, for example, Ward Anseeuw, Mathieu Boche, Thomas Breu, Markus Giger, Jann Lay, Peter Messerli and Kerstin Nolte, *Transnational Land Deals for Agriculture in the Global South: Analytical Report based on the Land Matrix Database* (International Land Coalition 2012); Saturnino M. Borras Jr., Jennifer C. Franco, S. Ryan Isakson, Les Levidow and Pietje Vervest, 'The Rise of Flex Crops and Commodities: Implications for Research' (2016) 43(1) *Journal of Peasant Studies* 93–15; Jampel Dell'Angelo, Paolo D'Odorico, Maria Cristina Rulli and Philippe Marchand, 'The Tragedy of the Grabbed Commons: Coercion and Dispossession in the Global Land Rush' (2017) 92 *World Development* 1–12; Lorenzo Cotula, ' "Land Grabbing" and International Investment Law: Toward a Global Reconfiguration of Property?' in Andrea K. Bjorklund (ed), *Yearbook on International Investment Law & Policy 2014–2015* (Oxford University Press 2016) 177–214; Kaitlin Y. Cordes and Anna Bulman, 'Corporate Agricultural Investment and the Right to Food: Addressing Disparate Protections and Promoting Rights-Consistent Outcomes' (2016) 20 *UCLA Journal of International Law and International Affairs* 87–161.

22 See, for example, Chaumtoli Huq, 'Charting Global Economic Inequalities and Emancipatory Human Rights Responses from the Ground Up: The Tea Workers' Movement of Bangladesh' (2020) 52(1) *Columbia Human Rights Law Review* 372–440.

23 Lorenzo Cotula, '(Dis)integration in Global Resource Governance: Extractivism, Human Rights, and Investment Treaties' (2020) 23(2) *Journal of International Economic Law* 431–54. See also GRAIN, *Asia's Agrarian Reform in Reverse: Laws Taking Land out of Small Farmers' Hands* (GRAIN 2015); Kevin Woods, 'A Political Anatomy of Land Grabs' *Myanmar Times* (3 March 2014) <www.mmtimes.com/national-news/9740-a-political-anatomy-of-land-grabs.html> accessed 9 June 2021.

24 UN Human Rights Council, 'Report of the Special Rapporteur on the Rights of Indigenous Peoples' (10 August 2018) UN Doc A/HRC/39/17. See also Global Witness, *Defending Tomorrow: The Climate Crisis and Threats against Land and Environmental Defenders* (Global Witness 2020).

25 Kojo S. Amanor, 'Global Resource Grabs, Agribusiness Concentration and the Smallholder: Two West African Case Studies' (2012) 39(3–4) *Journal of Peasant Studies* 731–49.

26 Ibid. For an oft-cited framework to describe rural differentiation in the context of commercial agriculture, see Andrew Dorward, 'Integrating Contested Aspirations, Processes and Policy: Development as Hanging In, Stepping Up and Stepping Out' (2009) 27(2) *Development Policy Review* 131–46.

27 See, for example, Comité Technique Foncier et Développement, *La Formalisation des Droits sur la Terre dans les Pays du Sud: Dépasser les Controverses et Alimenter les Strategies* (Ministère des Affaires Étrangères et du Développement International / Agence Française de Développement 2015). See also Patrick McAuslan, 'Only the Name of the Country Changes: The Diaspora of European Land Law in Commonwealth Africa,' in Camilla Toulmin and Julian Quan (eds), *Evolving Land Rights, Policy and Tenure in Africa* (Department for International Development/International Institute for Environment and Development / Natural Resource Institute 2000) *75–96*; Jean-Pierre Chauveau and Philippe Lavigne Delville, 'Quelles Politiques Foncières Intermédiaires en Afrique Rurale Francophone?' in Marc Lévy (ed), *Comment Réduire Pauvreté et Inégalités: Pour une Méthodologie des Politiques Publiques* (Karthala 2002) 211–39; Ambreena Manji, *The Politics of Land Reform in Africa: From Communal Tenure to Free Markets* (Zed Books 2006); Ben Cousins, 'Characterising "Communal" Tenure: Nested Systems and Flexible Boundaries' in Aninka Claasens and Ben Cousins (eds), *Land, Power and Custom: Controversies Generated by South Africa's Communal Land Rights Act* (UCT Press and Ohio University Press 2008) 109–37; Oussouby Touré, *Sécurisation du Foncier Pastoral en Afrique de l'Ouest: Des Modèles Divers et Riches d'Enseignements* (International Institute for Environment and Development 2018); Camilla Toulmin, *Land, Investment and Migration: Thirty-five Years of Village Life in Mali* (Oxford University Press 2020).

28 See, for example, Admos Chimhowu, 'The "New" African Customary Land Tenure. Characteristic, Features and Policy Implications of a New Paradigm' (2019) 81 *Land Use Policy* 897–903. See also Lorenzo Cotula (ed), *Changes in "Customary" Land Tenure Systems in Africa* (International Institute for Environment and Development 2006).
29 See, for example, Moussa Djiré, *Land Registration in Mali – No Land Ownership for Farmers? Observations from Peri-Urban Bamako* (International Institute for Environment and Development 2007).
30 Edelman and James, 'Peasants' Rights and the UN System,' supra note 18, at 83–92. See also Nora McKeon, 'One Does Not Sell the Land upon Which the People Walk: Land Grabbing, Transnational Rural Social Movements, and Global Governance' (2013) 10(1) *Globalizations*, 105–22.
31 See Mansour Fakih, Toto Rahardjo, and Michel Pimbert, *Community Integrated Pest Management in Indonesia: Institutionalizing Participation and People Centered Approaches* (particularly at 70–5 and Annex 5 containing a draft Peasant's Rights Charter, International Institute for Environment and Development 2003). On the role of participatory action research supported by the International Institute for Environment and Development (IIED) in the early stages of the development of peasants' rights, see Claeys, 'The Right to Land and Territory,' supra note 4, at 136.
32 Draft Peasants' Rights Charter, published as Annex 5 to Fakih, Rahardjo and Pimbert, *Community Integrated Pest Management in Indonesia*, supra note 31, art. A.7. See also art. B.16.
33 Ibid., art. B.20.
34 Fakih, Rahardjo and Pimbert, *Community Integrated Pest Management in Indonesia*, supra note 31, at 159.
35 Claeys, 'Food Sovereignty and the Recognition of New Rights for Peasants at the UN,' supra note 18, 5–6.
36 La Via Campesina, *Declaration of Rights of Peasants – Women and Men* (2009).
37 Ibid., art. 1.
38 Ibid., art. 4.
39 UN Human Rights Council 'Promotion and Protection of the Human Rights of Peasants and other People Working in Rural Areas' UN Doc A/HRC/RES/21/19 (11 October 2012). See the materials, <www.ohchr.org/EN/HRBodies/HRC/RuralAreas/Pages/WGRuralAreasIndex.aspx>, including: Christophe Golay, *Legal Reflections on the Rights of Peasants and Other People Working in Rural Areas* (Geneva Academy of International Humanitarian Law and Human Rights 2013); Geneva Academy of International Humanitarian Law and Human Rights, *The Right to Land and Other Natural Resources* (Geneva Academy of International Humanitarian Law and Human Rights 2017).
40 Human Rights Council, 'Promotion and Protection of all Human Rights, Civil, Political, Economic, Social and Cultural Rights, Including the Right to Development' UN Doc A/HRC/39/L.16 (26 September 2018).
41 UN General Assembly, 'United Nations Declaration on the Rights of Peasants and Other People Working in Rural Areas,' supra note 1.
42 UNDROP, supra note 4, sixth preambular para.
43 UNDROP, supra note 4, 11th preambular para.
44 UNDROP, supra note 4, 15th preambular para.
45 UNDROP, supra note 4, 13th preambular para.
46 UNDROP, supra note 4, 16th preambular para.
47 UNDROP, supra note 4, 19th preambular para.
48 UNDROP, supra note 4, art. 1.1.
49 UNDROP, supra note 4, artt. 1.2–1.4, emphasis added.
50 UNDROP, supra note 4, fifth preambular para.
51 UNDROP, supra note 4, art. 5.
52 UNDROP, supra note 4, art. 6.
53 UNDROP, supra note 4, art. 17.1.

54 Ibid.
55 In realising the 'fundamental right of everyone to be free from hunger,' for example, states are required to take a range of necessary measures, as relevant, including 'reforming agrarian systems in such a way as to achieve the most efficient development and utilization of natural resources.' See International Covenant on Economic, Social and Cultural Rights (adopted 16 December 1966, entered into force 3 January 1976) 93 UNTS 171 (ICESCR), art. 11.2.
56 UNDROP, supra note 4, art. 16.
57 For example, the UNDROP, supra note 4, artt. 17.1, 17.3, and 17.5
58 UNDROP, supra note 4, art. 17.3. See also FAO CFS VGGT, supra note 13, paras. 4.4, 53, and 7.1.
59 UNDROP, supra note 4, art. 17.3. See also art. 17.4, which reads: 'Peasants and other people working in rural areas have the right to be protected against arbitrary and unlawful displacement from their land or place of habitual residence, or from other natural resources used in their activities and necessary for the enjoyment of adequate living conditions. States shall incorporate protections against displacement into domestic legislation that are consistent with international human rights and humanitarian law. States shall prohibit arbitrary and unlawful forced eviction, the destruction of agricultural areas and the confiscation or expropriation of land and other natural resources, including as a punitive measure or as a means or method of war.'
60 UNDROP, supra note 4, art. 17.5.
61 UNDROP, supra note 4, art. 12.5.
62 See, for example, FAO CFS VGGT, supra note 13, paras. 3.1, 3.2, 4.4, 4.5, 5.3, 7.1, 8.2, 8.7, 9.4, 9.5, 11.6, 12.4, 12.6, 12.10, 12.15, 14.1, and 16.1.
63 For a fuller discussion, see Cotula, 'Between Hope and Critique,' supra note 6.
64 UNDROP, supra note 4, artt. 1.3 and 1.4. On the interface between human rights and redistributive action, see, for example, Julia Dehm, Ben Golder and Jessica Whyte, 'Introduction: 'Redistributive Human Rights?' Symposium' (2020) 8(2) *London Review of International Law* 225–32.
65 UNDROP, supra note 4, art. 17.6.
66 UNDROP, supra note 4, art. 28.
67 UNDROP, supra note 4, art. 17.6.
68 ICESCR, supra note 55, art. 11.2.
69 FAO CFS VGGT, supra note 13, para. 15.
70 UNDROP, supra note 4, art. 17.6.
71 Léon Duguit, *Les Transformations Générales du Droit Privé depuis le Code Napoléon* (2nd edn, Félix Alcan 1920).
72 For a fuller discussion, see Lorenzo Cotula, *Tenure Rights and Obligations: Towards a More Holistic Approach to Land Governance* (Food and Agriculture Organization of the United Nations, 2021).
73 Such as article 17 of the Universal Declaration of Human Rights (UDHR), which states that '[e]veryone has the right to own property alone as well as in association with others' and that '[n]o one shall be arbitrarily deprived of his property.'
74 UNDROP, supra note 4, art. 17.2; art. 1.2 clarifies that the UNDROP applies to peasants' 'dependent family members.'
75 UNDROP, supra note 4, artt. 3.1 and 3.3.
76 UNDROP, supra note 4, artt. 2.3 and 10.
77 UNDROP, supra note 4, art. 17.1 says 'Peasants and other people living in rural areas have the right to land, individually and/or collectively, in accordance with article 28 of the present Declaration.'
78 UNDROP, supra note 4, art. 28.2.
79 For example, while the La Via Campesina Declaration refers to the 'right to land and territory,' the UNDROP makes no explicit mention of territory, partly due to political sensitivities – though it does affirm the collective dimensions of the right to land.

80 See also Sérgio Sauer, 'Land and Territory: Meanings of Land between Modernity and Tradition' (2012) 1(1) *Agrarian South: Journal of Political Economy* 85–107.
81 UNDROP, supra note 4, art. 1.1.
82 UNDROP, supra note 4, artt. 1.1 and 1.2.
83 UNDROP, supra note 4, art. 13.
84 See, for example, Huq, 'Charting Global Economic Inequalities,' supra note 22, discussing the land struggles of tea plantation workers in Bangladesh. In her article, Huq reports that, based on arrangements dating back to colonial times, farm workers in the relevant site sublease small plots from plantation management to grow food crops and complement their meagre wages. The article retraces how farm workers mobilised against the compulsory acquisition of their plots for the creation of a special economic zone.
85 See, for example, Maya Hernández, 'Interview: Despite Escalating Assassinations, Colombian Farmworker Union Fights for the Right to Land' (2021) *Labour Notes* <www.labornotes.org/2021/01/interview-despite-escalating-assassinations-colombian-farmworker-union-fights-right-land> accessed 25 April 2021, containing an interview with Elsa Nury Martínez Silva, a farmworker activist and the president of the National Federation of Agricultural Unions (FENSUAGRO).
86 Rosset, 'Re-thinking Agrarian Reform,' supra note 19, at 727.
87 For a fuller discussion, see Cotula, '(Dis)integration in Global Resource Governance,' supra note 23.
88 Julia Dehm, 'Highlighting Inequalities in the Histories of Human Rights: Contestations Over Justice, Needs and Rights in the 1970s' (2018) 31(4) *Leiden Journal of International Law* 871–95.
89 Human Rights Committee, *Views Adopted by the Committee under Article 5(4) of the Optional Protocol, Concerning Communication No. 2751/2016* ('*Portillo Cáceres and Others v. Paraguay*'; 20 September 2019), CCPR/C/126/D/2751/2016. The complaint was filed by a group of peasants in connection with the impacts of agrochemical fumigations from large-scale soybean plantations on public health, waterways, crops and farm animals, and daily lives. The Committee found violations of the rights to life with dignity and to privacy, family life, and home (para. 8). In discussing the concept of 'home,' the Committee noted the petitioners' 'special attachment to and dependency on the land,' referencing the UNDROP (para. 7.8). The Inter-American Court of Human Rights cited the UNDROP in *Asociación Lhaka Honhat (Nuestra Tierra) v. Argentina* (Judgment 6 February 2020). The case concerns a long-running dispute over the recognition of the communal land rights of communities belonging to several Indigenous Peoples. The Court held that it could use the UNDROP when interpreting Argentina's obligations based on binding treaties (para. 136). For a discussion of these cases, see Golay, 'Research Brief: The Right to Land and Other Natural Resources,' supra note 3, at 4–5; Christophe Golay, *The Right to Land and the UNDROP: How Can We Use the United Nations Declaration on the Rights of Peasants and Other People Working in Rural Areas to Protect the Right to Land?* (Geneva Academy of International Humanitarian Law and Human Rights 2021) 52–3; Stéphanie de Moerloose and C. Ignacio de Casas, 'The Lhaka Honhat Case of The Inter-American Court of Human Rights: The Long-Awaited Granting of 400,000 Hectares under Communal Property Rights' (2020) *EJIL:Talk!*, <www.ejiltalk.org/the-lhaka-honhat-case-of-the-inter-american-court-of-human-rights-the-long-awaited-granting-of-400000-hectares-under-communal-property-rights/> accessed 28 April 2021.
90 See UN Human Rights Council, 'Report of the Special Rapporteur on the Rights of Indigenous Peoples,' supra note 23; Global Witness, *Defending Tomorrow,* supra note 23.
91 Alan Hunt, 'Rights and Social Movements: Counter-Hegemonic Strategies' (1990) 17(3), *Journal of Law and Society* 309–28; César Rodríguez-Garavito, 'Beyond Enforcement: Assessing and Enhancing Judicial Impact,' in Malcolm Langford, César Rodríguez-Garavito and Julieta Rossi (eds), *Social Rights Judgments and the Politics of Compliance: Making it Stick* (Cambridge University Press 2017) 75–108.
92 See also Sofía Monsalve Suárez, 'The Human Rights Framework in Contemporary Agrarian Struggles' (2013) 40(1) *Journal of Peasant Studies* 241–2.

Selected Bibliography

Claeys P, 'The Rise of New Rights for Peasants: From Reliance to NGO Intermediaries to Direct Representation' (2018) 9(3–4) *Transnational Legal Theory*.

Cotula L, 'Between Hope and Critique: Human Rights, Social Justice and Re-Imagining International Law from the Bottom Up' (2020) 48(2) *Georgia Journal of International and Comparative Law*.

De Schutter O and Rajagopal B (eds), *Property Rights from Below: Commodification of Land and the Counter-Movement* (Routledge 2020).

Suárez SM, *The Right to Land and Other Natural Resources in the United Nations Declaration on the Rights of Peasants and Other People Working in Rural Areas* (FIAN International 2015).

7

THE RIGHTS TO WATER AND SANITATION AND THE UNDROP

The implementation challenges of a comprehensive approach

Mara Tignino, Francesca Restifo, and Tadesse Kebebew

Introduction

The rights to water and sanitation are explicitly recognised under Article 21 of the United Nations Declaration on the Rights of Peasants and Other People Working in Rural Areas (UNDROP), stating that peasants and rural workers have the right to water for personal and domestic use, farming, fishing, and livestock. Moreover, the Declaration explicitly states their right to equitable access to water and to be free from arbitrary disconnections and the contamination of water supplies. It also emphasises that states have to prioritise water necessary for human needs, small-scale food production, ecosystem needs, and cultural use.

Research and evidence show that people in rural areas are disproportionately affected by poverty and lack of access to water and sanitation services.[1] According to the Food and Agriculture Organization (FAO), globally, the poverty rate in rural areas is more than three times higher than in urban areas, with rural areas accounting for over half of the world's population and 79% of the total poor.[2] Out of the two billion people without basic sanitation services, 70% live in rural areas. Furthermore, girls and women are significantly affected by lack of access to water and sanitation since they often have to collect water from far away, preventing them from attending school and often exposing them to sexual violence.[3]

Article 21 includes the obligation to prevent third parties from impairing peasants' and rural communities' enjoyment of the right to water. This dimension is particularly relevant in the context of the impacts of farmland investments on the human rights of peasants and rural communities. The UNDROP generally recognises that 'States shall take all necessary measures to ensure that non-state actors . . . such as . . . transnational corporations and other business enterprises, respect and strengthen the rights of peasants and other people working in rural areas.'[4]

DOI: 10.4324/9781003139874-10

The chapter first provides an overview of the content of the rights to water and sanitation and their relationship to the UNDROP. It then delves into the nexus between land and water rights, taking into account the case law of human rights and investment mechanisms.

The linkages between the rights to water and sanitation and the UNDROP

In 2002, the United Nations (UN) Committee on Economic, Social and Cultural Rights (CESCR) adopted General Comment No. 15 on the Right to Water, which states that the right to water is implicitly recognised in Article 11.1 on the right to an adequate standard of living and Article 12.1 on the right to health of the International Covenant on Economic, Social and Cultural Rights (ICESCR). Emphasising the interdependence of human rights, the General Comment demands that priority has to be given to the uses of water required to prevent starvation and underscores the need to consider the right to water beyond personal and domestic use. This right also covers water for securing livelihoods, including food production, which is pivotal for ensuring priority and sustainability of access to water for use in agriculture and pastoralism.[5]

The General Comment further enunciates the importance of providing equitable access to water for disadvantaged and marginalised farmers and indigenous communities, including women farmers. The General Comment also emphasises the need for sustainable rain harvesting and irrigation technology, as states are under obligation to ensure that local communities are not deprived of their means of subsistence.[6] Concerning sanitation, it stresses that access to adequate sanitation is not only fundamental for human dignity and privacy but one of the principal mechanisms for protecting the quality of drinking water supplies and resources.[7] Accordingly, states should progressively extend safe sanitation services, particularly to rural and deprived urban areas, taking into account the needs of women and children.

In 2010, the UN General Assembly (UNGA) and the Human Rights Council adopted two resolutions signalling the recognition of the rights to water and sanitation in international law.[8] Since then, both bodies regularly adopt resolutions on the rights to water and sanitation. In particular, the resolution of 2015 adopted by the General Assembly relies on the General Comment No. 15 to define the right to water as entitling 'everyone, without discrimination, to have access to sufficient, safe, acceptable, physically accessible and affordable water for personal and domestic use.' Therefore, the normative contents of the right include availability, quality, accessibility, and affordability of water for personal and domestic use.

The human right to sanitation is considered as entitling

> everyone, without discrimination, to have physical and affordable access to sanitation, in all spheres of life, that is safe, hygienic, secure, socially and culturally acceptable and that provides privacy and ensures dignity while reaffirming that both rights are components of the right to an adequate standard of living.[9]

In 2012, the UNGA adopted by consensus the UN Guiding Principles on Extreme Poverty and Human Rights. The UN Guiding Principles are the first global policy guidelines focused specifically on the human rights of people living in poverty, including people living in rural areas, with a strong emphasis on the importance of the right to water as a means to eradicate extreme poverty in rural areas.[10] Although non-legally binding per se, they stem from legally binding provisions included in international human rights instruments. The UN Guiding Principles are instrumental for concrete steps to be taken to ensure meaningful implementation of the UNDROP. In implementing the UNDROP as regards water provision, governments can be inspired by the guidance enshrined in its principles to ensure that public policies, including poverty eradication efforts, reach the poorest members of the society.

The right to water for peasants and rural communities

The UNDROP takes into account the aspects of the rights to water and sanitation, which have been analysed earlier. Indeed, it explicitly states under Article 21.1 that:

> Peasants and other people working in rural areas have the human rights to safe and clean drinking water and to sanitation, which are essential for the full enjoyment of life and all human rights and human dignity. These rights include water supply systems and sanitation facilities that are of good quality, affordable and physically accessible, and non-discriminatory and acceptable in cultural and gender terms.

When we compare the scope of the human rights to safe drinking water and sanitation under the UN resolutions specifically devoted to these rights and the content of Article 21 of the UNDROP, we can generally note that the Declaration broadens the extent of protection. First, although the rights to water and sanitation are developed as self-standing rights, they are also inextricably connected to the right to an adequate standard of living, the right to health, and the right to life. Thus, the recognition of these freedoms and entitlements under Article 21 with similar normative content is reasonable.

Second, in human rights discourse, though the right to water has been recognised as a 'prerequisite for the realisation of other human rights,' the scope is principally limited to personal and domestic uses.[11] The UNDROP recognises these linkages and broadens the right to water to include water necessary to ensure food security as well as the duty of states to ensure environmental conservation and restoration of water resources. The recognition of sustainable access to water resources for productive and agricultural use, which naturally exceeds what is traditionally understood as 'domestic use of water,' is pivotal as people in 'rural and peri-urban areas require water not only for personal and domestic uses but also for other uses relating to the broader goals of poverty reduction and livelihood security.'[12]

Third, the UNDROP supports a comprehensive approach to the reading of the human rights to safe drinking water and sanitation. The different Articles of the Declaration confirm this assertion. For example, its preamble states that 'peasants and other people working in rural areas suffer from the burdens caused by environmental degradation and climate change,' and Article 18 is devoted to environmental protection, also covering water resources. This provision explicitly recognises the right to a safe, clean and healthy environment for peasants and rural communities, as well as the conservation and protection of the productive capacity of the resources they use and manage, including water.[13]

The resolutions on the rights to safe drinking water and sanitation adopted by both the General Assembly and the Human Rights Council often mention the disparities in the access to water in rural areas and the necessity to improve their water services.[14] Among persons living in extreme poverty, rural communities' right to an adequate standard of living is heavily based on their access to land, water, and forests, not only as a source of food and shelter but also as a 'basis for social, cultural and religious practices and a central factor for economic growth.'[15] For example, the 2021 report of the mission of the UN Special Rapporteur on Extreme Poverty and Human Rights in Timor Leste highlights that:

> Access to adequate water and sanitation facilities remains poor, particularly in rural areas, where approximately 43% of households do not have access to clean water and 74% of households do not have access to improved sanitation. . . . Income inequality has risen significantly. Disaggregated data reveals that the 75% of Timorese people who live in rural areas suffer disproportionately from poverty and disadvantage.[16]

This aspect is directly related to Article 21 of the UNDROP, which emphasises the obligation of states to ensure the respect and protection of access to water, in particular, 'for rural women and girls, and persons belonging to disadvantaged or marginalised groups, such as nomadic pastoralists, workers on plantations, all migrants regardless of their migration status, and persons living in irregular or informal settlements.'[17] States also have the duty to promote appropriate and affordable technologies to reuse treated wastewater and for water collection and storage.[18]

The reports of the former Special Rapporteur on the rights to safe drinking water and sanitation, Mr Léo Heller, pointed out the importance of ensuring access to water for rural communities. In some of his reports, he identifies some programs targeting rural areas, including Mexico and Ethiopia.[19] Such programs may help to realise the objectives of the UNDROP, notably the rights of peasants and other people working in rural areas to safe, clean drinking water, and sanitation.[20]

In many regions, rural communities are deprived of their right to access water. In Mexico, indigenous groups live in rural areas where access to water is often sporadic, its quality is unsafe, and the cost is high. Indigenous communities complain about the pollution of water resources due to unregulated mining and hydrocarbon extraction.[21] Concessions concluded with private companies have been adopted by the government without taking into account the availability and the impact on the quality of water resources.[22] Although the Constitution of Mexico explicitly includes the rights to water and sanitation, there are still significant gaps in domestic water laws. National legal frameworks should be updated to be consistent with the rights enshrined in the Constitution. The right to water includes the States' obligation to ensure universal access to water for rural and indigenous communities. Moreover, the government should support the management of water at a community level and promote eco-technologies.[23] This last demand is included in UNDROP Article 21.3 as it determines that states shall respect and protect 'customary and community-based water management systems' on a non-discriminatory basis and promote technologies for collection and storage of water as well as treatment of wastewater.

Many African countries attract a large number of investments in land and, since land without water has no value and it is a critical resource for agriculture, water's role in investments in land is often underestimated. For example, in Ethiopia, considered the 'water tower' of East Africa,[24] investments in land have significantly increased. From 2009 to 2015, the government leased around 3.3 million hectares of land to private investors.[25] Indigenous communities are not involved in the negotiation of contracts between the government and investors, as their land is often considered 'wasteland' without prior users.[26] As has been rightly noted, weak domestic laws have curtailed the rights of indigenous communities in the planning and approval of large-scale agricultural projects.[27]

The UNDROP has addressed the critical normative gap concerning the interplay between water and land as well as the need for special protection of peasants and other people working in rural areas in this respect.

The nexus between water and land under the UNDROP

Under the UNDROP, the rights to water and sanitation cannot be fully understood without considering land rights. In its preamble, the Declaration recognises 'the special relationship and interaction between peasants and other people working in rural areas, and the land, water and nature to which they are attached and on which they depend for their livelihood.' Moreover, Article 17.1 recognises the individual and collective dimension of the right to land in accordance with Article 28 of the Declaration and specifies that 'the right to have access to, sustainably use and manage land' includes the use of water bodies.

The right to land has an increasingly important role in international human rights law, especially when one considers farmland investments.[28] Land rights are reflected in property rights owned by individuals, collective groups, or legal persons. They may also find expression in customary land tenure systems. Land rights may be connected to the right to food, the right to water, or the right to property. The enjoyment of these human rights may be dependent on the right to land. International human rights law recognises the particular attachment specific groups, such as indigenous peoples, have to their land. The right to land of indigenous peoples is duly recognised in the United Nations Labour Organisation (ILO) Convention No. 169 Concerning Indigenous Tribal Peoples in Independent Countries and the UNDRIP.[29]

Moreover, the United Nations FAO Voluntary Guidelines on the Responsible Governance of Tenure of Land, Fisheries, and Forests in the Context of National Food Security deal with the tenure rights of local communities and provide guidance on how to ensure equitable access to land, fisheries, and forests. Since their adoption in 2012, the implementation of the Voluntary Guidelines has been largely encouraged in interstate meetings and summits, e.g., the G-20, Rio+20 as well as the UNGA. International corporations and businesses have equally supported the document. Transnational companies such as Coca-Cola, Nestlé, and Unilever have announced they would incorporate the Voluntary Guidelines in their corporate policies.[30]

Although the Guidelines do not explicitly mention water, they note that the governance of land, fisheries, and forests includes the management of water resources.[31] Moreover, the document sheds particular light on procedural rights relevant to examining the nexus between land and water rights. They recommend seeking 'the support of those who, having legitimate tenure rights, could be affected by decisions, prior to decisions being taken' and they note that 'power imbalances between different parties' should be taken into consideration to ensure 'active, free, effective, meaningful and informed participation of individuals and groups in associated decision-making processes.'[32] Lastly, the Guidelines emphasise the linkages between civil and political rights and social, economic, and cultural rights.[33]

Other sets of human rights-relevant principles to be considered in the context of private investments in land are the 2014 Principles for Responsible Investment in Agriculture and Food Systems endorsed by the Committee on Food Security (CFS). Contrary to the 2012 Guidelines, this document explicitly takes into account the protection of water. The Principles aim to safeguard 'legitimate tenure rights,' including water rights and the environment, from risks that could arise from investments. Among other measures, states should take the necessary corrective action in addressing water use by farmland investments to ensure coherence with obligations under environmental and human rights law.[34]

While it might be argued that the Voluntary Guidelines, the UNDRIP, and the UNDROP have a limited legal impact given their soft-law nature, universal and regional human rights bodies have developed consistent jurisprudence on the protection of the rights of indigenous communities.

The UN Human Rights Committee (HRC)

In the landmark decision, *Portillo Cáceres and Others v. Paraguay case*,[35] the HRC drew upon the UNDROP. The Committee found Paraguay in violation of its international obligations to protect the rights to life and respect for private and family life and the home. The Committee points out that Paraguay failed to protect individuals from severe environmental contamination from large-scale farms' use of illegal chemicals. The Committee explained that Portillo Cáceres and his family members depend on their crops, fruit trees, livestock, fishing, and water resources for their livelihoods. It added that they have a 'special attachment to and dependency on the land' using the words of, and referring to, Article 1 of the UNDROP (paragraph 7.8). As the pollution has had direct repercussions on the Cáceres family's livelihood, their right to privacy, family, and their home had been violated.

The Inter-American human rights system

The Inter-American human rights system has rich jurisprudence on protecting indigenous rights to land and indirectly to the right to water. In *Mayagna (Sumo) Community of Awas Tingni v. Nicaragua*, the Inter-American Court of Human Rights (IACtHR) interpreted the protection of 'property' as entrenched in Article 21 of the American Convention on Human Rights transcended the right of possession under state domestic law so as to guarantee maximum protection of indigenous communities as individuals and peoples. The court affirmed that indigenous peoples' right to make use of natural resources, including water, is based on their special relationship with the territory and their dependence on nature for their physical and cultural survival.[36] In the words of the court in the *Saramaka* case against Suriname, indigenous people's land rights would be rendered meaningless 'if not connected to the natural resources that lie on and within the land.'[37] In *Kakmok Kasek Indigenous Community v. Paraguay*, the court detailed the connection between indigenous people's use of natural resources and their particular cultural identity.[38]

The IACtHR was the first regional human rights body to refer to the UNDROP to protect the right to land and other natural resources. In the case *Indigenous Communities of the Lhaka Honhat (Our Land) Association v. Argentina*, the court recognised for the first time autonomous violations of the rights to a healthy environment, adequate food, water, and cultural identity.[39] The case involves a claim from indigenous communities in Argentina composed of over ten thousand members and united under the association *Lhaka Honhat* ('our land'). They claimed Argentina violated their right to communal property by failing to provide legal security to their territory and allowing Creole settlers to reside on their lands. In doing so, their rights to a healthy environment, adequate food, participation in cultural life, and judicial protection were also breached. The IACtHR ordered Argentina, among other things, to grant titles to the indigenous communities, to refrain from doing anything on the indigenous property that might affect the property's value or use without prior consultation. Moreover, the decision also requests to create a fund for community development, and to prepare a study on actions to be taken for water conservation and to remedy contamination and to avoid loss of forest resources and recover lost food sources.

African human rights system

A similar set of concerns and relationships have been emphasised in the African human rights system. The African Commission on Human and People's Rights (the African Commission) has repeatedly used Article 21 of the African Charter on Human and People's Rights (African Charter) and other provisions to recognise and protect indigenous peoples' rights to natural resources.

Relying heavily on the jurisprudence of the IACtHR, the African Commission has interpreted Article 21's right to make use of wealth and resources in conjunction with Article 14's protection of property so as to provide vital protection for indigenous peoples' rights to resources found in their traditional lands.[40] In *Centre for Minority Rights Development v. Kenya*, the African Commission linked land and water rights with freedom of religion and the right to development.[41] In addition, access to information and public participation has been found to be means for exercising the right to use natural resources. In this sense, the African Commission found Kenya in violation of Article 21 of the African Charter for failing to properly consult with the concerned indigenous community and not having obtained their consent.[42]

The duty to consult placed on the state implies that the land and resource rights vested in indigenous groups are not absolute. Indeed, the African Commission has been clear that, so long as appropriate consultations occur, restrictions on these rights can be justified if they are in the 'general interest of the community.'[43] However, there are strict limits on the state's power to grant concessions against indigenous resources rights. Much in line with the structure of Article 21, and its analogue in Article 1 of the United Nations Covenants on Civil and Political Rights and Economic Social and Cultural Rights of 1966, no restrictions to or violations of rights can be allowed when the subsistence and thus the survival of the indigenous people are at risk. The landmark decision of the African Commission on Human and Peoples' Rights in the *Social and Economic Rights Action Center and the Center for Economic and Social Rights v. Nigeria* case confirms this assertion. The Commission found that the government's complicity and participation in the destruction and contamination of water, soil, and crops as well as the destruction of food sources as a result of oil company activities constituted a breach of Article 21 of the African Charter by both destroying the community land and 'devastatingly affecting' the well-being of the Ogoni people.[44]

The role of international investment law in farmland disputes and the right to water

In the discussion on the nexus between water and land rights, investment law has to be considered. This area of international law often governs disputes around land rights, including the uses of water resources. For example, foreign investors may conclude a contract with a country regarding a farmland investment that includes water rights. Moreover, private investors may seek to protect their land and water rights necessary to operate their contract on farmland investment through an investment treaty, mainly bilateral investment treaties (BITs). These treaties secure investors' rights, including the right to arbitration claims against the host state, potentially impacting its ability to manage water resources and securing associated individual and collective rights.

The reasons for disputes around land and water rights may be various. For example, in the case of long-term drought or a political change, the host state may decide to expropriate the land or reallocate water rights.[45] On the other hand, among the main standards applicable to protecting investors' rights is the fair and equitable treatment standard and the prohibition against expropriation without compensation.[46] According to customary law and many investment treaties, a state has the right to expropriate landholdings under the condition it pays compensation. This also includes the cases of land reform programs that may have the objective to protect rural communities' rights. This could lead to investment disputes under the investment treaty on the basis, for example, of the clauses on expropriation. In this context, farmland investments raise complex issues on how to reconcile the investments with peasants' and local communities' human right to water and the protection of the environment.[47]

Naturally, the protection of foreign investment cannot be viewed in isolation from human rights law. The recognition of the human right to water has legal implications for host states: they have to ensure that farmland investments do not interfere with peasants' and rural communities' right to water. In realising the human right to water, the Human Rights Council calls upon states to 'pay particular attention to persons belonging to vulnerable and marginalized groups.'[48] Farmland investments often take place in rural areas where local communities have customary rights, which are vulnerable compared to the rights of investors. This means that host states must pay particular attention in this context to ensure that peasants and rural communities have adequate access to water as required by international law. Specific to this point, Article 16.4 of the UNDROP enunciates that states' investment policies should 'contribute effectively to protecting and strengthening local livelihood options and to the transition to sustainable modes of agricultural production.' Similarly, Article 21.5 refers to states' obligation to 'prevent third parties from impairing the enjoyment of the right to water of peasants and other people working in rural areas.' Host states must ensure that water use by foreign investors does not interfere with the realisation of the right to water for local communities. In our view, the main challenge in such cases remains the short-sided take of the human right to water according to which this right focuses solely on water for drinking and sanitation services and ignores the inextricable link between water, land, and agricultural production.[49] The time is ripe for expanding this understanding: the protection of the right to water should encompass states' duty to guarantee individual and groups' access to water of sufficient quantity and good quality for domestic and small-scale agricultural activities.[50]

The legal obligations owed to investors under investment law must be viewed in connection with the states' wider international obligations. International human rights law is as equally binding on host states as investment law.[51] The arbitral jurisprudence has recognised the linkages between human rights law and investment law. In the *Urbaser S.A. and Consorcio de Aguas Bilbao Bizkaia, Bilbao Biskaia Ur Partzuergoa v. The Argentine Republic* case, the tribunal recognised that the BIT was not an isolated or closed system, since the BIT itself allows reference to other sources of international law, including 'general principles of international law.'[52] It stated that the reference to the 'general principles of international law' would become meaningless if

> the BIT is to be construed as an isolated set of rules of international law for the sole purpose of protecting investments through rights exclusively granted to investors. Such a view, which claimants favor, is not correct for more than one reason.[53]

Concerning international human rights law, the tribunal did not accept the argument that 'the human right to water is a duty that may be born solely by the State, and never borne also by private companies.'[54] For the tribunal, '[w]hile such principle had its importance in the past, it has lost its impact and relevance in similar terms and conditions as this applies to individuals.'[55] It further explained its position by arguing that:

> International law accepts corporate social responsibility as a standard of crucial importance for companies operating in the field of international commerce. This standard includes commitments to comply with human rights in the framework of those entities' operations conducted in countries other than the country of their seat or incorporation.[56]

Concluding remarks

The right to water has been traditionally understood as the right to access water of good quality for personal and domestic purposes. While this understanding may well reflect the challenges of individuals living and working in urban settings, it falls short of addressing the needs of rural workers and communities. We can affirm that, one of the main contributions of the UNDROP with respect to the right to water is to broaden the scope of application of the right to water so as to include water uses for food production, especially farming, fishing, and livestock. Furthermore, in tune with recent developments on the rights of indigenous peoples and local communities, the UNDROP draws special attention to the duty of states to ensure the conservation and restoration of water resources.

Notes

1 See, e.g., Ana Paula de la O Campos, Chiara Villani, Benjamin Davis, and Maya Takagi, *Ending Extreme Poverty in Rural Areas – Sustaining Livelihoods to Leave No One Behind* (FAO 2018) 12–29.

2 FAO, 'FAO hails landmark UN resolution that enshrines rights of peasants and rural workers' (2018) <www.fao.org/news/story/pt/item/1175208/icode/> accessed 25 April 2021.

3 Alison Graham (main), Francesca Restifo, and Janet Nelson, *Making Human Rights Work for People Living in Extreme Poverty: A Handbook for Implementing the UN Guiding Principles on Extreme Poverty and Human Rights* (International Movement ATD Fourth World and Franciscans International 2015).

4 UN Human Rights Council, 'United Nations Declaration on the Rights of Peasants and Other People Working in Rural Areas' (8 October 2018) UN Doc A/HRC/RES/39/12, art. 2(5).

5 UN Committee on Economic, Social and Cultural Rights (CESCR), 'General Comment No. 15: The Right to Water (Arts. 11 and 12 of the Covenant)' (20 January 2003) UN Doc E/C.12/2002/11, para. 6.

6 Ibid., para.7.

7 Ibid., para. 29.

8 UN General Assembly, 'The Human Right to Water and Sanitation' (28 July 2010) UN Doc A/RES/64/292; UN Human Rights Council, 'Human rights and access to safe drinking water and sanitation' (6 October 2010) UN Doc A/HRC/RES/15/9.

9 UN General Assembly, 'The Human Rights to Safe Drinking Water and Sanitation' (22 February 2016) UN Doc A/RES/70/169, para. 2.

10 UN Human Rights Council, 'Final Draft of the Guiding Principles on Extreme Poverty and Human Rights, submitted by the Special Rapporteur on extreme poverty and human rights, Magdalena Sepúlveda Carmona' (18 July 2021) UN Doc UN A/HRC/21/39.

11 CESCR, supra note 5, para. 1.

12 Inga Winkler, 'Water for Food' in Malcom Langford and Anne A. Russel (eds), *The Right to Water: Theory, Practice and Prospects* (Cambridge University Press 2017) 84.

13 UNDROP, supra note 4, art. 18(1).

14 UN General Assembly, 'The Human Rights to Safe Drinking Water and Sanitation,' supra note 9, para. 5.

15 UN Human Rights Council, 'Final Draft of the Guiding Principles on Extreme Poverty and Human Rights, submitted by the Special Rapporteur on extreme poverty and human rights, Magdalena Sepúlveda Carmona,' supra note 10, para. 71.

16 UN Human Rights Council, 'Report of the Special Rapporteur on Extreme Poverty and Human Rights, Magdalena Sepúlveda Carmona' (24 May 2021) UN Doc A/HRC/20/25/Add.1, para. 8.

17 UNDROP, supra note 4, art. 21(3).

18 Ibid.

19 UN Human Rights Council, 'Progress Towards the Realization of the Human Rights to Water and Sanitation (2010–2020), Report of the Special Rapporteur on the Human Rights to Safe Drinking Water and Sanitation' (6 August 2020) UN Doc A/HRC/45/11.

20 UNDROP, supra note 4, art. 21(1).

21 UN Human Rights Council, 'Report of the Special Rapporteur on the Human Rights to Safe Drinking Water and Sanitation on his Mission to Mexico' (2 August 2017) UN Doc A/HRC/36/45/add.2, paras. 45 and 48.

22 Ibid., para. 19.

23 Ibid., para. 45.

24 Emily Ingebrettsen, 'A Thirsty Third World: How Land Grabs are Leaving Ethiopia in the Dust' (2017) 4(1) *The Journal of Gender and Water* 97.

25 Ibid., 98.

26 Lorenzo Cotula, Sonja Vermeulen, Rebecca Leonard and James Keeley, *Land Grab or Development Opportunity? Agricultural Investment and International Land Deals in Africa* (IIED, FAO and IFAD 2009) 62 and 70.

27 Ibid.

28 Lorenzo Cotula, 'Land, Property and Sovereignty in International Law' (2007) 25(2) *Cardozo Journal of International and Comparative Law* 219–86.

29 International Labour Organization (ILO), Indigenous and Tribal Peoples Convention C169 (27 June 1989).
30 Makane M. Mbengue and Susanna Waltman, *Farmland Investments and Water Rights in Africa: The Legal Regimes at Stake* (International Institute for Sustainable Development 2015).
31 Subsequent to 2012 Guidelines, FAO developed the concept of 'water tenure.' This concept is defined as 'the relationship, whether legally or customarily defined between people, as individuals or groups, with respect to water resources.' See FAO, 'Exploring the Concept of Water Tenure' (2016) <www.fao.org/3/a-i5435e.pdf> accessed 27 May 2021 and Waterlex and FAO, 'Revisiting the Concept of Water Tenure: Filling the Gap between Water Rights and Water Governance' (2019) <http://humanright2water.org/wp-content/uploads/2020/04/191130-Water-Tenure-Paper-1.pdf> accessed 27 May 2021.
32 FAO, Voluntary Guidelines on the Responsible Governance of Tenure of land, fisheries and forests in the Context of National Food Security (FAO 2012) (Adopted at the 38th (Special) Session of the Committee on World Food Security, in Rome, on 22 May 2012), Principles of Implementation, para. 6.
33 Ibid., Rights and responsibilities related to tenure, para. 4.8.
34 Committee on World Food Security, 'Principles for Responsible Investment in Agriculture and Food Systems' (2014), Principle 5.
35 Human Rights Committee, *Views Adopted by the Committee under Article 5(4) of the Optional Protocol, Concerning Communication No. 2751/2016*, 'Portillo Cáceres v. Paraguay' (20 September 2019) CCPR/C/126/D/2751/2016.
36 *Indigenous Community Sawhoyamaxa v. Paraguay,* Inter-American Court of Human Rights Series C No 146 (26 March 2007), para. 118; *Indigenous Community Yakye Axa v. Paraguay* Inter-American Court of Human Rights Series C No 125 (17 June 2005), para. 137.
37 *Saramaka People v. Suriname,* Inter-American Court of Human Rights Series C No 172 (28 November 2007), para. 122.
38 *Kakmok Kasek Indigenous Community v. Paraguay,* Inter-American Court of Human Rights Series C No 214 (24 August 2010), para. 174.
39 Ibid., para. 289.
40 *Centre for Minority Rights Development (Kenya) and Minority Rights Group International on behalf of Endorois Welfare Council v. Kenya,* African Commission on Human and Peoples' Rights 276/2003(4 February 2010), para. 267.
41 Ibid., para. 173.
42 Ibid., paras. 266–8.
43 Ibid., para. 267.
44 *Social and Economic Rights Action Center for Economic and Social Rights v. Nigeria,* African Commission on Human and Peoples' Rights Communication 15/96 (27 October 2001), paras. 55–8.
45 Carine Smaller, 'A Global Thirst: How Water is Driving the New Wave of Foreign Investment in Farmland' (2010) *International Institute for Sustainable Development* <www.iisd.org/itn/en/2010/12/16/a-global-thirst-how-water-is-driving-the-new-wave-of-foreign-investment-in-farmland/> accessed 25 April 2021.
46 See Mbengue and Waltman, *Farmland Investments and Water Rights in Africa: The Legal Regimes at Stake*, supra note 30.
47 See Roberta Greco, *The Human Right to Water and International Economic Law* (Routledge-Giappichelli Studies in Law 2020) 123–94.
48 UN Human Rights Council, 'The Human Right to Water and Sanitation,' supra note 8, para. 8(c).
49 Will Schreiber, 'Realizing the Right to Water in International Investment Law: An Interdisciplinary Approach to BIT Obligations' (2008) 48(2) *Natural Resources Journal* 431–78.

50 See Edith Brown Weiss, *International Law for a Water-scarce World* (Martinus Nijhoff Publishers 2013).
51 Ibid.
52 Article 10(5) of the Spain-Argentina Bilateral Investment Treaty states that 'The arbitral tribunal shall make its decision on the basis of this Agreement and, where appropriate, on the basis of other treaties in force between the Parties . . . and the general principles of international law.'
53 *Urbaser S.A. and Consorcio de Aguas Bilbao Bizkaia, Bilbao Biskaia Ur Partzuergoa v. The Argentine Republic,* International Centre for Settlement of Investment Disputes Case No ARB/07/26(8 December 2016), para. 1189.
54 Ibid., para. 1193.
55 Ibid., para. 1194.
56 Ibid., paras. 1194–5.

Selected bibliography

Cotula L, 'Land, Property and Sovereignty in International Law' (2007) 25(2) *Cardozo Journal of International and Comparative Law* 219–86.

Golay C and Cismas I, *Legal Opinion: The Right to Property from a Human Rights Perspective* (Geneva Academy of International Humanitarian Law and Human Rights Law and International Centre for Human Rights and Democratic Development 2009).

Greco R, *The Human Right to Water and International Economic Law* (Routledge-Giappichelli Studies in Law 2020).

Heri C, 'The Human Right to Land, for Peasants and for All: Tracing the Social Function of Property to 1948' (2020) 20 *Human Rights Law Review* 433–52.

Mbengue MM and Waltman S, *Farmland Investments and Water Rights in Africa: The Legal Regimes at Stake* (International Institute for Sustainable Development 2015).

Winkler I, 'Water for food' in Malcom Langford and Anne Russel (eds), *The Right to Water: Theory, Practice and Prospects* (Cambridge University Press, 2017) 84–118.

8

THE RIGHT TO SEEDS

Using the commons as a sustainable governance scheme to implement peasants' rights?

*Sandrine Le Teno**, *Christine Frison, and Samuel Cogolati*

Introduction

In the end of 2020, Indian peasants protested against a new agricultural legislation feared to further favour agribusinesses over small-scale peasants [1] and the economic value of agrobiodiversity resources over their social and biological ones. Indeed, the development of biotechnologies and their accompanying private property systems of rights protection exacerbated the marketisation of agriculture.[2] In 2017, 72% of the global commercial seed market was controlled by six companies,[3] whereas rural populations accounted for an overwhelming 79% of the extremely poor.[4]

At the heart of this system lies a specific, crucial issue for peasants: access to, and control and management of seeds. Modernisation of agriculture and the related commodification of crops has led to the enclosure of seed varieties within an international regime complex.[5] This has caused the marginalisation of smallholder farmers and peasants,[6] despite their valuable contribution to the conservation and sustainable use objectives of the International Treaty on Plant Genetic Resources for Food and Agriculture[7] (ITPGRFA) and the Convention on Biological Diversity (CBD).[8] It also has led to the outlawing of peasants' seed systems,[9] depriving peasants of their customary rights over seeds.[10] In addition, it has increased seed diversity erosion, endangering local and global food security,[11] by encouraging homogeneity, when agrobiodiversity is necessary to increase resilience against climate change (only nine plant species produce 66% of the world crops[12]).[13] The needed diversification in seed varieties[14] could be achieved by supporting informal/peasants seed systems,[15] which substantially contribute to the distribution of more diverse and adaptable varieties.[16]

For a long time, civil societies and peasants' organisations fought for their recognition, developing claims for their food and seed sovereignty and autonomy.[17]

* Sandrine Le Teno carried out the research to draft this chapter based on Christine Frison's previous work, with few specific inputs from Samuel Cogolati. Dr Frison acted as a contributing supervisor for this publication.

DOI: 10.4324/9781003139874-11

Self-organised grassroots community networks[18] and participatory plant breeding initiatives based on shared access to seeds emerged. These contributed to revalue the commons,[19] although this concept does not benefit from strong legal protection in international law and only rarely in domestic legal systems.[20] Indeed, the current agricultural system remains dominated by 'monocultures of the mind,'[21] based on a growing commercial value of plant genetic resources for food and agriculture (PGRFA) and their individual privatisation. This is especially problematic as new threats to the agricultural system and world food security are emerging.[22]

However, with the recent adoption of the UN Declaration of the Rights of Peasants and Other People Living in Rural Area (UNDROP), especially its right to seeds,[23] this might change. This right can be defined as the right for peasants and peasant communities to, *inter alia*, access seeds, to protect their related traditional knowledge, to perpetuate their practices of saving, exchanging, selling, and using the seeds they grow, and to participate in decision-making.

The following analysis results from a literature review and the observatory participation in the developments regarding Farmers' Rights (FRs) and the right to seeds under the ITPGRFA and the FAO Committee on World Food Security (CFS).[24] Rather than restricting it to Article 19 of the UNDROP (section 3), we suggest broadening it to the ITPGRFA's FRs (section 2) and analysing the limits of these rights (section 4). Then, we propose anchoring their implementation into a commons governance approach[25] (section 5).

Farmers' rights under the international treaty for plant genetic resources for food and agriculture: a narrow attempt to recognise the right to seeds

The concept of FRs was first used in FAO Resolutions 4/89 and 5/89[26] attached to the International Undertaking on Plant Genetic Resources, a non-binding text. In 2001, the ITPGRFA recognised FRs in a legally binding treaty, thereby formally creating a counter claim to breeders' rights.[27]

Article 9 recognises the 'enormous contribution' of farmers to the conservation and development of PGRFA, from which their rights flow. However, it is only a recommendation, not a legal obligation, as it states: 'each Contracting Party should, as appropriate, and subject to its national legislation, take measures to protect and promote Farmers' Rights.' This includes the right to: protect their traditional knowledge related to PGRFA, participate in the equitable benefit sharing arising from their use, and participate in decision-making at the national level on related matters.[28] Additionally, Article 9 states that '[n]othing in this Article shall be interpreted to limit any rights that farmers have to save, use, exchange and sell farm-saved seed/propagating material, subject to national law and as appropriate.'[29] Article 9's provisions are therefore vague, not well-defined[30] and are subject to the political will of each contracting party, as the responsibility for realising FRs rests with national governments.

On the contrary, Intellectual Property Rights (IPRs) are legally binding (ITPGRFA, Article 12(f)) and are well implemented at the national level (see section 4)[31], creating an imbalance between FRs and breeders' rights.[32] In conclusion, the ITPGRFA does not challenge IPRs,[33] does not offer clearly defined obligations for states regarding FRs implementation,[34] nor reconcile the opposing interests at stake regarding PGRFA governance.[35] We argue, however, that FRs should be viewed as an integral part of ITPGRFA's core issues and obligations.[36] Combined with the UNDROP's Article 19, FRs and the peasants' right to seeds might become an effective reality.

Definition of the right to seeds: what is at stake?

The need to explicitly recognise peasants' rights and their central role in world food security urged the adoption of the UNDROP' (and its right to seeds). According to peasants themselves, the right to seeds is 'self-evident' and crucial for the right to food of all and for conservation purposes.[37] The UNDROP arose from a peasants' initiative[38] and included rights conflicting with the interest of the dominant agrochemical industry.[39]

It builds on other international law provisions such as the ITPGRFA, the Nagoya Protocol to the CBD, the UN Declaration on the Rights of Indigenous Peoples (UNDRIP),[40] as well as other human rights instruments.[41] But the right to seeds remained controversial during UNDROP negotiations.[42]

First, the scope of Article 19 expands to 'peasants and other people living in rural areas,'[43] i.e., 'any person . . . alone, or in association with others or as a community.'[44] Consequently, the UNDROP grants an individual and a collective right to seeds. Moreover, it allows local communities to formally emerge as a legitimate stakeholder,[45] especially at the CFS negotiating table.[46]

Second, Article 19 is divided into five more specific rights with five co-related state obligations.[47] Peasants and people working in rural areas have the right to:

1 *Enjoy the protection of their traditional knowledge associated with seeds,*[48] which extends to the protection of their innovation and practices[49] (already protected under the CBD[50]);
2 *Participate equitably in the sharing of benefits arising from seed utilisation;*[51]
3 *Participate in the decision-making on matters related to conservation and sustainable use of seeds,*[52] and more largely in the preparation and implementation of policies, programmes, and projects regarding seeds, that may affect them.[53]

The level at which participation is required is not specified. As the seed market is global, limiting participation in decision-making at the national level will not effectively rebalance peasants' and breeders' rights. Hence, we argue that peasants' participation in decision-making should also take place at a global level. Finally, this task might be easier if the participation of peasants is made through not only already-existing, inclusive institutions, such as the CFS, but also the governing body of the ITPGRFA.

4 *Save, use, exchange, and sell farm-saved seeds or propagating materials,*[54] which is the recognition of their customary right to do so.

This right was mentioned under the ITPGRFA,[55] but is completed, in the UNDROP, with the following right to:

5 'maintain, control, protect and develop their own seeds and the related traditional knowledge.'[56]

Read together with ITPGRFA's obligations, states shall support peasants' initiatives aimed at achieving the control, protection, or development of their seeds, similar to how they support formal seed systems.

Therefore, states' co-related obligations are to:

1 *Ensure consistency between this human right, IPRs, breeders' rights, and seed laws*, within their national, regional, and international laws and policies;[57]
2 *Respect, protect, and fulfil the right to seeds in taking appropriate steps to achieve the full implementation of FRs*[58], including ensuring to peasants access to seeds of sufficient quality, at a suitable time and an affordable price;[59]
3 *Ensure active and free participation of peasants in decision-making processes in relation to seeds*, directly or through their representatives, including in decision-making on evaluation and certification systems.[60] According to the FAO, participation is a prerequisite for FRs' realisation.[61]
4 *Support peasant seed systems and promote the use of peasant seeds and agrobiodiversity*;[62]
5 *Ensure that agricultural research and development integrate the needs of peasants and peasants themselves*,[63] notably through participatory research programs.[64]

Therefore, the UNDROP recognises the right of peasants to choose seeds and crops they want to grow.[65] This is crucial because small-scale peasants rely, in majority, on local peasant markets for their seed security.[66]

Finally, Article 19 is also directly linked to UNDROP's Article 15 on the right to food; Article 17 on the individual and collective right to land and land tenure systems; Article 18 on the right to a safe, clean, healthy, and protected environment; Article 20 on the right to biological diversity, its sustainable use and conservation; Article 21 on the human right to water and sanitation;[67] as well as peasants' right to culture, granted under classical human rights instruments and more specifically protected under the UNDROP.[68] All these rights are necessary for the realisation of peasants' right to seeds and should therefore be grasped together when interpreting and implementing Article 19. Interestingly, the UNDROP also makes specific reference to the rights of rural women. Their importance in the agricultural economy is recognised,[69] especially in low-income countries,[70] and they benefit from specific protection regarding seeds.[71] Consequently, their rights should also be considered when analysing Article 19.

In conclusion, the UNDROP remains a positive development, pushing FRs one step further. Yet, its possible primacy over IPRs remains debated[72] but extends beyond the limits of this chapter. We note, however, that peasants' right to seeds is recognised as a collective right in the African Union's Model Law for the Protection of the Rights of Local Communities, Farmers, and Breeders and for the Regulation of Access to Biological Resources, and in India,[73] which creates a balance between stakeholders' rights in spite of its strong right to (intellectual) property.[74] In most countries, strong IPRs and seed laws still hinder peasants rights' implementation.

Limits of the right to seeds under the United Nations Declaration on the Rights of Peasants and Other Peoples Working in Rural Area

With the adoption of the UNDROP, the right to seeds could become 'a standalone human right,'[75] but some limits impede its full bloom. Rather than focusing on the debate over the UNDROP's legal value,[76] we highlight two major limits we found necessary to address in order to enable the realisation of the right to seeds.

1 *In practice, the right to seeds is hindered by IPRs and seed law-related provisions.* States adopted them pursuant to the International Convention for the Protection of New Varieties of Plants 1991 (UPOV),[77] the Trade-Related Aspects of Intellectual Property Rights (TRIPS) agreement,[78] or IPRs clauses contained in trade agreements between countries.[79] They impose IPRs over plant varieties, hence, protecting breeders' and seed companies' interests. These are reinforced by a myriad of national seed laws aimed at controlling the seeds that are sold. Indeed, these laws often impose certification and commercialisation criteria for seeds to access the market (e.g., Distinctive, Stable and Uniform – DUS – criteria) that peasants' seeds frequently do not meet.[80] By strengthening control over seeds, these laws favour commercial and homogenised varieties over traditional and diverse ones,[81] in contradiction with FRs. Consequently, they limit the role of small-scale peasants within their national agricultural system. In China, for instance, the seed law went through a four-stage process, increasingly supporting seed marketisation at the expense of peasants and their own seeds.[82] Europe also struggles to implement Article 19.1(d) notably.[83] Therefore, peasant seed systems are impacted by these laws and in some cases criminalised.[84] Additionally, IPRs benefit from a strong implementation monitoring body: the WTO's Appellate Body.[85] On the contrary, FRs implementation remains poor and mainly depends on the state's will, as there is no international compliance mechanism for FRs. However, some improvement can be observed in a number of domestic legislations.[86] India and Norway, for instance,[87] adopted *sui generis* national laws protecting FRs.[88] The European Union increased its flexibility, heading towards better FRs realisation (i.e., through the progressive opening of its system to conservation seeds, peasants varieties, and heterogeneous reproductive material),[89] such as China, with its 2016 new seed law including FRs and their interests.[90] Yet, these improvements remain insufficient to counterbalance the global hegemony of IPRs and seed laws.

2 *The ITPGRFA framework, in which is embedded the right to seeds, constitutes another drag*, even though peasants' movements are changing the lines and increasingly transforming agrarian systems (e.g., the adoption of the UNDROP).[91] Indeed, by partly building its Multilateral System (MLS) on the foundations of IPRs,[92] the ITPGRFA only grants a narrow legal space for farmers to access seeds, as it fails to formally recognise peasants' seed systems. Being a traditional international law instrument, the ITPGRFA favours formal institutions models, *de jure* not recognising peasants seed systems and *de facto* reinforcing the formal global one. This lack of recognition within the MLS system should be reversed in order to enhance the right to seeds' capacities of realisation under the UNDROP. In line with the attempts to expand the scope of the MLS to all PGRFA, we argue that the Third Party Beneficiary[93] mechanism of the ITPGRFA's Standard Material Transfer Agreement could be extended to include FRs and the right to seeds in its dispute settlement mandate (i.e., applied to all PGRFA, regardless of a specific access contract). If the MLS was expanded, our proposition could become feasible and enable peasants' communities to monitor their rights' enforcement within its scope. Thereby, the Third Party Beneficiary, currently representing the collective interests of the MLS community,[94] could become a defensive tool for local to global peasant communities, which in practice benefit poorly from the system.[95]

In conclusion, although the system might appear incoherent,[96] we agree that the UNDROP 'can create positive synergies between farmers' rights, . . . and standard IPRs legislation, while maintaining predictability for breeders and life science companies'[97] and fill a gap in international law.[98] Nonetheless, peasants need to be formally recognised as effective stakeholders in the governing scheme, instead of remaining passive beneficiaries.[99] Consequently, we argue that a shift in the approach used to address FRs and their right to seeds, might overcome the limitations highlighted earlier and accelerate the realisation of the right to seeds.

The commons: a governance approach to strengthen peasant communities' right to seeds

International law implements state and top-down formal governance, leaving little room for alternative, decentralised models, such as community governance. However, over the last 40 years, scholars have demonstrated how local collective forms of governance can be more efficient than private or public ones in managing natural resources. Elinor Ostrom studied numerous communities implementing what she called 'common-pool resources' (i.e., common governance mechanism following rules designed by the community itself to manage resources equitably and sustainably).[100] Following that, the commons trend has expanded beyond economics to philosophy, political science, legal studies, etc. Bollier popularised the concept of 'commoning,' i.e., a political act to enhance communities' power in collectively governing resources in a sustainable manner.[101]

Previous attempts to use the dominant private or public property models in order to promote FRs failed and, in some cases, damaged the social order of communities.[102] However, considering the strong dependence of poor rural households on community-based land tenure system for their survival,[103] these systems' poor legal recognition,[104] and the importance of local access to and control over seeds for world food security,[105] there is an urgent need to recognise alternative commons governance systems. We, therefore, argue that governing seeds as a commons developed as a political tool,[106] and recognising peasant communities' practice of commoning at all levels of governance, might facilitate FRs and the right to seeds' implementation. Indeed, supporting a commons-based system where PGRFA belong collectively to the communities who use, save, exchange, and sell them following self-defined (social and cultural) rules[107] may overcome two major barriers: the rigidity of centralised, formalised (inter)governmental organisations and the case-by-case contractual relation of the current private property system.[108] Today, however, seeds as a commons are not protected.[109] Therefore, traditional practices of sharing and exchanging seeds also lack recognition and face threats of regulation by the formal sector.

In order to support a seed commons under the ITPGRFA, several improvements must be made, among which the recognition of:

1) The diversity of farmers operating at different scales, with different production objectives or under different sets of conditions, as well as the importance of supporting their interconnections;[110]
2) The importance of local control and self-sufficiency in the attainment of food security and sovereignty;[111]
3) Peasant communities as an actor[112] from the local to the global governance scale, which was initiated by the UNDROP.[113]

The inclusion of the 'commons' in the UNDROP's text did not provoke much debate during the last round of negotiations. However, it represents a significant breakthrough in the process of recognising it as a self-governance system of its own, separate from both public and private property, in international law. Interestingly, the UNDROP is the first international human rights instrument expressly putting forward a stand-alone 'standard' for states to recognise and protect the commons as an autonomous governance model (Article 17.3, last sentence). Indeed, we argue that the 'natural commons,' as understood in Article 17, is not limited to 'land' but includes 'other natural resources,' in particular, seeds.[114] We believe that various seed governance models can coexist,[115] pursuant to the UNDRIP's recognition of indigenous land commons, in parallel to the dominant private property land tenure system.[116] Applying this logic to the ITPGRFA, combined with the international recognition of FRs (i.e., no longer leaving this recognition up to states) and with the adaptation of the ITPGRFA governing body rules of procedures in light of the functioning of the CFS,[117] could enable the recognition of farming communities as active stakeholders in the governance of seeds at a global level. Commoning practices implementing peasants' right to seeds would hence take place from local to national levels thanks to their recognition under the UNDROP, and at a global level through the ITPGRFA (if reviewed in that direction). All levels of governance would therefore be 'commonized', thereby rebalancing the powers and interest at stake at all seed governance levels.

Conclusion: the right to seeds under the UNDROP as an essential collective human right

In this chapter, we analysed the right to seeds, its collective nature, its links with ITPGRFA's FRs, and their limits. We also emphasised the need, regarding PGRFA, to change the 'basic categories of ownership' already existing in law (i.e., private and public) to include a collective one: the commons.[118] In our opinion, it is necessary to recognise seeds' non-market values and to envisage a decentralised governance system relying on the community at various scales. This would enable the legal and social practices for PGRFA to be perceived as a commons[119] and engage with peasants' community rights. We believe that it would also displace the focus on the role of local communities in the governance of PGRFA, following the peasants' empowerment logic behind the adoption of the UNDROP. Moreover, these changes are key to further facilitate access by small-scale peasants to PGRFA in order to effectively face climate change and preserve world food security.[120] Finally, following the ongoing review process of the ITPGRFA, we are convinced that adapting the MLS[121] to include FRs and fully recognise these rights at both the international and national levels could overcome the lack of enforcement mechanism on the right to seeds. Indeed, even if it has its flaws,[122] the MLS constitutes progress towards the recognition of a global seed commons.[123] By enlarging the Third Party Beneficiary' scope, it could become a tool for peasant communities to enforce their (collective) right to seeds.

To conclude, the UNDROP constitutes an important step towards the realisation of FRs' and peasants' essential right to seeds. Using the commons to address the structural obstacles of international law might effectively facilitate their implementation.

We need to recognise the value of small-scale peasants' social and ecological sustainable contribution to food production in a concrete manner as well as their community rights to manage resources collectively.[124] As Kloppenburg highlighted,

> Resistance must be complemented by creative actions that are not just reactions to corporate/neoliberal depredations, but which are offensive, affirmative, positive, proactive undertakings designed to repossess and maintain alternative, (relatively) autonomous spaces.[125]

However, we admit that political will remains key to unlocking the full potential of the right to seeds, as an individual and collective human right and as a commons.

Notes

1 La Via Campensina, 'Les paysan·ne·s du monde entiers s'unissent! Ce 26 janvier, allumez une bougie en solidarité avec les paysan·ne·s de l'Inde' (2021) <https://viacampesina.org/fr/les-paysan·ne·s-du-monde-entier-sunissent-ce-26-janvier-allumez-une-bougie-en-solidarite-avec-les-paysan·ne·s-de-linde/> accessed 10 February 2021.
2 Christine Frison, *Redesigning the Global Seed Commons: Law and Policy for Agrobiodiversity and Food Security* (Routledge 2018) 62.
3 Hope Shand and Katie Jo Wetter, 'Plate Tech-Tonics: Mapping Corporate Power in Big Food' (2019) *ETC Group*; Philip H Howard, 'Intellectual Property and Consolidation in the Seed Industry' (2015) 55 *Crop Science* 2489–95.
4 UN General Assembly, 'Interim Report of the Special Rapporteur on the Right to Food' (17 July 2019) UN Doc A/74/164, para. 32.
5 Stefan Jungcurt, *Institutional Interplay in International Environmental Governance: Policy Interdependence and Strategic Interaction in the Regime Complex on Plant Genetic Resources for Food and Agriculture* (Landwirtschaftlich – Gärtnerischen Fakultät, Humboldt Universät 2007).
6 Philip McMichael, 'Peasants Make Their Own History, But Not Just as They Please . . .' (2008) 8 *Journal of Agrarian Change* 209.
7 International Treaty on Plant Genetic Resources for Food and Agriculture, (adopted 23 November 2001, entered into force 29 June 2004), 2400 UNTS 303, (ITPGRFA).
8 Convention on Biological Diversity (adopted 5 June 1992, entered into force 29 December 1993) 1760 UNTS 79 (CBD).
9 Tammara Wattnem, 'Seed Laws, Certification and Standardization: Outlawing Informal Seed Systems in the Global South' (2016) 43 *The Journal of Peasant Studies*.
10 Christophe Golay, 'The Implementation of the United Nations Declaration on the Rights of Peasants and Other People Working in Rural Areas' (2019) *Geneva Academy Research Brief* <www.geneva-academy.ch/joomlatools-files/docman-files/The%20implementation%20of%20the%20UN%20Declaration%20on%20the%20rights%20of%20peasants%20and%20other%20people%20w.pdf> accessed 20 January 2021, 16–7.
11 FAO, 'The second report on the state of the world's Plant Genetic Resources for Food and Agriculture' SoW2-PGRFA (26 October 2010) 186.
12 Julie Bélanger and Dafydd Pilling (eds), *The State of the World's Biodiversity for Food and Agriculture* (FAO Commission on Genetic Resources for Food and Agriculture Assessments 2019) 9.
13 UN General Assembly, 'Report of the Special Rapporteur on the Right to Food' (23 July 2009) UN Doc A/64/170, para. 39; IPES-Food, *From Uniformity to Diversity: A Paradigm Shift from Industrial Agriculture to Diversified Agroecological Systems* (International Panel of Experts on Sustainable Food systems 2016) 9–11.
14 Christoph Leibing, 'Crop and Forage Genetic Resources: International Interdependence in the Face of Climate Change' in Michael Halewood et al (eds), *Crop Genetic Resources as a Global Commons: Challenges in International Law and Governance* (Earthscan Publications 2013) 78–81.
15 International Treaty on Plant Genetic Resources for Food and Agriculture Secretariat, *The Global Consultation on Farmers' Rights 2016. Summary of presentations and discussions. International Treaty on Plant Genetic Resources for Food and Agriculture (ITPGRFA)* (Rome 2016) 13–5.
16 Leibing, 'Crop and Forage Genetic Resources: International Interdependence in the Face of Climate Change,' supra note 14, at 82.
17 Elise Demeulenaere, 'A Political Ontology of Seeds: The Transformative Frictions of a Farmers' Movement in Europe' (2014) 69 *Focaal – Journal of Global and Historical Anthropology* 8.
18 Elsa Tsioumani, 'Reimagining Agrobiodiversity Conservation and Agricultural Innovation from the Grassroots Up: The Case of the Peliti Seed Network in Greece' in Mariagrazia Alabrese et al (eds), *The United Nations Declaration on the Rights of Peasants and other*

People Working in Rural Areas in Perspective. New international Standards on Natural Resources and Food Systems (Routledge 2021) 216–227.

19 Stefanie Sievers-Glotzbach et al, 'Diverse Seeds – Shared Practices: Conceptualizing Seed Commons' (2020) 14 *International Journal of the Commons on Participatory Plant Breeding Initiatives*.

20 Samuel Cogolati, *International Law to Save the Commons. The International Legal Protection of the Commons in Development* (PhD thesis, KU Leuven 2021) 15.

21 See Vandana Shiva, *Monocultures of the Mind: Perspectives on Biodiversity and Biotechnology* (Palgrave Macmillan 1993).

22 Johannes Kotschi and Bernd Horneburg, 'The Open Source Seed Licence: A Novel Approach to Safeguarding Access to Plant Germplasm' (2018) 16 *PLOSBiology* <https://pubmed.ncbi.nlm.nih.gov/30352056/> accessed 30 January 2021, 2. See also Louise Sperling, 'Seed Security Response during COVID-19: Building on Evidence and Orienting to the Future' (2020) 12 *Food Security* 886.

23 UN General Assembly, 'United Nations Declaration on the Rights of Peasants and Other People Working in Rural Areas' (17 December 2018) UN Doc. A/RES/73/165 (UNDROP), art. 19.

24 Participatory observation of Dr. C. Frison to all the Governing Body meetings of the ITPGRFA, except the 2013 GB5 meeting in Oman. Participatory observation of Dr. S. Cogolati to the fifth round of the UNDROP's negotiation in 2018 at the Human Rights Council in Geneva.

25 See Elinor Ostrom, *Governing the Commons: The Evolution of Institutions for Collective Action* (Cambridge University Press 1990). See also the 2013 proposal by Burns H. Weston and David Bollier to adopt a 'Universal Covenant affirming a human right to commons- and rights-based governance of Earth's natural wealth and resources.'

26 FAO, 'Agreed Interpretation of the International Undertaking,' Conference Resolution 4/89 (29 November 1989); FAO, 'Farmers' Rights,' Conference Resolution 5/89 (29 November 1989).

27 Carlos M. Correa, 'Implementing Farmers' Rights Relating to Seeds' (2017) 75 *South Center Research Paper* 1.

28 ITPGRFA, supra note 7, artt. 9.2(a), (b) and (c).

29 ITPGRFA, supra note 7, art. 9.3.

30 Hans Morten Haugen, 'The UN Declaration on Peasants' Rights (UNDROP): Is Article 19 on Seed Rights Adequately Balancing Intellectual Property Rights and the Right to Food?' (2020) 23 *The Journal of World Intellectual Property* 297–9.

31 Christophe Golay, 'Research Brief: The right to seeds and Intellectual Property Rights' (2020) *Geneva Academy* 2.

32 Frison, *Redesigning the Global Seed Commons: Law and Policy for Agrobiodiversity and Food Security*, supra note 2, at 90ff.

33 Correa, 'Implementing Farmers' Rights Relating to Seeds,' supra note 27, at 6ff.

34 See Christine Frison, et al, *Plant Genetic Resources and Food Security Stakeholders Perspectives on the International Treaty on Plant Genetic Resources for Food and Agriculture* (The Food and Agriculture Organization of the United Nations and Bioversity International with Earthscan 2011).

35 Chiara Pontillo, *Implementing Farmers Rights through seed legislation: A Case Study in the Difficulty of Enforcing Policy Coherence at Multiple Levels* (Master thesis, University Roma Tre 2018) 10.

36 Anja Christinck, *Farmers' Rights: Views, Experiences, Lessons Learned and Proposed National Measures for the Implementation of Article 9* (FAO – ITPGRFA 2018) 19.

37 Christophe Golay, 'Legal Analysis on the Rights of Peasants and other People Working in Rural Areas. The Right to Seeds and Intellectual Property Rights' (2016) *Geneva Academy* 8–9.

38 The UNDROP emerged especially from the activism of La Vía Campesina (LVC), a transnational peasants organisation, that advocates for the recognition and protection of peasants' seed sovereignty over the seeds they cultivate.

39 Coline Hubert, 'The United Nations Declaration on the Rights of Peasants: A Tool in the Struggle for our Common Future' (2019) 42 *PubliCETIM* 20.
40 UN General Assembly, United Nations Declaration on the Rights of Indigenous Peoples, UN Doc A/RES/61/295, (2 October 2007) (UNDRIP), art. 31.1.
41 See for instance the Right to Food Guidelines, see Golay, 'The implementation of the United Nations Declaration on the Rights of Peasants and Other People Working in Rural Areas,' supra note 10; the International Covenant on Economic, Social and Cultural Rights (adopted 16 December 1966, entered into force 3 January 1976) 93 UNTS 171 (ICESCR) or the Convention on the Elimination of All Forms of Discrimination against Women (adopted 18 December 1979, entered into force 3 September 1981) 1249 UNTS 13 (CEDAW); see Haugen, 'The UN Declaration on Peasants' Rights (UNDROP): Is Article 19 on Seed Rights Adequately Balancing Intellectual Property Rights and the Right to Food?,' supra note 30.
42 Karine Peschard and Shalini Randeria, 'Keeping Seeds in Our Hands': The Rise of Seed Activism' (2020) 47 *The Journal of Peasant Studies* 623.
43 UNDROP, supra note 23, art. 19.1.
44 Ibid., art. 1.1.
45 Peschard and Randeria, 'Keeping Seeds in Our Hands': The Rise of Seed Activism,' supra note 42, at 622.
46 The CFS is an intergovernmental and highly inclusive platform based at the FAO, including peasants' organizations in its decision-making processes. Today, it is authoritative in the field of world food governance.
47 Golay, 'The Implementation of the United Nations Declaration on the Rights of Peasants and Other People Working in Rural Areas,' supra note 10, at 31–42.
48 UNDROP, supra note 23, art. 19.1(a).
49 UNDOP, supra note 23, art. 20.2.
50 CBD, supra note 8, art. 8(j).
51 UNDROP, supra note 23, art. 19.1(b).
52 UNDROP, supra note 23, art. 19.1(c).
53 UNDROP, supra note 23, art. 10.1.
54 UNDROP, supra note 23, art. 19.1(d).
55 ITPGRFA, supra note 7, preamble, art. 9.3.
56 UNDROP, supra note 23, art. 19.2.
57 UNDROP, supra note 23, artt. 19.8 and 2.4.
58 UNDROP, supra note 23, artt. 19.3 and 2.1.
59 UNDROP, supra note 23, art. 19.4.
60 UNDROP, supra note 23, artt. 2.4, 10.1, 11.3 and 19.8.
61 International Treaty on Plant Genetic Resources for Food and Agriculture Secretariat, *The Global Consultation on Farmers' Rights 2016. Summary of Presentations and Discussions. International Treaty on Plant Genetic Resources for Food and Agriculture*, supra note 15, at 2.
62 UNDROP, supra note 23, art. 19.6.
63 UNDROP, supra note 23, art. 19.7.
64 See Jacqueline A. Ashby and Nina Lilja, 'Participatory Research: Does it Work? Evidence from Participatory Plant Breeding' (International Crop Science Congress, Brisbane, Australia, 26 Sept to 1 Oct 2004).
65 UNDROP, supra note 23, art. 19.5.
66 Sperling, 'Seed Security Response during COVID-19: Building on Evidence and Orienting to the Future,' supra note 23, at 888
67 Stefania Errico and Priscilla Claeys, 'Human Rights and the Commons: Bridging Gaps and Exploring Complementary Approaches to the Governance of Land and Natural Resources' (XVI Biennial IASC Conference, Utrecht, Netherlands 2017) <www.iasc2017.org/> accessed 28 January 2021, 10–1.
68 UNDROP, supra note 23, art. 26.1. See also the UN General Assembly, 'Universal Declaration of Human Rights' (10 December 1948) UN Doc A/RES/217(III), art. 27; the International Covenant on Civil and Political Rights (adopted 16 December 1966,

entered into force 23 March 1976) 999 UNTS 171 (ICCPR), art. 27; ICESCR, supra note 41, art. 15.

69 UNDROP, supra note 23, preamble.

70 Bélanger and Pilling, *The State of the World's Biodiversity for Food and Agriculture*, supra note 12, at 98–101.

71 CEDAW, supra note 41, art. 14.2(g); UN Committee on the Elimination of Discrimination against Women, 'General Recommendation No. 34 on the Rights of Rural Women' (7 March 2016) UN Doc CEDAW/C/GC/34.

72 See Golay, 'Legal Analysis on the Rights of Peasants and other People Working in Rural Areas. The Right to Seeds and Intellectual Property Rights,' supra note 37, at 23–6; Philippe Cullet, 'Human Rights and Intellectual Property Protection in the TRIPS Era' (2007) 29 *Human Right Quarterly* 418.

73 Golay, 'The Implementation of the United Nations Declaration on the Rights of Peasants and Other People Working in Rural Areas,' supra note 10, at 36; Peschard and Randeria, 'Keeping Seeds in Our Hands': The Rise of Seed Activism,' supra note 42, at 627.

74 Cullet, 'Human Rights and Intellectual Property Protection in the TRIPS Era,' supra note 72, at 411.

75 Mohamed Coulibaly et al, 'The Right to Seeds and Legal Mobilization for the Protection of Peasant Seed Systems in Mali' (2020) *Journal of Human Rights Practice* 1.

76 Haugen, 'The UN Declaration on Peasants' Rights (UNDROP): Is Article 19 on Seed Rights Adequately Balancing Intellectual Property Rights and the Right to Food?,' supra note 30, at 289.

77 International Union for the Protection of New Varieties of Plants (UPOV), 'International Convention for the Protection of New Varieties of Plants' (18 March 1991) DC/91/130, art. 15.2. There are 59 contracting parties to the 1991 version, under which plant breeders' rights restrict the farmers' exemption to almost no possibility to carry out farmers' usual seed practices, thereby closing the opportunity to grant effective FRs' protection. Therefore, it is argued that TRIPS and UPOV are not compatible with the peasant seed system needs. See International Treaty on Plant Genetic Resources for Food and Agriculture Secretariat, *The Global Consultation on Farmers' Rights 2016. Summary of presentations and discussions. International Treaty on Plant Genetic Resources for Food and Agriculture*, supra note 15, at 20–2.

78 The Agreement on Trade Related Aspects of Intellectual Property Rights (adopted 15 April 1994, entered into force on 1 January 1995) 1869 UNTS 299 (TRIPS), Annex 1C, art. 27. It was adopted by all the 164 World Trade Organization members.

79 Peter Drahos, 'BITs and BIPS: Bilateralism in Intellectual Property' (2001) 4 *Journal of World Intellectual Property* 794; Fabien Girard and Christine Frison, 'From farmers' rights to the rights of peasants: seeds and the biocultural turn' (2021) *Review of Agricultural, Food and Environmental Studies* 1–16.

80 International Treaty on Plant Genetic Resources for Food and Agriculture Secretariat, *The Global Consultation on Farmers' Rights 2016. Summary of Presentations and Discussions. International Treaty on Plant Genetic Resources for Food and Agriculture*, supra note 15, at 21.

81 See Fabien Girard and Christine Noiville, 'Propriété Industrielle et Biotechnologies végétales: La Nova Atlantis. À propos de la recommandation du Haut Conseil de Biotechnologies' (2014) 1 *Revue Internationale de Droit Économique* 59.

82 See Song Yiching et al, *Sowing Diversity, Harvesting Security, Sustaining Future. An Impact Assessment of China's Seed Policy* (OXFAM – UNEP-IEMP 2020).

83 Golay, 'The Implementation of the United Nations Declaration on the Rights of Peasants and Other People Working in Rural Areas,' supra note 10, at 45ff.

84 La Via Campesina, 'Seed Laws that Criminalise Farmers: Resistance and Fightback' (2015) *GRAIN* <www.grain.org/article/entries/5142-seed-laws-that-criminalise-farmers-resistance-and-fightback> accessed 28 January 2021.

85 Christopher Golay and Adriana Bessa, 'The Right to Seeds in Europe: The United Nations Declaration on the Rights of Peasants and Other Peoples Working in Rural Areas and the Protection of the Right to Seeds in Europe' (2019) *The Geneva Academy*, Academy Briefing N°15.

86 See ITPGRFA, Ad Hoc Technical Expert Group on Farmers' Rights, 'Draft Inventory of National Measures, Best Practices and Lessons Learned from the Realization of Farmers' Rights as Set Out in Article 9 of the International Treaty' (23 May 2019) IT/GB-8/AHTEG-FR-2/19/3 Add.1.
87 Golay and Bessa, 'The Right to Seeds in Europe: The United Nations Declaration on the Rights of Peasants and Other Peoples Working in Rural Areas and the Protection of the Right to Seeds in Europe,' supra note 85, at 18–9; Haugen, 'The UN Declaration on Peasants' Rights (UNDROP): Is Article 19 on Seed Rights Adequately Balancing Intellectual Property Rights and the Right to Food?,' supra note 30, at 294–7.
88 TRIPS, supra note 78, art. 27.3(b).
89 Tsioumani, 'Reimagining Agrobiodiversity Conservation and Agricultural Innovation from the Grassroots Up: The Case of the Peliti Seed Network in Greece,' supra note 18, at 4ff.
90 Yiching et al, *Sowing Diversity, Havesting Security, Sustaining Future. An Impact Assessment of China's Seed Policy,* supra note 82, at 2.
91 Katie Sandwell, et al, 'A View from the Countryside: Contesting and Constructing Human Rights in an Age of Converging Crises' (2019) *TNI Issue Brief* 21–2.
92 See Carlos M. Correa, 'An Innovative Option for Benefit-Sharing Payment under the International Treaty on Plant Genetic Resources for Food and Agriculture. Implementing Article 6.11 Crop-Related Modality of SMTA' in Christine Frison, et al (eds), *Plant Genetic Resources and Food Security Stakeholders Perspectives on the International Treaty on Plant Genetic Resources for Food and Agriculture* (The Food and Agriculture Organization of the United Nations and Biodiversity International with Earthscan 2011) 250.
93 ITPGRFA, supra note 7, art. 22.
94 Frison, *Redesigning the Global Seed Commons: Law and Policy for Agrobiodiversity and Food Security,* supra note 2, at 117–21.
95 Beverly Dianne McIntyre et al, *Global Report. Agriculture at a Crossroads* (International Assessment of Agricultural Knowledge, Science and Technology for Development 2009); International Treaty on Plant Genetic Resources for Food and Agriculture Secretariat, *The Global Consultation on Farmers' Rights 2016. Summary of Presentations and Discussions. International Treaty on Plant Genetic Resources for Food and Agriculture*, supra note 15, at 20.
96 Ulrich Brand et al, *Conflicts in Environmental Regulation and the Internationalization of the State: Contested terrains* (Routledge 2008) 226.
97 Haugen, 'The UN Declaration on Peasants' Rights (UNDROP): Is Article 19 on seed rights adequately balancing intellectual property rights and the right to food?,' supra note 30, at 305.
98 Golay and Bessa, 'The Right to seeds in Europe: the United Nations Declaration on the Rights of Peasants and Other Peoples Working in Rural Areas and the Protection of the Right to seeds in Europe,' supra note 85, at 28.
99 Frison, *Redesigning the Global Seed Commons: Law and Policy for Agrobiodiversity and Food Security,* supra note 2, at 202.
100 See Ostrom, *Governing the Commons: The Evolution of Institutions for Collective Action*, supra note 25.
101 See David Bollier, *Think Like a Commoner: A Short Introduction to the Life of the Commons* (New Society Publishers 2014).
102 Jack R. Kloppenburg, 'Impeding Dispossession, Enabling Repossession: Biological Open Source and the Recovery of Seed Sovereignty' (2010) 10 *Journal of Agrarian Change* 373.
103 Oxfam (Rights and Resources Initiative (RRI)), *Common Ground: Securing land rights and safeguarding the earth* (Oxfam, International Land Coalition 2016) 7.
104 Liz Alden Wily, '"The Law Is to Blame": The Vulnerable Status of Common Property Rights in Sub-Saharan Africa' (2011) 42 *Development and Change* 735ff.

105 Kloppenburg, 'Impeding Dispossession, Enabling Repossession: Biological Open Source and the Recovery of Seed Sovereignty,' supra note 102, at 368–9; Sperling, 'Seed Security Response during COVID-19: Building on Evidence and Orienting to the Future,' supra note 22, at 886.
106 See Pierre Sauvêtre, 'Forget Ostrom: From the Development Commons to the Common as Social Sovereignty' in Samuel Cogolati and Jan Wouters (eds), *The Commons and A New Global Governance* (Edward Elgar 2019).
107 Olivier T. Coomes, 'Farmer Seed Networks Make a Limited Contribution to Agriculture? Four Common Misconceptions' (2015) 56 *Food Policy* 44–5.
108 Tom Dedeurwaerdere, 'Institutionalizing Global Genetic Resource Commons for Food and Agriculture' in Michael Halewood, et al (eds), *Crop Genetic Resources as a Global Commons: Challenges in International Law and Governance* (Earthscan Publications 2013) 370.
109 Cogolati, *International Law to Save the Commons. The International Legal Protection of the Commons in Development*, supra note 20, at 15.
110 Christinck, *Farmers' Rights: Views, Experiences, Lessons Learned and Proposed National Measures for the Implementation of Article 9*, supra note 36, at 72–3.
111 McIntyre, *Global Report. Agriculture at a Crossroads*, supra note 95.
112 Brand et al, *Conflicts in Environmental Regulation and the Internationalization of the State: Contested Terrains*, supra note 96, at 220.
113 Peschard and Randeria, 'Keeping Seeds in Our Hands': The Rise of Seed Activism, supra note 42, at 622.
114 UN Human Rights Council, 'Study on the Normative Sources and Rationale Underlying the Draft Declaration on the Rights of Peasants and Other People Working in Rural Areas' (4 May 2017) UN Doc A/HRC/WG.15/4/3, para. 240.
115 See Bollier, *Think Like a Commoner: A Short Introduction to the Life of the Commons*, supra note 101; Frison, *Redesigning the Global Seed Commons: Law and Policy for Agrobiodiversity and Food Security*, supra note 2, at 178.
116 See UNDRIP, supra note 40, art. 26.
117 Olivier De Schutter, 'The Reform of the Committee on World Food Security: The Quest for Governance in Global Governance,' in Nadia C.S. Lambeket al (eds), *Rethinking Food Systems. Structural Challenges, New Strategies and the Law* (Springer 2013) 219–38.
118 Cogolati, *International Law to Save the Commons. The International Legal Protection of the Commons in Development*, supra note 20, at 110.
119 Dedeurwaerdere, 'Institutionalizing Global Genetic Resource Commons for Food and Agriculture,' supra note 108, at 382ff.
120 Tone Winge, 'Linking Access and Benefit-sharing for Crop Genetic Resources to Climate Change Adaptation' (2016) 14 *Plant Genetic Resources: Characterization and Utilization* 23.
121 ITPGRFA, supra note 7, artt. 10 and 13.
122 Frédéric Thomas, 'Traité international de la FAO sur les ressources phytogénétiques pour l'agriculture et l'alimentation' in *Dictionnaire des biens communs* (Dir. Presses Universitaires de France 2017); Frison et al, *Plant Genetic Resources and Food Security Stakeholders Perspectives on the International Treaty on Plant Genetic Resources for Food and Agriculture*, supra note 34.
123 Frison, *Redesigning the Global Seed Commons: Law and Policy for Agrobiodiversity and Food Security*, supra note 2, at 163ff.
124 McMichael, 'Peasants Make Their Own History, But Not Just as They Please . . .,' supra note 6, at 217–20.
125 Kloppenburg, 'Impeding Dispossession, Enabling Repossession: Biological Open Source and the Recovery of Seed Sovereignty,' supra note 105, at 385.

Selected bibliography

Demeulenaere E, 'A Political Ontology of Seeds: The Transformative Frictions of a Farmers' Movement in Europe' (2014) 69 *Focaal – Journal of Global and Historical Anthropology* 45.

Frison C, *Redesigning the Global Seed Commons: Law and Policy for Agrobiodiversity and Food Security* (Routledge 2018).

Golay C and Bessa A, 'The Right to Seeds in Europe: The United Nations Declaration on the Rights of Peasants and Other Peoples Working in Rural Areas and the Protection of the Right to Seeds in Europe' (2019) *The Geneva Academy*, Academy Briefing N°15.

Sievers-Glotzbach S et al, 'Diverse Seeds – Shared Practices: Conceptualizing Seed Commons' (2020) 14 *International Journal of the Commons on Participatory Plant Breeding Initiatives* 418.

Wattnem T, 'Seed laws, certification and standardization: outlawing informal seed systems in the Global South' (2016) 43 *The Journal of Peasant Studies* 850.

9

THE RIGHTS TO FOOD AND FOOD SOVEREIGNTY IN THE UNDROP

Christophe Golay

Introduction

According to figures from the UN FAO and other UN agencies, throughout the last 50 years, there have always been more than 800 million undernourished people worldwide, primarily in rural areas in developing countries.[1] Of that total, 6 million children die every year, directly or indirectly, from the consequences of malnutrition – that is, 1 child every 5 seconds.[2]

These figures have not decreased since the first world conference on hunger in 1974, in which States committed to eradicating hunger in 10 years.[3] Hunger, food insecurity, and malnutrition have even been increasing since 2015, when the UNGA adopted the 2030 Agenda for Sustainable Development, in which all UN Member States committed to eradicating them by 2030.[4]

It is in this context that the UNDROP was adopted in 2018 and that its implementation began.[5]

Since 2020, the COVID-19 crisis has added another set of problems, with detrimental impacts on the right to food.[6]

At the core of the UNDROP, and of the fight against structural discrimination suffered by the most vulnerable people working in rural areas, are the rights to food and food sovereignty enshrined in UNDROP Article 15.

The right to food

There are two components to the right to food under international human rights law: the right to adequate food and the fundamental right to be free from hunger.[7] These two elements have been enshrined in Article 11 of the International Covenant on Economic, Social and Cultural Rights (ICESCR), and in UNDROP Article 15.1, which provides that 'Peasants and other people working in rural areas

DOI: 10.4324/9781003139874-12

have the right to adequate food and the fundamental right to be free from hunger.' In the context of the UNDROP, one element of the right to food defined in Article 15.1 – the right to produce food – is particularly important. Building on the most recent developments in the definition of the right to food,[8] UNDROP Article 15.1 also provides that this right includes the right to adequate nutrition.

The right to adequate food and nutrition

Defining the right to adequate food, UNDROP Article 15.2 provides that

> States shall ensure that peasants and other people working in rural areas enjoy physical and economic access at all times to sufficient and adequate food that is produced and consumed sustainably and equitably, respecting their cultures, preserving access to food for future generations, and that ensures a physically and mentally fulfilling and dignified life for them, individually and/or collectively, responding to their needs.[9]

The right to adequate food enshrined in the UNDROP and in other international human rights instruments includes four main elements: adequacy, availability, and sustainability of food, and permanent access to food with dignity.[10]

Adequate food requires that it be sufficient and adequate in quantity and quality. Food must also be healthy, free of toxins and contaminants, and culturally acceptable.[11] Adequate food should also be available, either directly from the land or other natural resources or through distribution systems capable of moving food to where it is needed.[12] Adequate food's availability must also be sustainable.[13]

Finally, adequate food must also be accessible. UNDROP Article 15.2 distinguishes two components connected to the accessibility to food: economic and physical.[14]

Economic accessibility implies that personal or household financial costs related to the acquisition of food required for an adequate diet should be at a level that does not threaten or compromise the exercise of other human rights (health, housing, education, among others). Physical accessibility implies that everyone, including physically vulnerable individuals, such as infants and young children, elderly people, disabled people, the terminally ill, and persons with persistent medical problems, including the mentally ill, should be ensured access to adequate food.[15]

There are multiple means for accessing food, the dynamics of which are complex. To be compatible with human dignity, the right to adequate food must be interpreted as the right to feed oneself and one's family with dignity.[16]

Based on this principle, UNDROP Article 15.1 provides that the right to adequate food of peasants, herders, pastoralists, and fisherfolk includes their 'right to produce food.'[17] The UNDROP also recognises other rights that are important for the enjoyment of the right to produce food, as well as for the right to food sovereignty, such as the rights to land and other natural resources (Articles 5 and 17), to participation and information (Articles 10 and 11), to facilitated access to the means of production, to engage in traditional ways of farming, fishing, livestock rearing, and forestry, and to develop community-based commercialisation systems (Article 16), to a safe, clean and healthy environment (Article 18), to seeds and biological diversity (Articles 19 and 20), to water for personal and domestic use, farming, fishing, and livestock keeping (Article 21), and to adequate training suited to the specific agroecological, sociocultural, and economic environments in which they find themselves (Article 25).

Other rights enshrined in the UNDROP are important for the enjoyment of the right to food, including the rights to decent work (Article 13), to an adequate standard of living (Article 16), and to social security (Article 22).

Finally, UNDROP Article 15.1 also recognises 'the right to adequate nutrition, which guarantee(s) the possibility of enjoying the highest degree of physical, emotional and intellectual development.'[18] In accordance with the Convention on the Elimination of Discrimination Against Women (CEDAW) and the CRC, UNDROP Article 15.3 provides that

> States shall take appropriate measures to combat malnutrition in rural children, including within the framework of primary health care through, inter alia, the application of readily available technology and the provision of adequate nutritious food and by ensuring that women have adequate nutrition during pregnancy and lactation.[19]

It also provides that 'States shall also ensure that all segments of society, in particular parents and children, are informed, have access to nutritional education and are supported in the use of basic knowledge on child nutrition and the advantages of breastfeeding.'[20]

As rightly stated by F. Kroff and A. Castañeda Flores, Article 15.3

> should be interpreted in a way of guaranteeing a global approach to nutrition through national strategies, in consultation with peasants, that enable them to overcome hunger and malnutrition through their own production and consumption of their own food as well as by strengthening their own food systems in the framework of food sovereignty.[21]

These authors also underline that Article 15.3

> lacks a clear gender approach and restricts women to a reproductive role. However, the right to food should be read together with the non-discrimination principle formulated in Article 4 of the UNDROP. . . . This interpretation recognises that firstly, the impact of hunger and malnutrition is higher among women and girls, and secondly, the contribution of women to realising the right to food for their families, their community and society in general must be recognised, protected and strengthened.[22]

The fundamental right to be free from hunger

The right to be free from hunger is the only right recognised as fundamental under international human rights law. For some authors, this recognition stems from the primacy conferred to the right to life.[23] The right to be free from hunger should therefore be interpreted as the core provision protecting individuals from hunger, defined as the insufficient or inadequate intake of food and low resistance to deadly diseases.

The right to be free from hunger was first established in Article 11.2 of the ICESCR. It should be interpreted, in its context, as a sub-norm to Article 11.1 of the ICESCR. For P. Alston, former Chair of the UN CESCR, the right to be free from hunger should be understood, consequently, as the minimum content of the right to food.[24] Other authors address the hard core or a minimum standard in respect to the right to food. As stated by P. Texier, another former Chair of the CESCR, it is 'the minimum threshold below which one should never, in principle, fall under any circumstances.'[25]

In its General Comment 12 on the right to food, the CESCR provided that:

> Every State is obliged to ensure for everyone under its jurisdiction access to the minimum essential food which is sufficient, nutritionally adequate and safe, to ensure their freedom from hunger. . . . Violations of the Covenant occur when a State fails to ensure the satisfaction of, at the very least, the minimum essential level required to be free from hunger.[26]

The right of peasants and other people working in rural areas to be free from hunger, therefore, must be defined as their right to have access to the minimum essential food that is sufficient and adequate to ensure that they are free from hunger and physical deterioration that would lead to death.

States' obligations

States have a key role to play in UNDROP's implementation. This role is underlined in the numerous provisions of the UNDROP that define their obligations.[27]

Building on international human rights law, the UNDROP provides that States shall, inter alia, respect, protect, and fulfil the rights of peasants and other people working in rural areas (Article 2.1), ensure the enjoyment of these rights without any discrimination (Articles 3 and 4), and engage in international cooperation for the same purposes (Article 2.6).

Non-discrimination

In its General Comment No 20 on non-discrimination, the CESCR defined formal and substantive discrimination in the following terms: eliminating formal discrimination requires 'ensuring that a State's constitution, laws and policy documents do not discriminate on prohibited grounds; for example, laws should not deny equal social security benefits to women on the basis of their marital status,' while eliminating discrimination in practice (substantive discrimination) requires 'paying sufficient attention to groups of individuals which suffer historical or persistent prejudice instead of merely comparing the formal treatment of individuals in similar situations.'[28]

The obligation to *de jure* non-discrimination is primarily fulfilled through legislative measures. In relation to the right to food, the obligation requires that States review their legislation to assure the absence of discriminatory provisions in the access to food or the means and entitlements for its procurement.[29] Further, States should adopt specific laws to combat discrimination and refrain from any discriminatory practices. The obligation to *de facto* non-discrimination implies that States should adopt positive measures to guarantee the equal realisation of the right to food by historically or socially discriminated persons or groups of persons.[30]

The UNDROP provides that States shall pay particular attention to the rights and special needs of individuals and groups who have historically been discriminated against, including older persons, women, youth, children, and persons with disabilities, taking into account the need to address multiple forms of discrimination (Article 2.2). It also provides that States shall take appropriate measures to eliminate conditions that cause or help to perpetuate discrimination, including multiple and intersecting forms of discrimination (Article 3.3).

The obligation to ensure the equal realisation of the right to food by women and men follows the same reasoning.[31] States should eliminate any *de jure* discrimination against women in the exercise of the right to food. For example, laws that discriminate against women in the access to productive resources, in particular land, or in the access to adequate remuneration should be amended. The same applies to any other laws that discriminate against women in the access to food. In the absence of legal protections against discrimination of women, such protections should be formulated and adopted. To ensure that women and girls have effective access to food on an equal basis to men and boys, States should implement positive measures to prevent discrimination abolished by law from being allowed to persist in practice.

The UNDROP provides that States shall take all appropriate measures to eliminate all forms of discrimination against peasant women and other women working in rural areas, to promote their empowerment, and to ensure that they enjoy without discrimination all the human rights and fundamental freedoms set out in the UNDROP (Article 4.2). It also provides that States shall, inter alia, ensure fair wages and equal remuneration for work of equal value, without distinction of any kind (Article 16.6), and remove and prohibit all forms of discrimination relating to the right to land, including those resulting from change of marital status, lack of legal capacity, or lack of access to economic resources (Article 17.2).

The obligations to respect, protect, and fulfil

A majority of peasants, herders, pastoralists, and fisherfolk who are forced to struggle to ensure their access to food must use their own resources, whether alone or collectively with others. This situation gave rise to the obligations of States to respect and protect their existing access to food, while, as regards assisting people and communities without access to these resources and who, in spite of their best efforts, are unable to access food, States have an obligation to fulfil their right to food.[32] The obligations of States to respect, protect, and fulfil the right to food were first developed by A. Eide[33] and were subsequently reaffirmed by the CESCR and by States in the FAO Right to Food Guidelines,[34] before being included in UNDROP Article 2.1.

The obligation to respect the right to food is essentially an obligation to refrain from action. It implies the obligation of States to refrain from taking measures that prevent people and communities from accessing food.[35] J. Ziegler underscored the arbitrary nature of measures that do not comply with this obligation. As he puts it:

> Violations of the obligation to respect [the right to food] would occur, for example, if the Government arbitrarily evicted or displaced people from their land, especially if the land was their primary means of feeding themselves, if the Government took away social security provisions without making sure that vulnerable people had alternative ways to feed themselves, or if the Government knowingly introduced toxic substances into the food chain, as the right to food entails access to food that is free from adverse substances.[36]

To ensure that States respect the rights of peasants and other people working in rural areas, the UNDROP provides that they shall, inter alia, respect their right not to be arbitrarily or unlawfully evicted or displaced from their land and other natural resources (Articles 17.3 and 17.4). States shall also recognise the natural commons and their related systems of collective use and management (Articles 17.3). It is also important that they choose policies that do not threaten local food production and quality.[37]

The obligation to protect the right to food requires that States ensure that enterprises and private individuals do not deprive individuals of their right to food.[38] This obligation corresponds to the obligation to protect civil and political rights. It requires that States implement a legislative and institutional framework and a legal system that are appropriate for the protection of the right to food.

The UNDROP provides that to protect the rights of peasants and other people working in rural areas, States shall take all necessary measures to ensure that non-State actors that they are in a position to regulate, such as private individuals and organisations, and transnational corporations and other business enterprises, respect and strengthen the rights enshrined in the UNDROP (Article 2.5). Such measures include the need to protect peasants and other people working in rural areas against arbitrary or unlawful evictions and displacements (Articles 17.3, 17.4, and 24.2), to ensure that no hazardous material, substance, or waste is stored or disposed of on their land (Article 18.4), and to prevent risks arising from the development, handling, transport, use, transfer, or release of any living modified organisms (Article 20(3)).

Finally, the obligation to fulfil the right to food implies that States should, first, facilitate the realisation of the right to food, creating an environment that enables people and communities to feed themselves by their own means, and, second, provide the right to food for those who are not capable of feeding themselves for reasons beyond their control, through the distribution of food and implementation of social security programs.[39]

The UNDROP provides that States shall fulfil the right of peasants and other people working in rural areas to food, inter alia, by carrying out agrarian reforms, giving priority to landless peasants, young people, small-scale fishers, and other rural workers in the allocation of public lands, fisheries, and forests (Article 17.6), by supporting peasant seed systems, and promoting the use of peasant seeds and agrobiodiversity (Article 19.6), and by establishing or maintaining social protection floors comprising basic social security guarantees (Article 22.3). When food distribution programmes are established, they shall give priority to food production that promotes peasant farming, artisanal fisheries, and local food production.[40]

Obligation to engage in international cooperation

In international human rights law, States have an obligation to international cooperation and assistance in order to respect, protect, and support the fulfilment of the right to food.[41]

UNDROP Article 2.6 provides useful guidance on how this shall be materialised. It underlines that States shall recognise the importance of international cooperation in support of national efforts aimed at implementing the UNDROP. It then explains that measures that they shall take in this regard include ensuring that international cooperation is inclusive, accessible, and pertinent to peasants and other people working in rural areas; facilitating and supporting capacity building; facilitating cooperation in research and in access to scientific and technical knowledge; providing technical and economic assistance; improving the functioning of markets on a global level; and facilitating timely access to market information, including on food reserves, in order to help to limit extreme food price volatility and the attractiveness of speculation.

The right to food sovereignty

The right to food sovereignty is at the heart of the rights-based struggles of communities of peasants and other people working in rural areas all over the world.[42] It is essential to ensuring that peasants and other people working in rural areas are able to exercise control over local food systems and natural and genetic resources.[43]

A short history of the right to food sovereignty

In the last decades, the right to food sovereignty has been recognised in an increasing number of instruments and documents at national, regional, and international levels.[44]

In a report presented in 2004, the first UN Special Rapporteur on the right to food, Jean Ziegler, argued that given the 'imbalances and inequities in the global trading system that can have profound negative effects on the right to food,' it was time 'to examine new and alternative models for agriculture and trade, such as that provided by the vision of food sovereignty, which places priority on food security and the right to food, for all people at all times.'[45]

In 2007, in a forum organised in Nyéléni, Mali, by LVC and partner organisations of peasants, family farmers, artisanal fisherfolk, indigenous peoples, landless peoples, rural workers, migrants, pastoralists, forest communities, women, youth, consumers, and environmental and urban movements, more than 500 representatives from more than 80 countries called for a world where 'all peoples, nations and states are able to determine their own food producing systems and policies that provide every one of us with good quality, adequate, affordable, healthy and culturally appropriate food.'[46] Food sovereignty was defined as 'the right of peoples to healthy and culturally appropriate food produced through ecologically sound and sustainable methods, and their right to define their own food and agriculture systems.'[47]

In his final report presented to the Human Rights Council in 2014, the second UN Special Rapporteur on the right to food, Olivier De Schutter, stated that

> [u]nderstood as a requirement for democracy in the food systems, which would imply the possibility for communities to choose which food systems to depend on and how to reshape those systems, food sovereignty is a condition for the full realization of the right to food.[48]

In its General Recommendation no. 34 adopted in 2016,[49] the CEDAW Committee stated that the 189 States parties to CEDAW should 'ensure the realization of the right to food and nutrition of rural women within the framework of food sovereignty and ensure that they have the authority to manage and control their natural resources.'[50]

On a regional level, food sovereignty has been recognised as one of the main objectives of the 2005 Agricultural Policy of the Economic Community of West African States (ECOWAS), and in the Declaration of Cochabamba on Food Security with Sovereignty in the Americas, adopted by the General Assembly of the Organization of American States (OAS) in 2012.[51] The Bolivarian Alliance for the Peoples of our America (ALBA) has also developed several instruments to promote food sovereignty. Food sovereignty has been recognised by a number of supranational parliamentary forums, including the Andean Parliament, the Central American Parliament, and the Latin American and Caribbean Parliament that adopted the Framework Law on the right to food, food security, and food sovereignty in 2012.[52]

The European Parliament has also acknowledged the right to food sovereignty in a number of resolutions in which it emphasised 'the need to respect the principle of food sovereignty' (2002), stressed that 'the right of developing countries to develop and implement agricultural policies guaranteeing them food sovereignty, in particular in WTO negotiations, must be respected and strengthened' (2006), called on the European Union (EU) 'to recognise developing countries' right to food sovereignty and to support this with targeted measures' (2009), and supported the promotion of food sovereignty in Europe (2017).[53] In 2017, French President Emmanuel Macron declared that one of the main objectives of the EU Common Agricultural Policy should be to protect Europe's food sovereignty.[54]

On a national level, food sovereignty has been recognised in a number of Constitutions and national framework laws.[55] It was enshrined in the Constitutions of Ecuador in 2008, Bolivia in 2009, and Nepal in 2015. It was further recognised by Venezuela in its 2008 Organic Law on Agricultural and Food Security and Sovereignty, by Nicaragua in its 2009 Law on Food and Nutrition Sovereignty and Security, by Ecuador in its 2009 and 2010 organic laws on the food sovereignty regime, by Bolivia in its 2013 Law on Sustainable Family Farming and Food Sovereignty, and by the Dominican Republic in its 2016 Law on Food Sovereignty and Food Security and Nutrition for the Right to Food.[56] Senegal also highlighted food sovereignty as one of the main objectives of its 2004 Agro-Sylvo-Pastoral Development Orientation Law, and Mali did the same in its 2006 Agricultural Orientation Law. Laws promoting food sovereignty have even been adopted at a local level. In Geneva, for example, regulations implementing the 2004 Law for the Promotion of Agriculture request the Canton of Geneva to promote agriculture following the principles of food sovereignty.[57]

The right to food sovereignty in the UNDROP

After fierce debate, the right to food sovereignty was included in UNDROP's Preamble and Article 15.[58] It is the first time that it has been recognised in an international instrument adopted by the UNGA.

The right to food sovereignty has been included in all drafts of the Declaration.[59] But during UNDROP negotiations, some States argued that it should not be included in the UNDROP, because doing so would breach World Trade Organization (WTO) rules, particularly the WTO Agreement on Agriculture.[60] Others asserted that national sovereignty is a State's prerogative that cannot be given to peoples or communities of peasants and other people working in rural areas.[61] The majority of experts and stakeholders involved in the negotiations, however, supported the idea that the right to food sovereignty should be included in the UNDROP.[62]

It is important to underline that the definition of food sovereignty included in the UNDROP does not provide peasants and other people working in rural areas, or their communities, with a degree of autonomy that is equivalent to national sovereignty. Like the rights to development, self-determination, and free disposal of natural resources, the right to food sovereignty can be described as having an internal dimension – the rights of individuals, communities, and peoples to define their local food and agricultural systems and to participate in decision-making processes around food and agricultural policies – and an external dimension – the right of states to define their food and agricultural policies and development.[63] It is this internal dimension that has been enshrined in UNDROP Article 15.

During the last year of negotiations, given that there was no precedent nor agreed language on the recognition and definition of food sovereignty at an international level, a compromise was found, with the indication that 'food sovereignty has been used in many States and regions,' i.e., not before 2018 at international level.[64]

With this indication, the right to food sovereignty is included in two parts of UNDROP. The Preamble of the UNDROP recognises that

> the concept of food sovereignty has been used in many States and regions to designate the right to define their food and agriculture systems and the right to healthy and culturally appropriate food produced through ecologically sound and sustainable methods that respect human rights.

UNDROP Article 15.4 recognises that:

> Peasants and other people working in rural areas have the right to determine their own food and agriculture systems, recognized by many States and regions as the right to food sovereignty. This includes the right to participate in decision-making processes around food and agriculture policy and the right to healthy and adequate food produced through ecologically sound and sustainable methods, respecting their cultures.

Article 15.5 then provides that:

> States shall formulate, in partnership with peasants and other people working in rural areas, public policies at the local, national, regional and international levels to advance and protect the right to adequate food, food security and food sovereignty and sustainable and equitable food systems that promote and protect the rights contained in the present Declaration. States shall establish mechanisms to ensure the coherence of their agricultural, economic, social, cultural and development policies with the realisation of the rights contained in the present Declaration.

Conclusion

Peasants and other people working in rural areas, including herders, pastoralists, and fisherfolk, represent 80% of the people suffering from hunger, food insecurity, and malnutrition in the world.

In a world in which there has been no progress in the fight against hunger in the last 50 years, the recognition of the rights to food and food sovereignty in the UNDROP was essential.

These two rights are at the heart of the rights-based struggles of communities of peasants and other people working in rural areas all over the world, and they were central in the call for the UNDROP. Now they need to be central in the UNDROP's implementation, to overcome structural discrimination suffered by the most vulnerable people in rural areas, and to ensure that they will be able to exercise control over local food systems and natural and genetic resources.

Notes

1 See, for example, FAO, International Fund for Agricultural Development (IFAD), United Nations Children's Fund (UNICEF), World Food Programme (WFP), and World Health Organization (WHO), *The State of Food Security and Nutrition in the World 2019: Safeguarding against Economic Slowdowns and Downturns* (FAO 2019).

2 Jean Ziegler et al., *The Fight for the Right to Food. Lessons Learned* (Palgrave Macmillan 2011) 1.

3 World Food Conference, 'Universal Declaration on the Eradication of Hunger and Malnutrition' (5–16 November 1974) UN Doc E/CONF.65/20, Part I, Chapter 2, para. 1.

4 UN Human Rights Council, 'Report of the Special Rapporteur on the Right to Food, Michael Fakhri' (24 December 2020) UN Doc A/HRC/46/33, para. 6; FAO, IFAD, UNICEF, WFP and WHO, *The State of Food Security and Nutrition in the World 2019,* supra note 1.

5 UN General Assembly, 'United Nations Declaration on the Rights of Peasants and Other People Working in Rural Areas' (17 December 2018) UN Doc A/RES/73/165. Christophe Golay, *The Implementation of the UN Declaration on the Rights of Peasants and Other People Working in Rural Areas* (Geneva Academy of International Humanitarian Law and Human Rights 2019).

6 Ibid.; see also FIAN International, *Monitoring Report on the Right to Food and Nutrition during COVID-19* (FIAN 2020); FIAN International, *Monitoring Toolkit on the Right to Food and Nutrition during COVID-19* (FIAN 2020).

7 Christophe Golay, *The Right to Food and Access to Justice. Examples at the National, Regional and International Levels* (FAO 2009) 11–4.

8 Florence Kroff and Angélica Castañeda Flores, 'Right to Adequate Food and Nutrition, and to Food Sovereignty' (2020) *UNDROP Series, FIAN International* 4–6.

9 See also UN Committee on Economic, Social and Cultural Rights (CESCR), 'General Comment No. 12: The right to food (Art. 11 of the Covenant)' (12 May 1999) UN Doc E/C.12/1999/5, paras. 6–8; UN Commission on Human Rights, 'Report of the Special Rapporteur on the right to food, Jean Ziegler' (7 February 2001) UN Doc E/CN.4/2001/53, para. 14.

10 CESCR, *General Comment No. 12*, supra note 9, paras. 6–8. See also UN Human Rights Council, 'Report of the Special Rapporteur on the Right to Food, Olivier De Schutter' (24 January 2014) UN Doc A/HRC/25/57, para. 2.

11 Ibid.; CESCR, *General Comment No. 12*, supra note 9, para. 9. The notion of adequacy and the dimension of nutrition could be further detailed by taking into account the FAO, *Voluntary Guidelines to Support the Progressive Realization of the Right to Adequate Food in the Context of National Food Security* (November 2004), guideline 10 on nutrition. See also UN General Assembly, 'Interim Report of the Special Rapporteur on the Right to Food' (3 August 2016) UN Doc A/71/282.

12 CESCR, *General Comment 12*, supra note 9, paras. 8 and 12.

13 As provided in the FAO Right to Food Guideline 8E: 'States should consider specific national policies, legal instruments and supporting mechanisms to protect

ecological sustainability and the carrying capacity of ecosystems to ensure the possibility for increased sustainable food production for present and future generations, prevent water pollution, protect the fertility of the soil and promote the sustainable management of fisheries and forestry.' FAO, *Voluntary Guidelines*, supra note 11, guideline 8.13.

14 See also CESCR, *General Comment 12*, supra note 9, para. 13.

15 CESCR, *General Comment 12*, supra note 9, para. 8.

16 UN Human Rights Council, 'Promotion and protection of all human rights, civil, political, economic, social and cultural rights, Including the right to development, Report of the Special Rapporteur on the right to food, Jean Ziegler' (10 January 2008) UN Doc A/HRC/7/5, para.18. FAO, *Voluntary Guidelines*, supra note 11, guideline 1.1 provides that 'States should create the conditions in which individuals can feed themselves and their families in freedom and dignity.'

17 See also Additional Protocol to the American Convention on Human Rights in the Area of Economic, Social and Cultural Rights (Protocol of San Salvador) (entered into force 16 November 1999) OAS Treaty Series No. 69, art. 12.

18 Ibid.

19 Convention on the Rights of the Child (adopted 20 November 1989, entered into force 2 September 1990) 1577 UNTS 3 (CRC), art. 27.1; Convention on the Elimination of All Forms of Discrimination against Women (adopted 18 December 1979, entered into force 3 September 1981) 1249 UNTS 13 (CEDAW), art. 12.2.

20 CRC, supra note 19, art. 24.2(c), (d), and (e).

21 Kroff and Flores, *Right to Adequate Food and Nutrition, and to Food Sovereignty*, supra note 8, at 5.

22 Ibid. See also UN Human Rights Council, 'Report of the Special Rapporteur on the right to food, Olivier De Schutter' (24 December 2012) UN Doc A/HRC/22/50, para. 6.

23 Golay, *The Right to Food and Access to Justice. Examples at the National, Regional and International Levels*, supra note 7, at 13–4.

24 Philip Alston, 'International Law and the Right to Food' in Asbjørn Eide et al (eds), *Food as a Human Right* (Second printing, United Nations University Press 1984) 167.

25 Philippe Texier, 'Commentaires sur l'observation générale du Comité des Droits Économiques. Sociaux et Culturels (DESC): le droit à une nourriture suffisante' in Roger Berthouzoz, Patrice Meyer-Bisch and Franck Nseka (eds), *Faim de vivre. La multidimensionnalité du droit à l'alimentation* (Fribourg 2000) 73. See also Emanuele Sommario, Francesca Spagnuolo, Mariagrazia Alabrese, '"Feeding the Hungry, Quenching the Thirsty": Shaping the Rights to Food and Water in Disaster Settings through Humanitarian Standards' (2019) 20(1) *Global Jurist.*

26 CESCR, *General Observation 12*, supra note 9, paras. 14 and 17.

27 Christophe Golay, *The Implementation of the UN Declaration on the Rights of Peasants and Other People Working in Rural Areas*, supra note 5, at 4–5.

28 UN Committee on Economic, Social and Cultural Rights (CESCR), 'General Comment No. 20: Non-Discrimination in Economic, Social and Cultural Rights (Art. 2 of the Covenant)' (2 July 2009) UN Doc E/C.12/GC/20, paras. 8–9.

29 CESCR, *General Comment 12*, supra note 9, para. 18.

30 Golay, *The Right to Food and Access to Justice. Examples at the National, Regional and International Levels*, supra note 7, at 15.

31 Ibid.

32 Ibid., at 17–9.

33 UN Commission on Human Rights, 'The right to adequate food and to be free from hunger. Updated study on the right to food, submitted by Asbjørn Eide' (28 June 1999) UN Doc E/CN.4/Sub.2/1999/12.

34 In the introduction to the FAO *Voluntary Guidelines*, supra note 11, States reiterated the exact same definition provided by the CESCR.

35 CESCR, *General Comment 12*, supra note 9, para. 15; Introduction to the FAO, *Voluntary Guidelines*, supra note 11.

36 UN Commission on Human Rights, 'Report of the Special Rapporteur for the right to food, Jean Ziegler' (16 March 2006) UN Doc E/CN.4/2006/44, para. 22.
37 Kroff and Flores, *Right to Adequate Food and Nutrition, and to Food Sovereignty*, supra note 8, at 9.
38 CESCR, *General Comment 12*, supra note 9, paras. 15 and 27.
39 FAO, *Voluntary Guidelines*, supra note 11, Preface and Introduction, para 17; CESCR, *General Comment 12*, supra note 9, at 15.
40 Kroff and Flores, *Right to Adequate Food and Nutrition, and to Food Sovereignty*, supra note 8, at 11.
41 CESCR, *General Comment 12*, supra note 9, at paras. 36–7; Ziegler, *The Fight for the Right to Food. Lessons Learned*, supra note 2, at 79–84.
42 Marc Edelman et al, 'Critical Perspectives on Food Sovereignty' (2014) 41(6) *Journal of Peasant Studies*, 911–1246. Kroff and Flores, *Right to Adequate Food and Nutrition, and to Food Sovereignty*, supra note 8.
43 Priscilla Claeys, *Human Rights and the Food Sovereignty Movement. Reclaiming Control* (Routledge 2015).
44 Christophe Golay, *The Rights to Food Sovereignty and to Free, Prior and Informed Consent* (Geneva Academy of International Humanitarian Law and Human Rights 2018) 2–4.
45 UN Commission on Human Rights, 'Report of the Special Rapporteur on the right to food, Jean Ziegler' (9 February 2004) UN Doc E/CN.4/2004/10, para. 53.
46 Declaration of the Forum for Food Sovereignty, Nyéléni (2007), para. 3.
47 Ibid.
48 UN Human Rights Council, 'Report of the Special Rapporteur on the right to food, Olivier De Schutter,' supra note 10, at para. 50.
49 CEDAW, supra note 19.
50 Ibid., para. 64.
51 Christophe Golay, *The Rights to Food Sovereignty and to Free, Prior and Informed Consent*, supra note 44, at 2.
52 FAO, *Parliamentary Fronts against Hunger, and Legislative Initiatives for the Right to Adequate Food and Nutrition. The Experience of Latin America and the Caribbean 2009–2016* (FAO 2017).
53 Golay, *The Rights to Food Sovereignty and to Free, Prior and Informed Consent*, supra note 44, at 2–3.
54 Élysée, *Discours d'Emmanuel Macron pour une Europe souveraine, unie et démocratique*, prononcé à la Sorbonne le 26 septembre 2017 <www.elysee.fr/front/pdf/elysee-module-795-fr.pdf>
55 Golay, *The Rights to Food Sovereignty and to Free, Prior and Informed Consent*, supra note 44, at 3.
56 FAO, *Parliamentary Fronts against Hunger, and Legislative Initiatives for the Right to Adequate Food and Nutrition*, supra note 52.
57 Golay, *The Rights to Food Sovereignty and to Free, Prior and Informed Consent*, supra note 44, at 3.
58 Christophe Golay, *Negotiation of a United Nations Declaration on the Rights of Peasants and Other People Working in Rural Areas* (Geneva Academy of International Humanitarian Law and Human Rights 2015) 23–4.
59 UN Human Rights Council, 'Draft Declaration on the Rights of Peasants and other People Working in Rural Areas' (6 March 2017) UN Doc A/HRC/WG.15/4/2, artt. 15.2 and 15.3; UN Human Rights Council, 'Draft Declaration on the Rights of Peasants and other People Working in Rural Areas' (8 March 2016) UN Doc A/HRC/WG.15/3/2, artt. 5.4 and 5.5; UN Human Rights Council, 'Draft Declaration on the Rights of Peasants and other People Working in Rural Areas' (20 June 2013) UN Doc A/HRC/WG.15/1/2, art. 2.5.
60 See, for example, UN Human Rights Council, 'Report of the Open-ended Intergovernmental Working Group on a Draft United Nations Declaration on the Rights of Peasants and other People Working in Rural Areas' (20 July 2016) UN Doc A/HRC/33/59, para. 28.

61 Ibid., para. 61.
62 Golay, *The Rights to Food Sovereignty and to Free, Prior and Informed Consent*, supra note 44, at 3.
63 Claeys, *Human Rights and the Food Sovereignty Movement. Reclaiming Control*, supra note 43.
64 UN Human Rights Council, 'Revised draft United Nation declaration on the rights of peasants and other people working in rural areas' (12 February 2018) UN Doc A/HRC/WG.15/5/2, Preamble and art. 5.4.

Selected bibliography

Claeys P, *Human Rights and the Food Sovereignty Movement. Reclaiming Control* (Routledge 2015).

Edelman M et al, 'Critical Perspectives on Food Sovereignty' (2014) 41(6) *Journal of Peasant Studies* 911–1246.

Golay C, *The Implementation of the UN Declaration on the Rights of Peasants and Other People Working in Rural Areas* (Geneva Academy of International Humanitarian Law and Human Rights 2019).

Kroff F and Castaneda Flores A, 'Right to Adequate Food and Nutrition, and to Food Sovereignty' (2020) *UNDROP Series, FIAN International* 4–6.

Ziegler J et al, *The Fight for the Right to Food. Lessons Learned* (Palgrave Macmillan 2011).

PART III

Food and agriculture governance

10

PEASANTS' RIGHTS AND FOOD SYSTEMS GOVERNANCE

*Smita Narula**

Introduction

In December 2018, following years of mobilisation by the transnational agrarian movement La Via Campesina (LVC) and its allies, the human rights of the one billion peasants who feed the world and safeguard the environment were enshrined in international law. Following six years of intense inter-governmental negotiations, the United Nations General Assembly adopted a new human rights declaration to protect the rights of peasants and other people working in rural areas. In so doing, the world's main policy-making body recognised the depth and breadth of human rights violations endured by rural communities. It also acknowledged the pivotal role that peasant farmers can play in producing food sustainably in the face of environmental disaster and ecological collapse.

This chapter comments on the significance of the United Nations Declaration on the Rights of Peasants and Other People Working in Rural Areas (UNDROP) as a tool for shifting the balance of power in the governance of food systems. The author argues that the evolving human rights agenda in relation to peasants' rights provides an important opportunity to fill critical gaps in international human rights law, address key environmental concerns, and build sustainable food pathways. The first part of the chapter looks at the political economy of our food system and the environmental and human rights impacts of industrial agriculture. The second part argues that the UNDROP's process, vision, and normative framework give voice to the needs and demands of an all but forgotten half of humanity and help chart a course for deep, transformative action in our food system.

The political economy of the industrial food system

The significance of the UNDROP to rural communities around the world cannot be overstated. Its landmark adoption must be viewed in light of the political

* The author thanks Cassandra Jurenci, Taylor Keselica, and Claire McLeod for their invaluable research support.

DOI: 10.4324/9781003139874-14

economy of our industrial food system – a system that generates hunger and malnutrition, impoverishes farmers, displaces rural communities and indigenous peoples, and wreaks havoc on the environment.[1]

Both the production of food and the governance of food systems have shifted dramatically over the past 60 or so years as a slew of agricultural reforms and trade policies have consolidated corporate power over the food chain, while significantly diminishing the role of the state. On the production side, the advent of the so-called Green Revolution in the 1960s ushered in a new era of food production that combined high-yielding plant varieties with increased irrigation, highly mechanised production processes, and the use of nitrogen-based fertilisers and pesticides.[2] These innovations not only increased the production volumes of major cereals and soybeans but also entailed significant environmental and social costs.[3]

In the decades that followed, structural adjustment programs mandated by international financial institutions encouraged developing countries to focus on the production and export of cash crops over basic food crops. These shifts in production, along with the liberalisation of trade in agricultural products in the 1990s, turned a number of food-exporting developing countries into net food importers,[4] making them particularly vulnerable to price volatility and supply chain disruptions. Moreover, as highly subsidised developed-country producers glutted markets, peasant farming was undermined across the Global South, fuelling waves of migration.[5] Globalisation and the financialisation of food systems have now enabled large agribusiness actors to exert tremendous control over the production, processing, and retailing of food,[6] which, as described below, has impoverished and decimated rural communities even further, while causing irreparable harm to the environment.

Agrarian distress

Peasant farmers, landless rural workers, indigenous peoples, pastoralists, fisherfolk, and their families comprise about half of the world's population and are the main or sole food providers for more than 70% of the planet.[7] They also comprise the majority of the world's food insecure. Of the nearly 700 million people hungry in the world today, 80% live in rural areas, particularly in developing countries, and no less than 50% are small-scale, peasant farmers who are particularly vulnerable to food insecurity, discrimination, and extreme exploitation.[8]

Small-scale farmers are hungry because the profits of their labour are being captured by agribusinesses along the food chain and because they cultivate plots of lands that are often very small, making the vast majority of them net food buyers.[9] They also face significant barriers to markets due to infrastructural deficiencies and technical trade requirements.[10] Most smallholder farmers are women, but few hold titles to land. The land they do own is often infertile or lacks water.[11] Many women also face wage inequities and sexual violence in the fields.[12]

In most countries, peasants and rural workers have no right to form unions, earn a living wage, or receive social security.[13] Peasant farmers struggle to access productive resources or save their seeds and are highly vulnerable to climate shocks.[14] Price volatility and a lack of proper supports for small-scale agriculture are also putting peasant farmers in a dire situation, as is evident by the high incidence of farmer suicides in multiple countries.[15] Rural communities throughout the world also bear the brunt of the large-scale land and water grabs that prop up the industrial food chain.[16] Protesting these acts of dispossession and environmental devastation is also proving to be a dangerous proposition. The very act of speaking out against land grabs or defending the environment from destructive industries, including agribusiness industries, can put people at risk of being threatened, arrested, sexually harassed, assaulted, and even killed.[17]

Environmental harms

The focus on integrating food producers into global markets that are now dominated by mega-sized agribusiness actors has also come at the expense of the environment and of sustainable and diversified local food systems. The natural resource base on which food production depends is being rapidly degraded. A hyper-focus on increasing agricultural production has led to a host of environmental problems including extension of monocultures, loss of biodiversity, and accelerated soil erosion.[18] The industrial food system – which relies on long petrochemical-based supply chains – also causes massive deforestation in order to make way for large-scale plantations to produce food commodities and animal feed and to secure land for animal grazing.[19]

According to some estimates, approximately one-fourth of human-made greenhouse gas emissions can be attributed to industrial modes of agricultural production when combined with the resource requirements to transport, package, and conserve food.[20] In turn, climate change patterns and soil degradation seriously constrain the potential productivity of current agricultural methods.[21] As industrial farming methods are further introduced into nonindustrial settings, soil depletion, petrochemical fertilisers, eutrophication, and coastal dead zones are also increasingly becoming commonplace in regions where such environmental degradation was previously unseen.[22]

The significance of the UNDROP for food systems governance

The UNDROP arises in response to the human rights abuses and ecological harms described above. Transnational agrarian movements – led by LVC [23] – have been advocating for such an instrument since at least 2001 on the premise that small-scale farmers, pastoralists, fisherfolk, and farmworkers are a vulnerable group increasingly subject to forced displacements from their lands, denied access to water and seeds, impoverished by trade and agricultural policies, and attacked and criminalised by state actors acting in collusion with the private sector.[24] The UNDROP is a historic landmark in a decades-long transnational agrarian movement to reclaim control over land and natural resources and to exercise sovereignty in the determination of food and agricultural systems. As discussed below, the Declaration's process, vision, and normative framework are all deeply significant for advancing this political project. The UNDROP also represents a significant milestone in the evolution of international human rights law.

A movement-driven process

In both its process and its substance, and as noted by Edelman and James, the Declaration 'represent[s] both the progressive extension of the existing human rights regime and a continuation of the democratization of rights-making.'[25] The extent and manner of civil society's involvement in the negotiations and drafting process marked a significant departure from global food systems governance processes where civil society actors rarely have a seat at the table, and from traditional human rights law-making processes wherein civil society groups can lobby delegates outside the halls of power but whose interventions during intergovernmental deliberations are frequently relegated to brief statements delivered at pre-set times.[26]

Because the UNDROP is the result of an iterative, bottom-up movement-driven process, it speaks more directly to the needs and realities of the communities it is meant to serve. As such, the Declaration makes visible widespread violations of peasants' rights that are often hidden from view. In its preambular paragraphs, for instance, the Declaration underscores that rural communities 'suffer disproportionately from poverty, hunger and malnutrition,' work in 'hazardous and exploitative conditions,' struggle to access 'land, water, seeds and natural resources,' and are subject to intimidation, violence, and forced evictions.[27]

Notably, the UNDROP underscores that 'the increasing concentration and unbalanced distribution of food systems and the uneven power relations along the value chains'[28] is a root cause of human rights abuses suffered by rural communities. The Declaration also makes visible peasants' profound contributions to humanity, noting that their 'past, present and future contributions' to 'conserving and improving biodiversity' and 'ensuring the right to adequate food and food security' are seen as fundamental to achieving the 2030 Agenda for Sustainable Development and its SDGs.[29]

But the intergovernmental negotiations for the Declaration were also rife with contention, including around the Declaration's reference to collective rights and food sovereignty.[30] Several delegations tried to downplay the language of rights altogether and to make human rights obligations and protections subservient to international trade and commercial interests. Each human rights gain, it seemed, was hotly contested, perhaps because of the seemingly radical worldview that the Declaration represents: that development projects and agricultural policies must be geared towards promoting the achievement of human rights, gender equality, public health, and environmental justice – and not simply towards the generation of profit or increasing the production of food commodities. Although the final draft is the end-product of much politicking and compromise,[31] read in its entirety, and as described below, the vision and scope of the Declaration is breathtaking.

Elevating the vision of an alternative food system

Agricultural policies are value-laden, as is all of law and policy. A key value informing agricultural policy is that of productivism – or the belief that the paramount goal of food systems is to increase the production of food, and food commodities in particular. The consolidation of corporate power in the food system described earlier is enabled, narratively and logically, by this productivist paradigm under which land and natural resources are directed to those who are seen as using these resources most competitively and 'efficiently' – where efficiency is defined as price efficiency rather than resource efficiency. The Declaration seeks to infuse the food system with a different set of values – that of human rights, environmental protection, and food sovereignty.

Agrarian movements have long framed their demands using the language and framework of food sovereignty as a means of reclaiming control over food and agricultural systems. Food sovereignty is defined as the right of people to determine their own food and agricultural systems, and to healthy food that is sustainably produced.[32] Long before the adoption of the UNDROP, the food sovereignty struggle had already reframed the human rights project. Most notably, it pushed the right to food framework to shift its focus from the question of 'How do we feed the world?' to the question: 'How do we ensure the capacity of peasants and rural communities to sustainably feed themselves (and the world), including through secure access to land and natural resources?'

Although trade and agricultural policies heavily favour large industrial farms, peasants are the main or sole food providers for more than 70% of the world's population and they produce this food using less than 25% of the world's agricultural land (and even less water and fossil fuels) while inflicting far less damage to soils and forests.[33] Peasant agriculture also often fortifies agro-biodiversity, carbon sequestration, and rural livelihoods. If provided complementary resources – such as extension services, credit, and irrigation – peasant agriculture can help alleviate rural poverty, address climate change, and achieve the right to food.[34]

But the struggle for food sovereignty is not just about ensuring the right to food – it is a transformative political project that seeks to decentralise power in the food system, restore ecological balance, and put forward an alternative development paradigm.[35] As stated by LVC,

> The alternative exists. . . . Small farmers and social movements from all over the world promote a model based on food sovereignty and orientated towards peasant-based agriculture and artisan fisheries, prioritising local markets and sustainable production methods. This model is based on the right to food and to the rights of peoples to define their own agricultural policies.[36]

The UNDROP elevates this alternative as a matter of international law. It does so by centring the agency and authority of peasants and other rural communities in decision-making processes, and by prescriptively putting a thumb on the scale in favour of more equitable land distribution, of food sovereignty and sustainable labour-intensive food systems, and by promoting agro-ecological methods, protecting seeds and biodiversity, and prioritising water for human needs.[37] In so doing, the Declaration challenges the orthodoxy of market-based mechanisms, large-scale development projects, industrial agriculture, and the attendant commodification of land and natural resources. Seen in this light, the UNDROP is a bold and essential bulwark against a destructive, extractive system. It is a repudiation of development models that are predicated on the destruction of the peasantry. It is a tool to help us correct course.

The UNDROP's normative framework

In addition to its process and vision, the UNDROP is significant for its normative framework in at least four key respects. First, it aims to correct a fundamental imbalance of power in the governance of food systems. Second, it delineates and elevates the category of peasants, tethering substantive rights to this group identity. Third, it fills critical gaps in rights protection, including in particular by affirming peasants' rights to land, seeds, and natural resources. And fourth, it reifies the interconnectedness of human rights and environmental concerns, modelling a holistic approach to food systems governance and advancing the right to a healthy environment. Each of these contributions is considered in turn below.

Correcting a fundamental imbalance of power

With its normative framework, the UNDROP aims to correct a fundamental imbalance of power between peasants and state and non-state actors in the food system. It does so first and foremost by asserting that peasants are rights holders who are entitled to specific rights guarantees that cannot be traded away in the context of investment and development projects. The Declaration also attempts to tip the scales in favour of peasants in more granular ways. Specifically, Article 2 of the Declaration calls on states to ensure the free, meaningful, and informed participation of peasants in decision-making processes that may affect their rights; interpret and apply relevant international agreements in line with their human rights obligations; and ensure that non-state actors that states are in a position to regulate (including transnational corporations) respect and strengthen the rights of peasants and other people working in rural areas.[38]

The free, meaningful, and informed participation standard borrows from but does not fully extend the United Nations Declaration on the Rights of Indigenous Peoples (UNDRIP)' 'free, prior and informed consent' standard.[39] The call to apply relevant agreements and standards in line with human rights obligations asserts the primacy of states' human rights obligations and addresses the problem of fragmentation in international law wherein states' human rights obligations routinely come into conflict with their obligations under other international agreements.[40] Finally, the call to regulate transnational corporations reinforces states' duty to protect against acts by non-state actors and is itself significant given the deeply negative impact agribusiness actors and other corporations can have on peasants' rights.

Reclaiming the category of 'peasants'

The term 'peasant' has long carried derogatory connotations indicative of 'both peasants' extreme subordination and of a ubiquitous elite practice of blaming peasants for a variety of economic and social ills.'[41] Over time, however, and as articulated by Marc Edelman, the term has become an 'identity marker, [which has] served to inspire the collective action of diverse kinds of rural movements.'[42]

Drawing from LVC's definition, and as explored further in this volume,[43] Article 1 of the Declaration cements this reclamation of the peasant identity and defines a 'peasant' as

> any person who engages or who seeks to engage alone, or in association with others or as a community, in small-scale agricultural production for subsistence and/or for the market, and who relies significantly, though not necessarily exclusively, on family or household labour and other non-monetized ways of organizing labour, and who has a special dependency on and attachment to the land.[44]

As is also reflected in the Preamble, the definition of peasants in the Declaration is centred on the recognition of the 'special relationship and interaction between peasants and other people working in rural areas, and the land, water and nature to which they are attached and on which they depend for their livelihood.'[45] Such recognition – which to date has been reserved for indigenous peoples under international law[46] – is notable because it sets the stage for the articulation of substantive rights not heretofore granted to peasants under international law, including, in particular, their right to land and natural resources.

In this view, the value of land and natural resources arises out of the relationship between a group of people and the resources that they care for, use, and occupy, rather than the relationship between these resources and the market. In essence, the UNDROP is signalling that those who have a connection with and know how to care for land and natural resources should have a greater right to access and use those resources, rather than those who see the commodification and exploitation of nature as the *sine qua non* of economic development and societal progress. This, in itself, is pathbreaking, both as a legal development and as a value narrative that further dismantles the productivist paradigm.

Filling gaps in rights protections

Although the UNDROP is not a legally binding instrument, the principles and rights enumerated therein are based on human rights standards enshrined in international instruments that are legally binding, such as the International Covenant on Economic,

Social and Cultural Rights (ICESCR), the International Covenant on Civil and Political Rights (ICCPR), the Convention on the Elimination of All Forms of Discrimination Against Women, as well as several International Labour Organization Conventions.[47]

But in key areas, the UNDROP goes a step further, extending specific substantive rights to peasants and other people working in rural areas on the reasoning that they have a special relationship to the land and environs to which they are attached and which they depend upon for their livelihood. The Declaration's articulation of the right to land is a case in point. To date, international human rights law has guaranteed only limited land rights. States cannot arbitrarily deprive people of property and cannot evict settled communities that rely on a piece of land but lack legal title to it, without first meeting certain conditions. The right to property, however, protects the rights of existing property owners, and the general bar against eviction can easily be circumvented, suggesting that greater normative development is needed to protect land rights.[48] Article 17 of the UNDROP helps fill this critical gap in international human rights law. It states:

> Peasants and other people living in rural areas have the right to land, individually and/or collectively . . . including the right to have access to, sustainably use and manage land and the water bodies, coastal seas, fisheries, pastures and forests therein, to achieve an adequate standard of living, to have a place to live in security, peace and dignity and to develop their cultures.[49]

Article 17 recognises that secure and stable access to land is essential to the realisation of a range of human rights, including the right to an adequate standard of living and the right to enjoy one's culture. Land is also intimately linked to community and identity and plays both a collective and a social function.[50] As reflected in Article 17, land is a common good that communities and people access, control, manage, and use to live a dignified life, according to both their social and cultural context.[51] Significantly, the right to land is recognised as both an individual and a collective right given the communal nature of land use and ownership in many parts of the world.[52]

In essence, Article 17 embodies a holistic understanding of the importance of land to rural communities and underscores the fact that for peasants and other people working in rural areas, the relationship to land and natural resources cannot simply be viewed in transactional terms. In acknowledging these myriad uses of land, the UNDROP implicitly challenges the dominant market narrative and its treatment of land as a commodity. In addition to articulating the nature of the right, the Declaration also defines states' obligations to respect, protect, and fulfil the right to land.[53] This includes undertaking redistributive agrarian reforms to ensure a more equitable distribution of land and to limit excessive land concentration,[54] ensuring the conservation and sustainable use of land, and supporting practices that are resilient and regenerative, such as agro-ecological methods.[55]

Article 5 of the UNDROP similarly extends to peasants and other people working in rural areas 'the right to have access to and to use in a sustainable manner the natural resources present in their communities that are required to enjoy adequate living conditions.' Article 5 also affirms peasants' rights to participate in the management of these resources and the obligation of states to conduct good-faith consultations and social and environmental impact assessments before exploiting the natural resources that peasants traditionally use or hold, and to share the benefits of such exploitation in a fair and equitable manner.

Like land, seeds are another site of contestation in the struggle for control over food systems.[56] Article 19 of the UNDROP asserts, *inter alia*, that peasants and other people working in rural areas have the right to seeds, including the right to 'maintain, control, protect and develop their own seeds and traditional knowledge.' States must also ensure that seed policies and intellectual property laws take peasants' rights into account. Although the UNDROP does not fundamentally seek to dismantle the architecture of global food systems governance, as articulated by Lorenzo Cotula, '[t]he provisions of the Declaration that deal with control over land, seeds, and market systems embody a challenge to prevailing economic structures, and their full implementation would change the way agriculture and food systems are presently organised.'[57]

Integrating environmental concerns

Finally, the UNDROP reifies the interdependence of human rights and environmental concerns and advances the right to a healthy environment. Although many of the rights enshrined in the 1948 Universal Declaration on Human Rights (UDHR) cannot be meaningfully achieved without a healthy natural environment, the word 'environment' is nowhere to be found in the UDHR, nor are environmental concerns reflected in the ICCPR or the ICESCR. Despite their obvious interdependence, the fields of international human rights law and international environmental law have largely developed along separate tracks. This has now begun to change as the human rights impacts of environmental degradation and climate change have become increasingly undeniable and as human rights mechanisms dedicate greater attention to these issues.

The UNDROP contributes markedly to this process of integration. In addition to human rights treaties, the language of the Declaration draws from multilateral environmental agreements to articulate states' obligations to maintain biological diversity, comply with their international obligations to combat climate change, and promote and protect peasants' traditional knowledge and practices, including in the design and implementation of climate change adaptation and mitigation policies.[58]

Article 18 of the UNDROP goes even further, stating that 'Peasants and other people working in rural areas have the right to the conservation and protection of the environment and the productive capacity of their lands, and of the resources that they use and manage.' Article 18 adds that 'States shall take appropriate measures to ensure that peasants and other people working in rural areas enjoy, without discrimination, a safe, clean and healthy environment' and 'shall cooperate to address the threats to the enjoyment of their rights that result from transboundary environmental harm.' In essence, the UNDROP recognises the right to a healthy environment for peasants and other people working in rural areas – a right that at this writing had not been recognised in international law, despite its presence in numerous constitutions and regional instruments.

In these respects, the UNDROP can be seen as sitting at the intersection of the fields of international human rights law and international environmental law, and in so doing, answers a call for greater coherence between environmental policies and states' human rights obligations. As such, the Declaration models a holistic approach to food systems governance, underscoring the interdependence of environmental and human rights concerns while calling on states to ensure protection from transboundary harms and advancing the right to a healthy environment. Indeed, the faithful implementation of the Declaration's directives would both reduce the enormous pressure of agriculture on our ecosystems and would enhance the ability of rural communities to contribute to environmental regeneration and recovery in line with human rights.[59]

Conclusion

Following nearly two decades of mobilisation, the adoption of the UNDROP represents a historic milestone for food systems governance and the first time that the plight of peasants and other rural communities has taken centre-stage in international human rights law. The effective implementation of the UNDROP has the potential to significantly alter human rights and environmental struggles, shift the narrative and vision for food systems, and decentralise power therein.

Following decades of neoliberal reforms that have removed the hand of the state from agriculture, the UNDROP powerfully re-affirms the role of the state in this consequential realm, grounded in states' obligations under international human rights and environmental law.[60] While the international human rights framework must still contend with significant power dynamics that undermine the implementation of rights guarantees, both the process and substance of the UNDROP represent an important win for participatory governance and signal a critical shift in consciousness as well as the elevation of an alternative development paradigm and a holistic worldview.

Notes

1 For more on the points raised in this section, see Smita Narula, 'Achieving Zero Hunger Using a Rights-Based Approach to Food Security and Sustainable Agriculture' in Narinder Kakar et al (eds), *Fulfilling the Sustainable Development Goals* (Routledge 2021). Some portions of 'Achieving Zero Hunger' are reproduced here with the permission of INFORMA UK LIMITED through PLS clear.

2 UN Human Rights Council, 'Report of the Special Rapporteur on the Right to Food, Olivier De Schutter' (24 January 2014) UN Doc A/HRC/25/57, para. 6; see also Bryan Newman, *Development Report No. 15: A Bitter Harvest: Farmer Suicide and the Unforeseen Social, Environmental and Economic Impacts of the Green Revolution in Punjab, India* (Institute for Food and Development Policy 2007) 1.

3 'Report of the Special Rapporteur on the Right to Food, Olivier De Schutter,' supra note 2, at para. 6.
4 Smita Narula, 'Reclaiming the Right to Food as a Normative Response to the Global Food Crisis' (2010) 13/2 *Yale Human Rights & Development Law Journal* 411.
5 'Report of the Special Rapporteur on the Right to Food, Olivier De Schutter,' supra note 2, at paras. 10–2.
6 See generally, IPES-Food, Too Big to Feed: Exploring the Impacts of Mega-mergers, Consolidation and Concentration of Power in the Agri-food Sector (IPES-Food 2017). Agribusiness companies have also flooded markets with processed, calorie-dense, and nutrient-poor foods. Inadequate diets are a major contributing factor to the increase of non-communicable diseases occurring now in all regions of the world. UN Human Rights Council, Special Rapporteur on the right to food, 'Critical Perspective on Food Systems, Food Crises and the Future of the Right to Food' (21 January 2020) UN Doc A/HRC/43/44, paras. 31–3.
7 ETC Group, 'Who Will Feed Us? The Peasant Food Web vs. The Industrial Food Chain' (3rd edn, ETC Group 2017) 17.
8 UN Human Rights Council, 'Critical Perspective on Food Systems, Food Crises and the Future of the Right to Food,' supra note 6, para. 14.
9 UN General Assembly, 'Report of the Special Rapporteur on the Right to Food' (11 August 2010) UN Doc A/65/281.
10 UN Human Rights Council, 'Critical Perspective on Food Systems, Food Crises and the Future of the Right to Food,' supra note 6, para. 14.
11 Maryellen Kennedy Duckett, 'Empowering Female Farmers to Feed the World' (2019) *National Geographic* <www.nationalgeographic.com/culture/article/partner-content-empowering-female-farmers> accessed 25 April 2021 (asserting that, among other barriers, women face barriers to farm land ownership that negatively affects their ability to seek 'higher earnings and reliable sources of income').
12 Sara Kominers, *Working in Fear: Sexual Violence Against Women Farmworkers in the United States* (Oxfam America 2015) 15–6.
13 Peter Hurst et al and Food and Agriculture Organization et al, *Agricultural Workers and Their Contribution to Sustainable Agriculture and Rural Development* (ILO 2007) at 19, Point 1.4 at 25, 32, 3.4 at 35, and 2.21 at 58.
14 UN High-Level Political Forum on Sustainable Development,' 2017 HLPF Thematic review of SDG2: End hunger, achieve food security and improved nutrition, and promote sustainable agriculture' (2017) 1.
15 See Pragati, 'Farmer Suicides: A Global Phenomenon' (2015) *The Indian National Interest Review* <http://pragati.nationalinterest.in/2015/05/farmer-suicides-a-global-phenomena/> accessed 27 May 2021.
16 Smita Narula, 'The Global Land Rush: Markets, Rights, and the Politics of Food' (2013) 49(1) *Stanford Journal of International Law* part 1.
17 Global Witness, *Defending Tomorrow: The Climate Crisis and Threats against Land and Environmental Defenders* (Global Witness 2020) 6.
18 'Report of the Special Rapporteur on the Right to Food, Olivier De Schutter,' supra note 2, para. 6.
19 See ETC Group, 'Who Will Feed Us?,' supra note 7, at 32 (noting that the industrial food chain 'dominates more than 75% of global agricultural land . . . and 80% of the Chain's agricultural land is used for livestock production').
20 Joseph Poore and Thomas Nemecek, 'Reducing Food's Environmental Impacts through Producers and Consumers' (2019) 360(6392) *Science* 987; see also '2017 HLPF Thematic review of SDG2,' supra note 14, at 6.
21 HLPF, '2017 HLPF Thematic review of SDG2,' supra note 14, at 6 (noting that 'increased climate variability and weather-related disasters affect agriculture and threaten the stability of food prices').
22 Eli D. Lazarus, 'Land Grabbing as a Driver of Environmental Change' (2014) 46(1) *Area* 79; see also Priyadarshi R. Shukla et al (eds), 'Summary for Policymakers' (2019),

Climate Change and Land: An IPCC special report on climate change, desertification, land degradation, sustainable land management, food security, and greenhouse gas fluxes in terrestrial ecosystems, Table 4.1 at 356–8, and Findings 4.2.1 at 354–6, 4.3.1 at 366, 4.4.1.3 at 372–3, and 4.7.2 at 379–80.

23 *La Via Campesina* is a transnational grassroots peasants' movement that promotes and defends food sovereignty and small-scale sustainable agriculture 'as a way to promote social justice and dignity.' La Via Campesina, 'The International Peasants' Voice' <https://viacampesina.org/en/international-peasants-voice/> accessed 7 May 2021. See also Annette Aurélie Desmarais, 'United in the Vía Campesina' (2006) *Food First* <https://viacampesina.org/en/united-in-the-vcampesina/> accessed 7 May 2021.

24 La Via Campesina, 'Declaration of the Rights of Peasants – Women and Men' (2009); Marc Edelman, 'Linking the Rights of Peasants to the Right to Food in the United Nations' (2014) 10(2) *Law, Culture and the Humanities* 196 (analysing *La Via Campesina's* efforts to have the UN adopt an instrument on peasants' rights).

25 Marc Edelman and Carwil James, 'Peasants' Rights and the UN System: Quixotic Struggle? Or Emancipatory Idea Whose Time Has Come?' (2011) 38(1) *The Journal of Peasant Studies* 82.

26 For more on the process of promulgating the UNDROP, see the Introduction to this volume.

27 UN Human Rights Council, 'United Nations Declaration on the Rights of Peasants and Other People Working in Rural Areas' (8 October 2018) UN Doc A/HRC/RES/39/12, preamble.

28 Ibid.

29 Ibid.

30 For more on the right to food sovereignty in the UNDROP, see Chapter 9 of this volume.

31 On this point, see Noha Shawki, 'New Rights Advocacy and the Human Rights of Peasants: La Via Campesina and the Evolution of New Human Rights Norms' (2014) 6(2) *Journal of Human Rights Practice* 320–2 (describing the ways in which '[i]nstitutionalizing human rights can have adverse consequences for grassroots-led and radical movements like the peasant movement').

32 See Declaration of the Forum for Food Sovereignty, Nyéléni (2007). See also UNDROP, supra note 27, preamble (defining food sovereignty as the right of people 'to define their food and agriculture systems and the right to healthy and culturally appropriate food produced through ecologically sound and sustainable methods that respect human rights'). See Chapter 9 for more on the right to food sovereignty in the UNDROP.

33 ETC Group, 'Who Will Feed Us?,' supra note 7, at 17.

34 Ibid., at 10, 48 (explaining that, '[a]ll the links in the [Industrial Food] Chain are connected within the financial and political system,' and that, '[w]ith the right policies, land and rights, peasant-led agroecological strategies could double or even triple rural employment, substantially reduce the pressure for urban migration, significantly improve nutritional quality and availability and eliminate hunger while slashing agriculture's GHG emissions by more than 90%.'). See also IPES-Food, *Breaking Away from Industrial Food and Farming Systems: Seven Case Studies of Agroecological Transition* (IPES-Food 2018) 2 (reviewing the results of seven case studies in agroecological transition and noting that with changes in 'production practices, in knowledge generation and dissemination, in social and economic relations, and in institutional frameworks,' 'the multiple "lock-ins" of industrial food systems can be overcome and new sustainable food systems can start to emerge').

35 See also Priscilla Claeys, *Human Rights and the Food Sovereignty Movement: Reclaiming Control* (Routledge 2015).

36 La Via Campesina, 'Farmers and Social Movements call for a Fundamental Restructuring of the Global Food System' (2009) <https://viacampesina.org/en/farmers-and-social-movements-call-for-a-fundamental-restructuring-of-the-global-food-system/> accessed 7 May 2021.

37 UNDROP, supra note 27, artt. 2, 3.2, 4.2, 5, 13.4, 15.4, 16.4, 17, 18, 19, 20, and 21.
38 UNDROP, supra note 27, artt. 2.3, 2.4, and 2.5. Art. 18.5 also calls on states to protect peasants from abuses by non-state actors, including by enforcing environmental laws.
39 See Chapter 3 of this volume.
40 Narula, 'Reclaiming the Right to Food as a Normative Response to the Global Food Crisis,' supra note 4, at 407.
41 Marc Edelman, 'What Is a Peasant? What Are Peasantries? A Briefing Paper on Issues of Definition' (2013) *First Session of the Intergovernmental Working Group on a United Nations Declaration on the Rights of Peasants and Other People Working in Rural Areas* 3.
42 Ibid., 9.
43 See Chapter 2 of this volume.
44 UNDROP, supra note 27, art. 1.1.
45 UNDROP, supra note 27, preamble.
46 See Chapter 3 of this volume.
47 The Declaration also draws on soft law instruments, such as the Voluntary Guidelines on the Responsible Governance of Tenure of Land, Fisheries and Forests in the Context of National Food Security. For more, see Chapter 1 of this volume.
48 Narula, 'The Global Land Rush: Markets, Rights, and the Politics of Food,' supra note 16, at 170–1.
49 UNDROP, supra note 27, art. 17.1.
50 UNDROP, supra note 27, art. 17.6.
51 UNDROP, supra note 27, art. 17.3.
52 See FIAN, *The Human Right to Land* (FIAN 2017) 25.
53 For more see Chapter 6 of this volume.
54 UNDROP, supra note 27, art. 17.6.
55 UNDROP, supra note 27, art. 17.7. For more on the right to land, see Christophe Golay, *The Right to Land and the UNDROP: How can we use the United Nations Declaration on the Rights of Peasants and Other People Working in Rural Areas to Protect the Right to Land?* (International Land Coalition et al 2020).
56 For more see Chapter 8 of this volume.
57 Lorenzo Cotula, 'Between Hope and Critique: Human Rights, Social Justice and Re-Imagining International Law from the Bottom Up' (2020) 48(2) *Georgia Journal of International and Comparative Law* 513.
58 UNDROP, supra note 27, artt. 18 and 20.
59 World Future Council, 'THE TIME IS NOW! Global Call for the UN Human Rights Council to Urgently Recognize the Right to a Safe, Clean, Healthy and Sustainable Environment' (2020) <www.ciel.org/wp-content/uploads/2020/09/Global-Call-for-the-UN-to-Recognize-the-Right-to-a-Healthy-Environment-English.pdf> accessed 13 May 2021, 3.
60 Hannan Mollah, 'Agriculture, Peasantry, Food Sovereignty and Peasants' Rights in India' <http://nfc.gov.np/downloadfile/Agriculture,%20Peasantry,%20Food%20Sovereignty%20and%20Peasants%20Rights%20in%20India_1552896341.pdf> accessed 13 May 2021 (remarks to delegates of the International Conference on Food Sovereignty and Peasant Rights).

Selected Bibliography

Claeys P, *Human Rights and the Food Sovereignty Movement: Reclaiming Control* (Routledge 2015).

Cotula L, 'Between Hope and Critique: Human Rights, Social Justice and Re-Imagining International Law from the Bottom Up' (2020) 48(2) *Georgia Journal of International and Comparative Law* 513.

Edelman M, 'Linking the Rights of Peasants to the Right to Food in the United Nations' (2014) 10(2) *Law, Culture and the Humanities* 196.

ETC Group, 'Who Will Feed Us? The Peasant Food Web vs. The Industrial Food Chain' (3rd edn, ETC Group 2017).

Narula S, 'Achieving Zero Hunger Using a Rights-Based Approach to Food Security and Sustainable Agriculture' in Narinder Kakar et al (eds), *Fulfilling the Sustainable Development Goals* (Routledge 2021).

11

THE UNDROP AND CLIMATE CHANGE

Squaring the circle?

Mariagrazia Alabrese and Annalisa Savaresi[*]

Introduction

Agriculture is a major source of emissions causing climate change, and, at the same time, one of the sectors most threatened by it.

On the one hand, land uses and their changes act both as sources of emissions – releasing carbon in the atmosphere, as a result of soil disturbances – and as sinks – by storing atmospheric carbon in soils. The Intergovernmental Panel on Climate Change (IPCC) estimates that Agriculture, Forestry, and Other Land Use activities – so-called AFOLU – account for 23% of global anthropogenic greenhouse gas emissions.[1] The staggering contribution of AFOLU activities to global emissions has led some experts to describe agriculture and animal farming as 'the new coal.'[2] Maintaining and enhancing the land carbon sink while reducing GHG emissions associated with land uses is key to achieving net zero emissions in the second half of this century, in line with the Paris Agreement.[3] However, mitigating emissions from agriculture requires significant adjustments to farming practices worldwide.

On the other hand, changes in weather patterns and extreme events disproportionately affect farming communities around the world, with significant adverse impacts on livelihood and human health. The IPCC notes how climate extremes will have immediate and long-term impacts on the livelihoods of poor and vulnerable communities and especially affect rural areas and smallholder farming systems.[4] Specifically, climate variability and extreme events diminish agricultural production and the availability of food, cause spikes in commodity prices, and, as a result, constrain access to food.[5] Rising CO_2 levels in the atmosphere are also expected to impair the nutritional quality of food, which is likely to become poorer in micronutrients, with particularly detrimental effects on the most vulnerable segments

* The authors are grateful to Roberto Talenti, graduate student of International Security Studies at the Sant'Anna School of Advanced Studies, for invaluable research assistance in the preparation of this piece.

DOI: 10.4324/9781003139874-15

of the world's population.[6] Finally, rising temperatures are expected to lead to the spread of pathogens, such as mycotoxins, thus threatening food safety.[7]

Against this backdrop, the UN Declaration on the Rights of Peasants and Other People Working in Rural Areas (UNDROP) recognises the right of peasants and other people working in rural areas to contribute to the design and implementation of national and local climate change adaptation and mitigation policies, including through traditional knowledge (Article 18). Furthermore, UNDROP recognises the right of peasants and other people working in rural areas to adequate training, including the ability to cope with climate change and weather-related events (Article 25).

This chapter considers the relationship between agriculture and climate change through the lens of the UNDROP. The first section considers how the international climate regime deals with the matter of agriculture. The second examines climate-related rights enshrined in the UNDROP and the extent to which they align with and influence international climate change law and practice. We consider how and why the UNDROP deals with climate change and explore the implications of the UNDROP for the international climate legal regime. The third section reviews developments occurred since the adoption of the Paris Agreement concerning local communities and traditional knowledge, and their relevance to the rights recognised by the UNDROP.

The international climate regime and agriculture

Both the United Nations Framework Convention on Climate Change (UNFCCC) and the Paris Agreement acknowledge the crucial interplay between climate change, food production, and food security and largely conceive agriculture as a sector to be protected from climate variability to ensure that food production 'is not threatened.'[8]

Furthermore, the Paris Agreement recognises 'the fundamental priority of safeguarding food security and ending hunger, and the particular vulnerabilities of food production systems to the adverse impacts of climate change.'[9] The Agreement, however, does not make express references to agriculture.[10]

Conversely, the UNFCCC includes agriculture among the 'relevant sectors' where Parties should control, reduce, or prevent anthropogenic GHG emissions.[11] In a similar fashion, the Convention asks Parties to develop and elaborate appropriate and integrated adaptation plans for agriculture.[12]

The Kyoto Protocol (KP) famously introduced emission reduction targets for a group of industrialised (so-called Annex I) countries.[13] The Protocol asks these countries also to develop policies and measures to promote 'sustainable forms of agriculture in light of climate change considerations.'[14] The KP also specifically mentions agriculture amongst the sectors that may be included in the adaptation and mitigation programmes that Annex I countries are required to periodically formulate, implement, publish, and regularly update.[15]

However, under the KP, it does not really matter in which sector the emissions reductions are carried out, as long as the targets set for Annex I countries are achieved. So, Annex I countries need not take any measures relating to agriculture, provided that they meet their emission reduction targets by taking action in other sectors. Indeed, emission reductions under the KP were largely achieved by reducing emissions in the energy sector.[16]

More generally, Parties to the climate regime have been reluctant to adopt specific commitments for AFOLU activities, which have historically been regarded as difficult to regulate.[17] This is largely due to uncertainty over how to quantify emissions and removals in this specific sector, and how to tackle the impacts of climate policies on global food production and prices.[18]

As a result, a 2016 review of the UNFCCC Parties' emission reduction plans that were submitted in the lead up to the adoption of the Paris Agreement – so-called Intended Nationally Determined Contributions – found that, even though 148 plans included agriculture, they rarely mentioned sector-specific targets for the sector.[19] Another survey furthermore shows how only 2 out of 20 OECD countries have introduced legally binding mitigation targets for the agriculture sector so far.[20]

At the same time, numerous attempts have been made to address 'agriculture' as a standalone negotiation item under the UNFCCC.[21] In 2017, Parties agreed to launch the so-called Koronivia Joint Work on Agriculture, with the aim of advancing cooperation on matters such as methods and approaches for assessing adaptation, adaptation co-benefits and resilience; improved soil carbon, soil health, and soil fertility under grassland and cropland as well as integrated systems; improved nutrient use and manure management; improved livestock management systems; and socio-economic and food security dimensions of climate change in the agricultural sector.[22] The Koronivia Joint Work on Agriculture represents a landmark step towards greater integration of agriculture into the international climate regime.[23] Yet, so far UNFCCC Parties have largely focused on the harmful effects of climate change on agri-food systems, rather than on the mitigation potential of the sector.[24] Thus, presently, international climate change law seems to be more concerned with the vulnerability of agriculture, rather than its role in accelerating human-induced climate change.

UNDROP and climate change

Peasants' dependence on agriculture makes them especially vulnerable to the impacts of climate change. It is therefore hardly a surprise that the UNDROP pays special attention to this matter, calling on States to 'comply with their respective international obligations to combat climate change.'[25]

This section looks at climate change through the lens of the UNDROP. It begins by explaining why the UNDROP deals with the matter of climate change and how. It then discusses the possible implications of the UNDROP for the climate regime.

The UNDROP's preamble highlights how peasants 'suffer from the burdens caused by environmental degradation and climate change.' The expression 'suffer from the burdens' emphasises the impact of climate change on the enjoyment of a wide range of human rights. It resonates with the increasing recognition of the relationship between climate change and human rights law in the literature,[26] by the parties to the climate regime,[27] and by human rights bodies.[28] Rather than articulating new obligations, these sources emphasise the relevance of human rights obligations to climate change action and the need to systemically interpret state obligations in this connection, both at national and at international levels.[29]

On the one hand, climate change has an impact on the enjoyment of a wide range of human rights,[30] particularly those associated with food security and the enjoyment of the right to food.[31] Climate change has already impaired the ability of some communities to feed themselves, posing particular threats for agriculture workers.[32]

On the other hand, some climate change response measures, like biofuel subsidies, have exacerbated pressure on land uses, contributing to spikes in food prices, riots, and hunger, affecting the enjoyment of human rights.[33] The United Nations Special Rapporteur on the right to food has specifically raised concerns over 'green-grabbing or land-grabbing' with the alleged purpose of mitigating climate change.[34] The United Nations Special Rapporteur on extreme poverty and human rights has furthermore cautioned of a 'climate apartheid scenario' in which the wealthy pay to escape overheating, hunger and conflict, while the rest of the world is left to suffer.[35]

Rights-based approaches for understanding climate change and shaping the development and implementation of climate decisions and actions have been increasingly advocated for to resolve these tensions.[36] Such approaches bring to the fore the implications of climate change for the pursuit of social and environmental justice.[37]

In this vein, UNDROP's provisions on climate change build on a wave of recent developments concerning the relationship between climate change and human rights law. The Declaration focuses on two specific aspects.

Article 25 recognises peasants' right to 'adequate' training. Training is deemed to be adequate when it fits with the specific agroecological, sociocultural, and economic environment in which peasants live and work.[38] The UNDROP notes that training 'should include, but not [be] limited to, improving productivity, marketing and *the ability to cope with* pests, pathogens, system shocks, the effects of chemicals, *climate change and weather-related events*' (emphasis added). This context-aware training is not supposed to provide guidance on eliminating or reducing greenhouse gas emissions from agriculture. Instead, training is expected to enable farmers to cope with the effects of climate change and weather-related events through adaptation. This reading of Article 25 maintains the focus on the vulnerability of agriculture, as depicted in section 1. However, the expression 'should include, but not [be] limited to' leaves open the possibility that training might also cover other matters. For example, training could raise peasants' awareness about the challenges posed both by climate change and by some climate change response measures, highlighting win-wins for agriculture and emission reductions.

Arguably, training programmes with a focus on mitigation are not centre stage in the UNDROP because the agriculture model brought about by the Declaration hinges on farming practices that are regarded as not detrimental for the environment and the climate.[39] Yet, even traditional ways of farming envisaged by the UNDROP[40] might need to be aligned with environmental and climate change objectives. For example, the IPCC has highlighted trade-offs for climate change mitigation associated with the use of fire in land clearance.[41] As a result, equipping peasants with the knowledge and tools to deliver emission reductions while maintaining profitability is essential. Therefore, the notion of training under Article 25 of the UNDROP should be interpreted to encompass both adaptation and mitigation. The Paris Agreement specifically mentions training in connection with capacity building, with the aim of implementing effective climate change adaptation and mitigation measures also in developing countries.[42]

Article 18 of the UNDROP recognises peasants' right to contribute to the design and implementation of national and local climate change policies at a national and a local level. This call for peasants' participation in climate decision-making aligns with the Paris Agreement's provisions on cooperation measures to enhance public participation.[43] Furthermore, it aligns with the widespread call for greater public participation in climate decision-making in the transition to net zero emissions.[44] Best practices here include the use of human rights impact and other due diligence assessments, of safeguards to address tensions between the protection of fundamental rights and climate action, and of rights-based approaches for developing, implementing, monitoring, and evaluating climate policies and programmes.[45] The latter specifically emphasise the capacity of citizens and communities to participate in climate decision-making, and their access to judicial or administrative proceedings and effective remedies for harm caused both by climate change and by climate change response measures.

These obligations take specific connotations in relation to peasants. Peasant participation in climate decision-making processes is especially important at a time when states are finally looking at how to reduce emissions from agriculture. The contribution of peasants to the design of climate legislation and policies could deliver more resilient and sustainable agriculture while addressing the socio-economic and food security dimensions of climate change. The participation of those affected could also result in more ambitious plans for accelerating mitigation and adaptation in the agriculture sector and in legal frameworks that are more inclusive and effective.

At least implicitly, therefore, the UNDROP draws attention to links between climate change and agriculture, which have been neglected for many years under the climate regime. Specifically, it draws attention to the benefits of peasants' contribution to the design and implementation of climate policies and measures, building on traditional knowledge, amongst other things. The next section considers this matter in further detail.

The rights of peasants and traditional knowledge

Climate science has increasingly acknowledged the role of traditional knowledge holders, like peasants, both in assessing changes in the climate[46] and in designing climate change response measures.[47] These developments have raised the question of how to create partnerships between traditional knowledge holders and scientists,[48] on the one hand, and how to create pathways to enable farmers to participate in the making of climate policy, on the other.

Since the adoption of the Paris Agreement, these matters have been at the centre of considerable developments. The Agreement is the first climate treaty to explicitly reference traditional knowledge, acknowledging the need to protect local communities and indigenous people, and the importance of their knowledge to adaptation.[49] Parties to the UNFCCC have since created a specific body, the Local Communities and Indigenous Peoples Platform.[50] As such, the Agreement has opened up a new frontier to address traditional knowledge in the climate regime, building on developments occurring under a range of diverse international law instruments concerning disparate matters, such as biodiversity, plant genetic resources, heritage, and human rights.[51]

The Local Communities and Indigenous Peoples Platform is aimed to strengthen indigenous peoples' and local communities' ability to cope with climate change, to engage in the fight against climate change, and to share their ideas and best practices.[52] The platform's workplan for 2020–2021 includes the development of recommendations concerning the engagement and input of indigenous peoples and local communities across the UNFCCC process.[53] The platform has also organised a series of training webinars and commissioned the creation of a specific portal. More generally, the platform seeks to facilitate the integration of diverse knowledge systems, practices, and innovations in designing and implementing international and national climate action, programmes, and policies, in a manner that respects and promotes the rights and interests of local communities and indigenous peoples. It is too early to say whether and to what extent the platform will realise these objectives and deal with agriculture specifically.

It nevertheless seems clear that, at least in principle, activities carried out under the Local Communities and Indigenous Peoples Platform resonate with the training and public participation provisions under the UNDROP, which we reviewed earlier.[54] There are clear overlaps between peasants, local communities, and indigenous peoples, although the groups have different characteristics and rights. Distinguishing between these groups is a complex endeavour, which has already been attempted under other processes. For example, the CBD makes reference to the traditional knowledge of 'indigenous and local communities embodying traditional lifestyles.'[55] While CBD Parties have not attempted to define these groups,[56] this terminology has prompted some scholars to argue that the notion of traditional knowledge holders under the CBD leaves out individuals, as well as groups who are neither indigenous nor local,[57] and possibly even indigenous peoples and local communities who live in urbanised/western societies.[58]

Similarly, other international treaties have attempted to deal with the matter of how to enable partnerships between traditional knowledge holders, scientists, and industry. The UN Convention to Combat Desertification in Countries Experiencing Serious Drought and/or Desertification, particularly in Africa requires state parties to collect and catalogue traditional knowledge.[59] The Nagoya Protocol to the CBD,[60] instead, adopts a contractual approach, restricting access to traditional knowledge. No consistent approach to preventing the misuse and misappropriation of traditional knowledge has been developed, however.[61]

With reference to peasants, traditional knowledge is especially sensitive, as it is associated with potential commercialisation, for example, of weather-resistant crop varieties, which have led to fierce controversies between farmers and industry.[62] At the same time, the practice of international scientific bodies, like the IPCC and the Intergovernmental Panel on Biodiversity and Ecosystem Services (IPBES),[63] has evidenced the need for enabling greater and better interaction between traditional knowledge holders and scientists.

In this connection, the UNDROP falls within the context of a growing body of international human rights practice requesting states to enable traditional knowledge holders' participation in decision-making and science creation.[64] State obligations associated with the right to culture and the right to access to justice are especially relevant here and suggest that effective remedies, including judicial ones, should be made available to individuals and groups who feel that their traditional knowledge has not been adequately protected and has been unduly appropriated.[65] A human rights-based approach provides a much-needed safety net against the misuse and misappropriation of traditional knowledge and the reporting of unauthorised access.[66]

Conclusions

The UNDROP deals with the matter of climate change in a rather specific, circumscribed way. It addresses the matters of farmer training and capacity building, on the one hand, and of farmer participation in climate-related policy and decision-making, on the other. It seemingly focuses on adaptation, rather than on mitigation. As such, the UNDROP suffers the same ambiguities that have characterised the approach to the highly contentious matter of emissions from agriculture under the climate regime.

The UNDROP's provisions, however, must not be read in isolation. Rather, the UNDROP is part of a wave of recent developments concerning the interplay between human rights and climate change law. These developments suggest that striking a balance between competing interests in the context of the race to decarbonise human societies is complex. In this context, rights-based approaches are a helpful tool to achieve environmental objectives, while safeguarding rights, especially those of the most vulnerable.

Notes

1 Priyadarshi R. Shukla et al (eds), 'Summary for Policymakers' (2019) *Climate Change and Land: An IPCC Special Report on Climate Change, Desertification, land Degradation, Sustainable Land Management, Food Security, and Greenhouse Gas Fluxes in Terrestrial Ecosystems* 8.

2 Cfr. Charlotte E. Blattner, 'Just Transition for Agriculture? A Critical Step in Tackling Climate Change' (2020) *Journal of Agriculture, Food Systems, and Community Development* 1.
3 Paris Agreement (adopted 12 December 2015, entered into force 4 November 2016) 55 ILM 740, art. 4.1.
4 IPCC, *Summary for Policymakers*, supra note 1, at 455.
5 Ibid., at 451.
6 The numerous scientific studies on the subject include: Jessica Fanzo, Claire Davis, Rebecca Mclaren and Jowel Choufani, 'The Effect of Climate Change Across Food Systems: Implications for Nutrition Outcomes' (2018) *Global Food Security* 12; Danielle E. Medek, Joel Schwartz and Samuel S. Myers, 'Estimated Effects of Future Atmospheric CO_2 Concentrations on Protein Intake and the Risk of Protein Deficiency by Country and Region' (2017) 125(8) *Environmental Health Perspectives*; Nate Seltenrich, 'Estimated Deficiencies Resulting from Reduced Protein Content of Staple Foods: Taking the Cream out of the Crop?' (2017) 125(9) *Environmental Health Perspective*; Matthew R. Smith, Christopher D. Golden, Samuel S. Myers, 'Potential Rise in Iron Deficiency Due to Future Anthropogenic Carbon Dioxide Emissions' (2017) *GeoHealth* 248; and Samuel S. Myers et al, 'Increasing CO_2 Threatens Human Nutrition' (2014) *Nature* 139. For a legal analysis of the topic, see Mariagrazia Alabrese, 'The Climate-related Nutrient Depletion Phenomenon and its Legal Implications,' in Alessandra Di Lauro (ed), *Les métamorphoses de l'aliment. Les frontières entre les aliments, les médicaments et les cosmétiques* (ETS 2019) 157.
7 Antonio Moretti, Michelangelo Pascale, Antonio F. Logrieco, 'Mycotoxin Risks Under a Climate Change Scenario in Europe' (2019) 84 *Trends in Food Science & Technology* 38. See also an interesting EFSA report on the topic, Angelo Maggiore, Ana Alfonso, Federica Barrucci, Giacomo De Sanctis, 'Climate Change as a Driver of Emerging Risks for Food and Feed Safety, Plant, Animal Health and Nutritional Quality' (2020) 17(6) *EFSA Supporting Publication*.
8 United Nations Framework Convention on Climate Change (adopted 9 May 1992, entered into force 21 March 1994) 1771 UNTS 107 (UNFCCC), art. 2; similarly PA, supra note 3, art. 2.
9 PA, supra note 3, preamble.
10 Jonathan Verschuuren, 'Climate Change and Agriculture under the United Nations Framework Convention on Climate Change and Related Documents' in Mary Jane Angelo, Anél Du Plessis (eds), *Research Handbook on Climate Change and Agricultural Law* (Edward Elgar 2017) 43.
11 UNFCCC, supra note 8, art. 4.1(d).
12 UNFCCC, supra note 8, art. 4.1(e).
13 Kyoto Protocol to the United Nations Framework Convention on Climate Change (adopted 11 December 1997, entered into force 16 February 2005) 2303 UNTS 14, Annex B.
14 Ibid., art. 3.
15 Ibid., art. 10.
16 Michael Grubb, 'Full Legal Compliance with the Kyoto Protocol's First Commitment Period – Some Lessons' (2016) 16 *Climate Policy* 673.
17 See e.g. Zoya E. Bailey, 'Sink That Sank the Hague: A Comment on the Kyoto Protocol' (2002) 16 *Temple International & Comparative Law Journal* 103; Kate Dooley and Aarti Gupta, 'Governing by Expertise: The Contested Politics of (Accounting for) Land-Based Mitigation in a New Climate Agreement' (2017) 17 *International Environmental Agreements: Politics, Law and Economics* 483; Joachim HA Krug, 'Accounting of Ghg Emissions and Removals from Forest Management: A Long Road from Kyoto to Paris' (2018) 13 *Carbon Balance and Management* 1; Annalisa Savaresi and Lucia Perugini, 'Article 5: Sinks, Reservoirs of GHGs and Forests' in Geert van Calster and Leonie Reins (eds), *The Paris Agreement on Climate Change: A Commentary* (Edward Elgar Publishing 2021).

18 Gabriela Steier, 'The Carbon Tax Vacuum and the Debate about Climate Change Impacts: Emission Taxation of Commodity Crop Production in Food System Regulation' (2018) *Pace Environmental Law Review* 346.
19 FAO, *The Agriculture Sectors in the Intended Nationally Determined Contributions: Analysis* (Roma 2016) <www.fao.org/3/a-i5687e.pdf>. 'In the context of mitigation, "Agriculture" – in accordance with IPCC terminology – includes emissions from enteric fermentation, manure management, rice cultivation, prescribed burning of savannas and grassland, and from soils (i.e. agricultural emissions).'
20 Ben Henderson, Clara Frezal, and Eimear Flynn, 'A Survey of GHG Mitigation Policies for the Agriculture, Forestry and other Land Use Sector' (2020) 145 *OECD Food, Agriculture and Fisheries Papers*. The 20 surveyed countries covered six global regions (Europe, North America, South America, Oceania, East and South East Asia, and Sub-Saharan Africa). Within each of the global regions, countries were selected based on a combination of their regional contribution to AFOLU emissions and the progress they have made in introducing policies or commitments to reduce these emissions. The two countries having adopted legally binding mitigation measures are New Zealand and Ireland.
21 Savaresi and Perugini, 'Article 5: Sinks, Reservoirs of GHGs and Forests,' supra note 17.
22 United Nation Framework Convention on Climate Change, 'Report of the Conference of the Parties on its twenty-third session, Dec 4/CP.23 Koronivia Joint Work on Agriculture' (8 February 2018) FCCC/CP/2017/11/Add.1.
23 Savaresi and Perugini, 'Article 5: Sinks, Reservoirs of GHGs and Forests,' supra note 17.
24 Mariagrazia Alabrese, 'Politiche climatiche, politiche agricole e il bisogno di coordinamento' (2020) 1 *Rivista di diritto agrario* 618.
25 UN Human Rights Council, 'United Nations Declaration on the Rights of Peasants and Other People Working in Rural Areas' (8 October 2018) UN Doc A/HRC/RES/39/12, art. 18.3.
26 See Daniel Bodansky, 'Introduction: Climate Change and Human Rights: Unpacking the Issues' (2009) 38 *Georgia Journal of International and Comparative Law* 511; Stephen Humphreys, *Human Rights and Climate Change* (Cambridge University Press 2010); Lavanya Rajamani, 'The Increasing Currency and Relevance of Rights-Based Perspectives in the International Negotiations on Climate Change' (2010) 22 *Journal of Environmental Law* 391; Siobhan McInerney-Lankford, Mac Darrow and Lavanya Rajamani, *Human Rights and Climate Change. A Review of the International Legal Dimensions* (World Bank 2011); Sébastien Duyck et al, 'Human Rights and the Paris Agreement's Implementation Guidelines: Opportunities to Develop a Rights-Based Approach' (2018) 12 *Carbon & Climate Law Review* 191; Annalisa Savaresi and Joanne Scott (eds), 'Implementing the Paris Agreement: Lessons from the Global Human Rights Regime' (2019) 9 *Climate Law* 159.
27 Cancun Agreements, Appendix I, 2; PA, supra note 3, preamble.
28 The Human Rights Council adopted ten resolutions on human rights and climate change between 2008 and 2020. See <www.ohchr.org/EN/Issues/HRAndClimateChange/Pages/Resolutions.aspx> A summary of the activities of the Office of the UN High Commissioner for Human Rights: <www.ohchr.org/en/issues/hrandclimatechange/pages/hrclimatechangeindex.aspx> See also: UN Human Rights Council, 'Report of the Special Rapporteur on the Issue of Human Rights Obligations Relating to the Enjoyment of a Safe, Clean, Healthy and Sustainable Environment' (1 February 2016) UN Doc A/HRC/31/52; UN Special Rapporteur on Human Rights and the Environment, 'Human Rights Obligations Relating to the Enjoyment of a Safe, Clean, Healthy and Sustainable Environment' (15 July 2019) UN Doc A/74/161.
29 Annalisa Savaresi, 'Climate Change and Human Rights: Fragmentation, Interplay and Institutional Linkages' in Sébastien Duyck, Sébastien Jodoin and Alyssa Johl (eds), *Routledge Handbook of Human Rights and Climate Governance* (Routledge, Taylor & Francis Group 2018) 42.

30 UN Human Rights Council, 'Report of the Special Rapporteur on the Issue of Human Rights Obligations Relating to the Enjoyment of a Safe, Clean, Healthy and Sustainable Environment,' supra note 28, paras. 2–3.
31 Astrud Lea Beringer, 'Environmental and Climate Justice' (2020) *UNDROP Series, FIAN International*, which, *inter alia*, highlights the importance of land tenure security for rural communities in the climate debate and the centrality of peasants' capacity to use their own seeds and adapt them to changing environmental and climatic conditions. See also Christophe Golay and Adriana Bessa, 'The Right to Seed in Europe. The United Nation Declaration on the Rights of Peasants and Other People Working in Rural Areas and the Protection of the Right of Seeds in Europe' (2019) *The Geneva Academy of International Humanitarian Law and Human Rights* 32.
32 UN Human Rights Council, 'Report of the Special Rapporteur on the Issue of Human Rights Obligations Relating to the Enjoyment of a Safe, Clean, Healthy and Sustainable Environment,' supra note 28, paras. 24–6.
33 UN Human Rights Council, 'Report of the Special Rapporteur on the Issue of Human Rights Obligations Relating to the Enjoyment of a Safe, Clean, Healthy and Sustainable Environment,' supra note 28, para. 69.
34 UN Human Rights Council, 'Special Rapporteur on the right to food, "Critical Perspective on Food Systems, Food Crises and the Future of the Right to Food"' (21 January 2020) UN Doc A/HRC/43/44, para. 21.
35 UN Human Rights Council, 'Special Rapporteur on extreme poverty and human rights, "Climate Change and Poverty"' (17 July 2019) UN Doc A/HRC/41/39, para. 51.
36 Duyck et al, *Routledge Handbook of Human Rights and Climate Governance*, supra note 29; Sébastien Jodoin, Annalisa Savaresi and Margaretha Wewerinke-Singh, 'Rights-Based Approaches to Climate Decision-Making' (2021) *Current Opinion in Environmental Sustainability* 45.
37 See, e.g., Annalisa Savaresi and Kim Bouwer, 'Equity and Justice in Climate Change Response Measures: Benefit Sharing as a Safeguard' in Tahseen Jafry (ed), *Routledge Handbook of Climate Justice* (Routledge 2018); Jodoin, Savaresi, and Wewerinke-Singh, 'Rights-Based Approaches to Climate Decision-Making' supra note 36.
38 UNDROP, supra note 25, art. 25.
39 UNDROP, supra note 25, art. 16.4.
40 UNDROP, supra note 25, art. 16.1.
41 IPCC, *Summary for Policymakers*, supra note 1, art. 7.
42 PA, supra note 3, art. 11
43 PA, supra note 3, art. 12
44 Duyck et al, *Routledge Handbook of Human Rights and Climate Governance*, supra note 29; Jodoin, Savaresi and Wewerinke-Singh, 'Rights-Based Approaches to Climate Decision-Making,' supra note 36.
45 Jodoin, Savaresi, and Wewerinke-Singh, 'Rights-Based Approaches to Climate Decision-Making,' supra note 36.
46 IPCC, *Climate Change 2007: Impacts, Adaptation and Vulnerability. Contribution of Working Group II to the Fourth Assessment Report of the Intergovernmental Panel on Climate Change Contribution* (Cambridge University Press 2007) 138 and 673. For a review, see Claudio Chiarolla and Annalisa Savaresi, 'Indigenous Challenges Under IPBES: Embracing Indigenous Knowledge and Beyond' in Marie Hrabanski and Denis Pesche (eds), *The Intergovernmental Platform on Biodiversity and Ecosystem Service (IPBES): Challenges, Knowledge and Actors* (Routledge 2016).
47 IPCC, *Summary for Policymakers*, supra note 1, at 4.1.
48 UN Framework Convention on Climate Change Secretariat, 'Best practices and available tools for the use of indigenous and traditional knowledge and practices for adaptation, and the application of gender-sensitive approaches and tools for understanding and assessing impacts, vulnerability and adaptation to climate change' (31 October 2013) FCCC/TP/2013/11, 8.
49 PA, supra note 3, art. 7.5.

50 Decision 1/CP.21, Adoption of the Paris Agreement, FCCC/CP/2015/10/Add.1 (2015), paras. 135–6
51 For a review, see Annalisa Savaresi, 'Traditional Knowledge and Climate Change: A New Legal Frontier?' (2018) 9 *Journal of Human Rights and the Environment* 32.
52 Decision 2/CP.23 Local communities and indigenous peoples platform, FCCC/CP/2017/11/Add.1 (2018), paras. 6.
53 UN Framework Convention on Climate Change Subsidiary Body of Scientific and Technological Advice, 'The 1st Meeting of the Facilitative Working Group of the Local Communities and Indigenous Peoples Platform' (18 September 2019) FCCC/SBSTA/2019/4.
54 UNDROP, supra note 25, artt. 18 and 25
55 Convention on Biological Diversity (adopted 5 June 1992, entered into force on 29 December 1993) 1760 UNTS 69, art. 8(j).
56 For a reflection on the interpretation of these terms, see Savaresi, 'The International Human Rights Law Implications of the Nagoya Protocol' in Elisa Morgera, Matthias Buck and Elsa Tsioumani (eds), *The 2010 Nagoya Protocol on Access and Benefit-sharing in Perspective* (Martinus Nijhoff 2013). In relation to the use of the term 'indigenous peoples' in the context of the CBD, see Decision XI/14, Article 8(j) and related provisions (5 December 2012) UN Doc UNEP/CBD/COP/DEC/XI/14, 18–9.
57 Graham Dutfield, 'Protecting the Rights of Indigenous Peoples: Can Prior Informed Consent Help?' in Rachel Wynberg, Doris Schroeder, and Roger Chennells (eds), *Indigenous Peoples, Consent and Benefit Sharing* (Springer 2009).
58 See Lyle Glowka, Francoise Burhenne-Guilmin, and Hugh Synge, *A Guide to the Convention on Biological Diversity* (IUCN 1994) 11. Contra, see Elisa Morgera, Elsa Tsioumani, and Matthias Buck, *Unraveling the Nagoya Protocol: A Commentary on the Nagoya Protocol on Access and Benefit-Sharing to the Convention on Biological Diversity* (Martinus Nijhoff 2014) 25.
59 United Nations Convention to Combat Desertification in Those Countries Experiencing Serious Drought and/or Desertification, Particularly in Africa (adopted 17 June 1994, entered into force on December 1996)(UNCCD), art. 18.2(a).
60 Nagoya Protocol on Access to Genetic Resources and the Fair and Equitable Sharing of Benefits Arising from their Utilization to the Convention on Biological Diversity (adopted 29 October 2010, entered into force 12 October 2014) 30619 UNTS 27, especially artt. 7 and 12. See, for example: Camena Guneratne, *Genetic Resources, Equity and International Law* (Edward Elgar Publishing 2012) 92; Christoph B. Graber, 'Institutionalization of Creativity in Traditional Societies and in International Trade Law' (2011) *SSRN Scholarly Paper.*
61 Savaresi, 'Traditional Knowledge and Climate Change: A New Legal Frontier,' supra note 51.
62 These tensions are explored, e.g., in Elsa Tsioumani, 'Exploring Benefit-Sharing from the Lab to the Land (Part I): Agricultural Research and Development in the Context of Conservation and Sustainable Use' (2014) *SSRN Scholarly Paper.*
63 The IPBES was established to strengthen the science-policy interface for biodiversity, carry out knowledge generation catalysis, policy support, and capacity building – IPBES, Functions, Operating Principles and Institutional Arrangements of IPBES, 21 April 2012, Panama City, Appendix 1, 1.
64 This practice is comprehensively reviewed in: UN Human Rights Council, 'Report of the Independent Expert on the issue of human rights obligations relating to the enjoyment of a safe, clean, healthy and sustainable environment, John H. Knox, Mapping Report'(30 December 2013) UN Doc A/HRC/25/53.
65 See, e.g., Concluding Observations of the Committee on Economic, Social and Cultural Rights: Russian Federation (1 June 2011) E/C.12/RUS/CO/5, 34.
66 As argued also in Annalisa Savaresi, 'The International Human Rights Law Implications of the Nagoya Protocol' in Elisa Morgera, Matthias Buck and Elsa Tsioumani (eds), *The*

2010 Nagoya Protocol on Access and Benefit-sharing in Perspective (Martinus Nijhoff 2013); Savaresi, 'Traditional Knowledge and Climate Change: A New Legal Frontier,' supra note 51.

Selected bibliography

Angelo MJ and Du Plessis A (eds), *Research Handbook on Climate Change and Agricultural Law* (Edward Elgar 2017).

Duyck S, Jodoin S and Johl A (eds), *Routledge Handbook of Human Rights and Climate Governance* (Routledge, Taylor & Francis Group 2018).

Humphreys S, *Human Rights and Climate Change* (Cambridge University Press 2010).

Steier G, 'The Carbon Tax Vacuum and the Debate about Climate Change Impacts: Emission Taxation of Commodity Crop Production in Food System Regulation' (2018) *Pace Environmental Law Review* 346.

Wynberg R, Schroeder D and Chennells R (eds), *Indigenous Peoples, Consent and Benefit Sharing* (Springer 2009).

12

FARMING MODELS AND PEASANTS' RIGHTS*

Eloisa Cristiani and Giuliana Strambi

Introduction

The preamble to the United Nations' *Declaration on the Rights of Peasants and Other People Working in Rural Areas* (UNDROP) portrays peasants and peasant communities as guardians of biodiversity due to their living in symbiosis with the natural elements they depend on for their livelihoods. In concrete, the UNDROP recognises:

> The past, present and future contributions of peasants and other people working in rural areas in all regions of the world to development and to conserving and improving biodiversity, which constitute the basis of food and agricultural production throughout the world, and their contribution in ensuring the right to adequate food and food security.

This assumption underpins peasants' rights to produce food according to traditional, culture-respectful, and sustainable methods. The close relationship between peasants and biodiversity also justifies the call for production practices free from toxic or polluting chemicals and the right of peasants to autonomously select and adopt food systems and agricultural production methods, by relying on natural processes and adapting to climate change.

The present chapter explores the particular characteristics of sustainable agriculture models as proposed by the UNDROP. We first briefly note pre-UNDROP documents to then discuss the extent to which this concept has been incorporated in the UNDROP's text and how it has been related to peasant's production methods and rights.

* While the article represents a joint effort, sections 1, 2, 3 and 4 are ascribed to Eloisa Cristiani, and sections 5 and 6 to Giuliana Strambi. The authors are grateful to the editors for their constructive comments.

DOI: 10.4324/9781003139874-16

Sustainable agriculture in pre-UNDROP documents

The peasants' charter

The 'Peasants' Charter' of 1979, mentioned in the UNDROP's Preamble, touched upon many of the themes developed in the Declaration; however, it does not directly mention sustainable agriculture methods.[1] The Peasants' Charter, notwithstanding the fact that it was conceived when the environmental concerns were only surfacing, referred to the importance of the equitable distribution of land and the 'efficient' use of natural resources in compliance with the ecological balance and environmental protection, which were considered as prerequisites for rural development. Rural development, in turn, was seen as essential to increase production and to stem poverty.[2] The Declaration, in turn, comes in the middle of the ecological crisis. It has thus taken a decisive step further by advocating for sustainable agricultural measures, and explicitly referring to agroecology, and by dismissing the idea that the rural development model based on monoculture and mechanisation is the silver bullet to reduce poverty and food insecurity.

The UN declaration on the rights of indigenous peoples

The UNDROP has borrowed from many of the normative developments already present in the UNDRIP.[3] First, the UNDRIP recognises in its preamble 'that respect for indigenous knowledge, cultures and traditional practices contributes to sustainable and equitable development and proper management of the environment.' Second, it recognises the 'collective' nature of the rights ascribed to indigenous peoples. Likewise, the UNDROP has also largely built upon the UNDRIP by giving traditional agricultural practices a central role in the document and by emphasising the importance of recognising the right to participation in its collective form, thus in line with the most advanced trends of international law concerning 'environmental democracy' and 'biocultural rights.'[4]

In this connection, the UNDROP repeatedly emphasises the right of peasants to maintain their cultural wealth and natural resources, as a corollary to the right to development. The recurrence throughout the text of the word 'their' also reinforces peasants' sense of belonging to land, to bodies of water, pastures, and forests.[5]

References to the sustainable use of resources and to agroecology are more explicit and clearer in the UNDROP. Peasants and other people working in rural areas play a public role or a role of notable public importance, since they contribute to guaranteeing food security in the world. This is discussed both in terms of justifying peasants' right to land and other natural resources necessary to conduct a safe and dignified life as well as in references to public policies aimed at encouraging the transition towards sustainable agri-food systems.

Sustainable agriculture in the UNDROP

Although the scientific community has no universally agreed definition of sustainable agriculture, a textual analysis of the preamble to the UNDROP introduces this notion in reference to sustainability in the agricultural sector. In its preamble, the UNDROP states:

> Peasants and other people working in rural areas should be supported in their efforts to promote and undertake sustainable practices of agricultural production that support and are in harmony with nature, . . . including by respecting the biological and natural ability of ecosystems to adapt and regenerate through natural processes and cycles.

In other passages, references are made to peasants' practices that conserve and improve biodiversity, agroecology, and respect for the environmental, social, and economic carrying capacity of ecosystems.[6] In this sense, the preamble suggests that sustainable agriculture models are intended not only to be ecologically safe but also respectful of the human rights of both producers and consumers, promoting healthy and culturally adequate food. This idea is clearly reinforced in the concept of food sovereignty, defined as the right of peasants 'to define their food and agriculture systems and the right to healthy and culturally appropriate food produced through ecologically sound and sustainable methods that respect human rights.'[7]

A number of articles of the UNDROP refer to the sustainable agricultural production. Article 16.4 of the UNDROP explicitly includes, within a non-exhaustive list, agroecology and organic agriculture under the umbrella concept of sustainable agriculture. It incorporates within the human rights catalogue specific norms related to modes of production, rural livelihood, and environmental protection. It thus clarifies state duties in this respect and literally demands states take all appropriate measures 'to ensure that their rural development, agricultural, environmental, trade and investment policies and programmes contribute effectively to protecting and strengthening local livelihood options and to the transition to sustainable modes of agricultural production.' The UNDROP goes on to state that it is the role of states to 'stimulate sustainable production, including agroecological and organic production.' Moreover, the UNDROP makes a strong case for the association of ecological farming and social justice in rural settings, by demanding States 'facilitate direct farmer-to-consumer sales,' whenever possible.

Article 20.2 of the UNDROP also makes express reference to agroecology as an agriculture model in line with conservation strategies and therefore to be encouraged, maintained, and further developed. It demands states take appropriate measures to promote and protect the traditional knowledge, innovation, and practices of peasants associated with 'agroecological systems relevant to the conservation and sustainable use of biological diversity.' In doing so, it links cultural heritage, biodiversity protection, and sustainability goals[8] and calls for the adoption of concerted laws, public policies, and programs.

Article 25.1 of the UNDROP, which focuses on the right of peasants to adequate training, prescribes that all training should be designed to be 'suited to the specific agroecological, sociocultural and economic environments' where peasants live and work. Training programmes should cover, among other topics, methods and techniques to improve 'productivity, marketing and the ability to cope with pests, pathogens, system shocks, the effects of chemicals, climate change and weather-related events.' In tune with recent findings and UN reports, the UNDROP connects agroecology to climate change by suggesting agroecological techniques may help reduce greenhouse gas emissions and increase peasants' produce resilience to climate variability.[9]

More nuanced references to sustainable agriculture are observed in other passages. These reinforce the idea of peasants' traditional farming systems as ecologically and socially sustainable and should therefore be maintained and promoted. Article 20.2 of the UNDROP, for instance, affirms that 'traditional agrarian, pastoral, forestry, fisheries, livestock and agroecological systems [are] relevant to the conservation and sustainable use of biological diversity.' Article 16.1 of the UNDROP in turn evokes the right of peasants to 'engage . . . in traditional ways of farming, fishing, livestock rearing and forestry and to develop community-based commercialization systems.'

It is worth noting that the right to access information and the right to participate in law-, policy-, and decision-making processes, as well as in the implementation of strategies that have an impact on peasants' lives and livelihoods have been expressly linked to the preservation of traditional farming methods in an international instrument for the first time. In this regard, Article 11.2 demands States ensure that peasants

> have access to relevant, transparent, timely and adequate information in a language and form and through means adequate to their cultural methods so as to promote their empowerment and to ensure their effective participation in decision-making in matters that may affect their lives, land and livelihoods.

What is more, in Article 11.3, the UNDROP determines that states must promote peasants' participation in the formulation of appropriate certification systems for the evaluation of the quality of their products.

The same is true of the right to education, which has been reaffirmed and elaborated upon with a view to valuing and preserving traditional farming methods – rather than assisting peasants in the transition to 'modern' agriculture. Article 25.1 clarifies that peasants' right to adequate training must be 'suited to the specific agroecological, sociocultural and economic environments in which they find themselves.' An example of training of this sort is suggested in Article 14.4(e), which prescribes educational programmes on the health and environmental effects of chemical inputs and pesticides and on alternatives.

Sustainable agriculture methods within the international scenario: the role of the FAO and the SDGs

This section will investigate whether there is an established definition of sustainable agriculture at an international level, from both a regulatory and a soft law perspective; then it will consider whether this model of sustainable agriculture corresponds to the one emerging from the UNDROP.

Goal 2 of the 2030 Agenda for Sustainable Development refers specifically to the need to 'End hunger, achieve food security and improved nutrition and promote sustainable agriculture.' However, it does not provide a specific definition of sustainable agriculture and it does not detail sustainable production techniques.

The FAO has done considerable work on sustainable agriculture.[10] In the 2018 document entitled 'The 10 Elements of Agroecology. Guiding the transition to sustainable food and agricultural systems,' sustainable agriculture is characterised under the model of agroecology.[11] This document lists the ten constitutive elements of agroecology, which are then analysed extensively. These highly interconnected elements include diversity; synergies; efficiency; resilience; recycling; co-creation and sharing of knowledge; human and social values; culture and food traditions; responsible governance; and circular and solidarity economy.[12]

A detailed study of these concepts goes beyond the scope of this chapter. Nevertheless, from a legal perspective, in order to outline the main points of the sustainable agricultural model envisioned by the UNDROP, it is important to highlight the similarities and differences between the 'sustainable practices of agricultural production that support and are in harmony with nature' enshrined in the UNDROP and the FAO's 'agroecological' approach.

Firstly, both the UNDROP and the FAO's document highlight the need to rethink the current food systems. Countries need to take political, administrative, and legal measures to support and implement more sustainable and resilient food systems, which make rational use of non-renewable resources and that resist and adapt to climate shocks. Production also needs to be in tune with environmental protection and social inclusion. This is based on the idea that it is possible and essential to create an agricultural system for the third millennium that is a 'win-win' for people and their livelihoods and the planet.[13]

In addition, as the next section highlights, the approaches of both the UNDROP and the FAO focus on the traditional, practical, and local knowledge of food producers. Traditional knowledge is key for diversification and better integration of food production systems into complex ecological processes. It builds on synergies with the natural habitat instead of depleting natural resources. The preservation of traditional knowledge is thus as fundamental as switching to sustainable and resilient production models.

The UNDROP and the FAO also identify small-scale farmers as the key agents of this change. In fact, the FAO refers to 'family farmers, including smallholder farmers, indigenous peoples, fisher folks, mountain farmers and pastoralists' who are all included in the UNDROP notion of 'peasants.'[14] It is evident that in both these scenarios, the transition to sustainability is based on bottom-up processes.

Notwithstanding the common elements between the UNDROP and the FAO to sustainable agriculture, it should be acknowledged that in complex agrifood systems, there is no single way to implement sustainable agricultural practices. In fact, the FAO often refers to the concept of 'sustainable intensification,' which involves considerable use of 'technical knowledge.'[15] The FAO emphasises the massive growth of the global population, which consumes non-renewable natural resources and which must be supported by all the possible means available in order to intensify production, including those provided by biotechnologies.[16] By not dismissing FAO's approach, the UNDROP envisions tensions in this respect and claims that states should protect peasants' rights from the risk of violation possibly arising from living modified organisms (Article 20.3). Such a violation risk can refer to the depletion of biodiversity, which, in turn, could endanger the fulfilment of peasants' rights.

The protection of traditional farming in the transition to sustainable models of agriculture

The protection of peasants' human rights in the UNDROP is not only an aim in itself, but also an instrument for achieving SDGs, such as food security and combating climate change, which are important for the whole of humanity. Accordingly, the Declaration protects the inalienable human rights of peasants to development, while, as mentioned earlier, promoting 'sustainable practices of agricultural production that support and are in harmony with nature' (see Preamble).

The right to the conservation and protection of the environment and the productive capacity of peasants' land and of the resources that they use and manage (Article 18.1.) is particularly interesting. This is closely linked to the adoption of appropriate measures to ensure that peasants and other people working in rural areas enjoy, without discrimination, a safe, clean, and healthy environment (Article 18.2). It also entails measures to guarantee 'the right to contribute to the design and implementation of national and local climate change adaptation and mitigation policies, including through the use of practices and traditional knowledge,' as 'States shall comply with their respective international obligations to combat climate change' (Article 18.3).[17] Similarly, the right to seed as formulated in the UNDROP[18] should be protected by States in order to maintain agrobiodiversity for future generations (Article 19).[19]

The authors acknowledge that this approach may entail a utilitarian view of peasants in a human rights instrument. That said, we would like to focus on the role of traditional knowledge and practices for the sustainable development the UNDROP aims to achieve, as a secondary and indirect objective.[20]

The contribution of traditional knowledge to combating climate change and food security is explicitly cited in the aforementioned provisions, as well as in Article 16, where States are called upon to take 'appropriate measures to ensure that their rural development, agricultural, environmental, trade and investment policies and programmes contribute effectively to . . . the transition to sustainable modes of agricultural production' (Article 16.4). States are also called upon to take into consideration the specific culture and traditional practices of the peasant communities involved and especially to guarantee their right to participate in decision-making processes.[21] Which practices, however, are based on traditional knowledge and what are traditional practices?

These practices vary depending on the specific peasant communities and on the particular link with the territory[22] and concern, for example, soil protection and fertilisation, terracing, irrigation, and water harvesting.[23] These practices have been handed down through the generations to tackle the challenges of the natural environment and to conserve the productive capacity of the land and the resources that peasants use and manage.[24]

Considering this legal framework, it is therefore fair to ask what role peasants' innovation may play in the transition to sustainability.[25] The term 'innovation' is mentioned only once in the UNDROP, in a context strictly connected to the right to biological diversity. Article 20, in fact, provides that States should take appropriate measures to promote and protect innovation as well as the traditional knowledge and practices of peasants (and other people working in rural areas) in order to conserve the sustainable use of biodiversity.[26] Traditional knowledge and innovation are thus put on the same footing. They are equally promoted and protected as a means to ensure the full enjoyment of the rights of peasants, even though no actual definition of the terms is provided.[27]

Article 20 in conjunction with other important Articles of the UNDROP, such as Article 25 on the right to adequate training and Article 15 on the right to participate in decision-making processes on agriculture policy, helps in understanding the purpose of the UNDROP to protect a dynamic reality, where peasants' development should be guaranteed, thus ensuring 'the right to determine and develop priorities and strategies to exercise their right to development' (Article 3), therefore also in terms of innovation policies.[28] However, it should not be forgotten that the aim of UNDROP is to protect alternative rather than industrial farming approaches. In this regard, as a human rights instrument, first and foremost, the UNDROP aims to guarantee the principle of non-discrimination, providing peasants with 'the right to maintain, express, control, protect and develop their traditional and local knowledge, such as ways of life, methods of production or technology, or customs and tradition' (Article 26).[29]

In other words, the innovation protected by the UNDROP should concern practices and agrarian systems that do not negatively impact the 'special dependency on and attachment to the land' as a key element of the notion of peasant (Article 1)[30] but instead improve its sustainability at a local level, in order to guarantee the preservation of land for future generations (and the resources they use) aimed at producing healthy and adequate food, the environment, landscape, and culture.[31]

Closing remarks

Our analyses suggest that agricultural systems are facing an epochal crossroads: deep transformations are required, which can only be achieved following the principle of sustainability. However, it is evident that within the framework of complex modern agricultural models, there is no single way to attain sustainability.

The UNDROP expressly mentions agroecological and organic methods of food production as examples of sustainable modes of agriculture to be promoted by states. Traditional ways of farming (fishing, livestock rearing, and forestry) are levelled to a human right, in the context of peasants and other people working in rural areas. They are also protected and supported by states as a model of sustainable agriculture when they involve the conservation and sustainable use of land and other natural resources used in food production and contribute to the adaptation to and mitigation of climate change.

The key role assigned by the UNDROP to traditional knowledge and practices in the transition towards sustainable agriculture should inform the FAO and other international organisations working to improve global food security and the living conditions of people in rural areas.

In particular, the FAO has to enhance the link between the UNDROP and the Global Action Plan 'United Nations Decade of Family Farming 2019–2028' as they are moving in the same direction.[32] The role of FAO could thus be decisive in creating international momentum, aimed at ensuring that the principles of the Declaration are acknowledged and implemented also by the most sceptical and least forthright States.

Notes

1 Declaration of Principles and Programme of Action of the World Conference on Agrarian Reform and Rural Development (adopted July 1979).

2 As expected, the term 'sustainable' never appears in the Peasants' Charter of 1979, whereas it is used 17 times in the Declaration, repeatedly associated with modes of agricultural production, use of land and other natural resources, or with the whole food system.

3 This concept can be applied to the UNDROP, which was effectively expressed by Simone Vezzani, 'The Protection of Traditional Knowledge of Agricultural Interest in International Law' in Antonietta Di Blase and Valentina Vadi (eds), *The Inherent Rights of Indigenous Peoples in International Law* (Roma TrE-Press 2020), with a specific reference to the UN General Assembly, 'United Nations Declaration on the Rights of Indigenous Peoples' (2 October 2007) UN Doc A/RES/61/295.

4 In this regard, see Lorenza Paoloni and Simone Vezzani, 'La Dichiarazione ONU sui diritti dei contadini e delle altre persone che lavorano nelle aree rurali: prime riflessioni' (2019) 1 *federalismi.it* <www.federalismi.it/focus/index_focus.cfm?FOCUS_ID=104&focus=humanrights> accessed 31 August 2020, 26–7.

5 See, for instance, UN Human Rights Council, 'United Nations Declaration on the Rights of Peasants and Other People Working in Rural Areas' (8 October 2018) UN Doc. A/HRC/RES/39/12, artt. 11.2, 15.4, 17.5.

6 On sustainable agriculture, see Terry Gips, 'What is a Sustainable Agriculture?' in Patricia Allen and Debra van Dusen (eds), *Global Perspectives on Agroecology and Sustainable Agricultural Systems* (Proceedings of the Sixth International Conference of the International Federation of Organic Agriculture Movements 1988) 63; Olaf Christen, 'Sustainable Agriculture: History, Concept and Consequences for Research, Education and Extension' (1996) 74(1) *Berichte Uber Landwirtschaft* 66; Eric Lichtfouse et al (eds), *Sustainable Agriculture* (Springer 2009); Paul B. Thompson, 'Agricultural Sustainability: What it is and What It Is Not' (2007) 5(1) *International Journal of Agricultural Sustainability* 8; Jules Pretty, 'Agricultural Sustainability: Concepts, Principles and Evidence' (2008) 363 *Philosophical Transactions of the Royal Society B* 447–51; Christine Jakobsson, 'Definitions of the Ecosystems Approach and Sustainability' in Christine Jakobsson (ed), *Sustainable Agriculture* (The Baltic University Programme – Uppsala University 2012) 13.

7 On the right to food and food sovereignty, see Chapter 9 in this volume.

8 On this point, see also Mariagrazia Alabrese, 'Politiche climatiche, politiche agricole e il bisogno di coordinamento' (2020) 1 *Rivista di diritto agrario* 618ff.

9 See, for example, the UN General Assembly, 'Interim Report of the Special Rapporteur on the right to food' (5 August 2015) UN Doc A/70/287, sec. 2 ('The impact of climate change on the right to food').
10 See FAO, *Scaling up Agroecology Initiative Transforming Food and Agricultural Systems in Support of the SDGS – A Proposal Prepared for the International Symposium on* Agroecology (FAO 2018) <www.fao.org/3/I9049EN/i9049en.pdf> or the FAO Report, *FAO'S Work on Agroecology. A Pathway to Achieving the SDGs* (FAO 2018) <www.fao.org/3/I9021EN/i9021en.pdf> accessed 1 August 2020.
11 The FAO talks about 'agro-ecology' by proposing the following definition: 'Agroecology is based on applying ecological concepts and principles to optimize interactions between plants, animals, humans and the environment while taking into consideration the social aspects that need to be addressed for a sustainable and fair food system. By building synergies, agroecology can support food production and food security and nutrition while restoring the ecosystem services and biodiversity that are essential for sustainable agriculture. Agroecology can play an important role in building resilience and adapting to climate change' <www.fao.org/agroecology/en/>. In addition, there is an analysis of the relationship between agro-ecology and sustainable agriculture in Paola Migliorini-Alexander Wezel, '*Converging and Diverging Principles* and Practices of Organic Agriculture Regulations and Agroecology' (2017) *Agronomy for Sustainable Development.* More generally, in Massimo Monteduro, 'Ius et rus: la rilevanza dell'agroecologia per il diritto' (2019) 2 *Rivista quadrimestrale di diritto dell'ambiente* 4.
12 As in note n. 4 of the FAO Dossier, these principles have both a political and a scientific foundation, in particular referring to Miguel A. Altieri, *Agroecology: The Science of Sustainable Agriculture* (CRC Press 1995) and Stephen R. Gliessman, *Agroecology: The Ecology of Sustainable Food Systems* (3rd edn, CRC Press 2015).
13 This is the usual and highly effective term used in all the FAO documents. See, for example, FAO, FAO'S Work on agroecology, supra note 10, at 6.
14 See Chapter 2.
15 For example, FAO, *Save and Grow* (FAO 2011), the FAO presents a 'new paradigm' based on 'sustainable crop production intensification (SCPI), which produces more from the same area of land while conserving resources, reducing negative impacts on the environment and enhancing natural capital and the flow of ecosystem services' (vii). Production systems for SCPI are 'knowledge-intensive and relatively complex to learn and implement' (26); therefore, the document recognizes the important role of farmers as innovators.
16 See HLPE, *Agroecological and other Innovative Approaches for Sustainable Agriculture and Food Systems that Enhance Food Security and Nutrition. A Report by the High Level Panel of Experts on Food Security and Nutrition of the Committee on World Food Security* (HLPE 2019).
17 See Chapter 11.
18 See Christophe Golay and Adriana Bessa, *The Right to Seed in Europe. The United Nation Declaration on the Rights of Peasants and Other People Working in Rural Areas and the Protection of the Right of Seeds in Europe* (The Geneva Academy of International Humanitarian Law and Human Rights 2019) 32ff.
19 See Chapter 8.
20 See Paoloni and Vezzani, 'La Dichiarazione ONU sui diritti dei contadini e delle altre persone che lavorano nelle aree rurali: prime riflessioni,' supra note 4, at 27.
21 The topic of the 'Traditional Knowledge and Sustainable Development' was already tackled by the World Bank Conference in 1993 with specific reference to the indigenous conditions. The core idea debated was that 'no contradiction exists between the maintenance of strong cultural traditions and identities and economic development. To the contrary, for development to be socially and environmentally sustainable, it must take into account and draw upon the values, traditions, and cultures of the people in the countries and societies that it serves' (see Davis H. Shelton and KatrinkaEbbe (eds), 'Traditional Knowledge and Sustainable Development. Conference held at The World Bank Washington' (1993) *Environmentally Sustainable Development Proceedings Series No. 4*).

On the broad body of international treaties and customary rules governing the safeguarding and promotion of knowledge associated with genetic resources, see, for example, Vezzani, 'The Protection of Traditional Knowledge of Agricultural Interest in International Law,' supra note 3, at 281ff.

22 On the absence of the explicit reference to the territory issue and to the right to territory in the UNDROP compared to La Via Campesina's Declaration on the Rights of Peasants – Women and Men and to the UNDRIP, see Priscilla Claeys and Marc Edelman, 'The United Nations Declaration on the Rights of Peasants and other People Working in Rural Areas' (2020) 47 *The Journal of Peasant Studies* 43.

23 For some agricultural practices, see the Traditional Knowledge World Bank's inventory, which is available at <www.tkwb.org/w/index.php/TKWB_-_Traditional_Knowledge_World_Bank> accessed 28 July 2020, mentioned by Vezzani 'The Protection of Traditional Knowledge of Agricultural Interest in International Law,' supra note 3, at 280. On the TKWB website, the inventory is described as 'a Wiki-like system set up as a platform for knowledge sharing and dissemination with full acknowledgement of the originators.' The inventory was created because 'traditional knowledge is in danger and its disappearance would cause not only the loss of its people's capability to keep and pass on the cultural and natural heritage, but also the loss of an extraordinary source of knowledge and cultural diversity from which appropriate innovative sustainable solutions can be derived both today and in the future.'

24 UNDROP, supra note 5, art 18.1. See Vezzani, 'The Protection of Traditional Knowledge of Agricultural Interest in International Law,' supra note 3, at 280.

25 For example, FAO, *Building a Common Vision for Sustainable Food and Agriculture. Principles and Approaches* (FAO 2014) 37. According to the FAO, 'innovation, in general and particularly in agriculture, is the central driving force for achieving a world free from hunger and malnutrition. Innovation is usually perceived as related to technology. In fact, innovation is broader than that. Agricultural innovation is the process whereby individuals or organizations bring new or existing products, processes or ways of organization into use for the first time in a specific context, to increase effectiveness, competitiveness and resilience with the goal of solving a problem.' In this sense and also for other information on FAO's role in promoting agricultural innovation to increase food security, sustainable development, and promote rural development, see <www.fao.org/innovation/en/> accessed 1 August 2020.

26 See Chapter 3.

27 On the concepts and definitions of innovation, see HLPE, *Agroecological and other Innovative Approaches for Sustainable Agriculture and Food Systems that Enhance Food Security and Nutrition*, supra note 16, at 53.

28 According to the FAO, 'The role of research, development and innovation in the transition to sustainable agriculture cannot be overemphasized. Countries will need to assign very high priority to strengthening their agricultural education and innovation capacities, and ensure the availability of affordable technologies that are locally adapted.' See FAO, *Building a Common Vision for Sustainable Food and Agriculture*, supra note 25, at 22.

29 On 'the need to define key dimensions to assess sustainable innovation' in agriculture, which 'should not be only about technology nor productivity,' but also have 'a holistic and multidisciplinary perspective that includes social, economic, cultural, environmental and policy processes, and seek a positive impact on the lives of small-scale food providers, workers and their communities,' see FOEI, 'Agroecology: Innovating for Sustainable Agriculture & Food Systems' (2018) *Who Benefits? Series 1*.

30 See Golay and Bessa, *The Right to Seed in Europe. The United Nation Declaration on the Rights of Peasants and Other People Working in Rural Areas and the Protection of the Right of Seeds in Europe*, supra note 24, at 22–3; Paoloni and Vezzani, 'La Dichiarazione ONU sui diritti dei contadini e delle altre persone che lavorano nelle aree rurali: prime riflessioni,' supra note 4, at 14.

31 In 2018, the FAO's Committee on Agriculture affirmed that 'It is also important to recognize that innovation is not a goal *per se*, recognizing that some forms of innovation

may contribute to environmental degradation, be disruptive of livelihoods or exacerbate inequalities. It is important to understand which kinds of innovation need to be encouraged, where and for whom and how innovative approaches and innovation in general are conducive to the achievement of SDGs' (FAO Committee on Agriculture, 'Sustainable pathways to engage food and agriculture for the achievement of the 2030 Agenda for Sustainable Development' (1–5 October 2018) COAG/2018/3/Rev.1, 7). In this regard, the FAO Director-General Josè Graziano da Silva was particularly pertinent when he wrote: 'The Future of Agriculture is not input-intensive but knowledge-intensive. We need the integrated approach that agroecology can offer' (FAO, *FAO'S Work on Agroecology,* supra note 10, at 3).

32 FAO and IFAD, *United Nations Decade of Family Farming 2019–2028. Global Action Plan* (FAO and IFAD 2019) 10.

Selected bibliography

Altieri MA, *Agroecology: The Science of Sustainable Agriculture* (CRC Press 1995).

Gliessman SR, *Agroecology: The Ecology of Sustainable Food Systems* (3rd edn, Boca Raton, FL, USA, CRC Press, Taylor & Francis Group 2015).

Golay C, *Negotiation of a United Nations Declaration on the Rights of Peasants and Other People Working in Rural Areas* (Geneva Academy of International Humanitarian Law and Human Rights, 2015).

Hubert C, *The United Nations Declaration on the Rights of Peasants: A Tool in the Struggle for Our Common Future* (Centre Europe – Tiers Monde 2019).

Migliorini P and Wezel A, 'Converging and diverging principles and practices of organic agriculture regulations and agroecology' (2017) *Agronomy for Sustainable Development.*

Paoloni L and Vezzani S, 'La Dichiarazione ONU sui diritti dei contadini e delle alter persone che lavorano nelle aree rurali: prime riflessioni' (2019) 1 *federalismi.it* <www.federalismi.it/focus/index_focus.cfm?FOCUS_ID=104&focus=humanrights> accessed 31 August 2020, 14.

Vezzani S, 'Protection of Traditional Knowledge of Agricultural Interest in International Law' in Antonietta Di Blase and Valentina Vadi (eds), *The Inherent Rights of Indigenous Peoples in International Law* (Roma Tre-Press 2020).

13

THE UNDROP, BIOCULTURAL RIGHTS, AND SUSTAINABILITY STANDARDS FOR AGRI-FOOD SYSTEMS

Alessandra Di Lauro

Introduction

The implementation of the UN Declaration on the Rights of Peasants and Other People Working in Rural Areas (UNDROP) requires, in the eyes of an agricultural lawyer, the development of new, sustainable production and consumption models which correspond to a systemic, complex, holistic sustainability, without which the rights covered by the UNDROP would be left unprotected or would be distorted in their most significant contents. These rights include individual rights, collective rights, and rights conditioned by sustainability values and standards that can be largely associated with biocultural rights.

The idea of biocultural rights proposes a rethinking of sustainable development models by adding a further layer to the classic pillars of sustainability, i.e., the economic, social, and environmental. In this sense, the fourth pillar would incorporate the cultural dimension. A number of international documents adopted in the last three decades, after the 1992 UN Conference on Environment and Development, have shed light on the cultural dimension of sustainability and on the fact that national and international strategies for environmental conservation, human rights protection, and the pursuit of development should take culture and cultural manifestation duly into account.

The UNESCO Universal Declaration on Cultural Diversity (UDCD), adopted in Paris in November 2001, laid the first stone to the consolidation of the idea of biocultural rights.[1] The preamble of the UDCD points out that

> culture should be regarded as the set of distinctive spiritual, material, intellectual and emotional features of society or a social group, and that it encompasses, in addition to art and literature, lifestyles, ways of living together, value systems, traditions and beliefs.

DOI: 10.4324/9781003139874-17

This declaration reaffirms the definition of culture as proposed in the Mexico City Declaration on Cultural Policies, adopted in 1982. This definition has been reaffirmed in the World Commission Report on Culture and Development, 'Our Creative Diversity,' 1990, and in the Action Plan of the Intergovernmental Conference on Cultural Policies for Development held in Stockholm in 1998. More recently, the UNESCO Convention on the Safeguarding of the Intangible Cultural Heritage (ICH Convention) elaborated upon this notion of culture by defining intangible cultural heritage as:

> the practices, representations, expressions, knowledge, skills – as well as the instruments, objects, artefacts and cultural spaces associated therewith – that communities, groups and, in some cases, individuals recognize as part of their cultural heritage . . . it is created and transformed continuously by communities and groups . . . and it provides to communities a sense of identity and continuity.

Further on, the ICH Convention connects culture and the environment, by identifying 'knowledge and practices concerning nature and the universe' as an example of intangible cultural manifestations to be protected under this instrument.

In 2005, the *UNESCO Convention on the Protection and Promotion of the Diversity of Cultural Expressions emphasises the collective dimension as well as the intra and intergenerational aspects of culture and cultural rights by stating that cultural diversity* 'refers to the manifold ways in which the cultures of groups and societies find expression. These expressions are passed on within and among groups and societies.'[2]

The IV UNESCO World Forum on 'Culture and Food: Innovative Strategies for Sustainable Development' discussed the 'linkages between food, culture and society, as well as the evolving landscape of food security and food systems, and their pivotal role for the implementation of the 2030 Agenda for Sustainable Development.'[3] This forum paved the way to the adoption of the Parma Declaration on culture, food, and sustainable development in September 2019.[4]

The notion of culture emerging from these documents is ample and includes places, people, assets, and resources, both tangible and intangible. This notion emphasises the role that culture plays in the development of approaches that can be deemed sustainable. It also shows how pivotal the cultural aspect should be in environmental strategies. This dimension needs to be held in consideration at least at the same level as the economic, environmental, and social aspects. In fact, it is precisely the cultural dimension of sustainable development that fully reveals the most significant elements of the rights recognised by the UNDROP, that is, their partly conditioned, collective and widespread, biocultural nature.

For this reason, we will consider the evolution of biocultural rights and the relevance of the examination in the context and then the evolution of sustainable legal models of production and consumption.

Individual, collective, and biocultural rights

The UNDROP reaffirms the recognition of rights such as the right to life and personal integrity (Article 6 paragraph 1), freedom of expression, thought, belief, conscience, religion (Article 8), recognition as a person and freedom of movement (Article 7), and the right not to be tortured (Article 6 Paragraph 2). Peasants and rural communities are already holders of these rights under international law.[5] However, this reaffirmation strengthens the existing protections.[6] In some cases, UNDROP does not provide a mere mechanical repetition of rights that have been formulated elsewhere, but it is possible to appreciate a different sensibility towards already recognised rights that in the context of the UNDROP receive specific attention. Examples are Article 11 on the right to information, which concerns information about the prices of the factors of production and about different aspects related to agricultural policies and Article 21 on the right to water, focusing on the use of water resources for agricultural purposes.[7]

Another distinctive feature of the approach taken by the UNDROP is the articulated relationship that it weaves between individual and collective rights, that is, between the rights of individuals and the rights that can be understood as belonging to a community. This invites us to look at these rights not only as 'new rights,' which have an individual scope but also as collective, diffuse rights in that they are recognised on the basis of the criterion of equity within the same generation and between generations.[8]

It is clear that such a system of rights is strongly structured around the rural community as a community that, through a sustainable relationship with their territory, should ensure that any form of production takes place in harmony with the natural environment and local culture, in an inclusive social context that does not discriminate against the vulnerable community members (women, children, and the old and disabled people). However, even this description fails to give a full idea of the scope of the protected subjects. The UNDROP identifies a category of rights that may not only be attributable to peasants and/or to a group of kindred people who somehow share farm work and/or conditions of vulnerability, but rather with a community that, by direct and indirect reference also to other international sources, is potentially the widest possible, and it may be seen as encompassing the present and future human generations as well as the environment. It is not a matter, therefore, of trying to understand just how extended the notion of peasants is as laid down by this Declaration of 2018[9] or how broad the notion of agriculture contained therein is,[10] given that the protection of the rights laid down by the UNDROP should be of interest to the whole community (and to the future generations) as collective and diffused (as well as individual) rights.

This idea is corroborated by the realisation of how heavily such rights are 'conditioned' by claims for sustainable development. In our opinion, the text of the UNDROP lacks clarity on this point, and its wording seems to go back to an outdated concept of community rights. In this sense, Article 3.2 provides that: 'Peasants and other people working in rural areas have the right to determine and develop priorities and strategies to exercise their right to development.' However, it is clear that sustainability must be considered a key pillar of the UNDROP even where it is not explicitly mentioned. The rights listed in the UNDROP, including access to resources, the right to work, the right to education, and the right to food, cannot but be regarded as dependent on the implementation of sustainable practices, not only for the individuals and/or the community involved but also for the achievement of a sustainable context, even beyond the local level and even beyond the present generations.

Also, we must underline that, since the nineties, Principle 3 of the Rio Declaration on Environment and Development has highlighted that 'The right to development must be fulfilled so as to equitably meet developmental and environmental needs of present and future generations.' And again in 2003, the UN Commission on Human Rights reaffirmed that the enjoyment of 'the right to development as well as the respect for cultural diversity are essential for achieving sustainable development and ensuring that sustainable development benefits all' (Resolution 2003/71).

The perspective that makes sustainability a prerequisite for the fulfilment of rights should inspire the interpretation of the right of access to and use of natural resources, while also being a benchmark for the assessment of the suitability of living conditions (Article 5), the possibility of access to justice (Article 12), working conditions (Article 13), participation in decision-making processes, and determination of one's own food systems (Article 15). Sustainability, therefore, is a dimension that should always be present, at the same level as when it is clearly expressed, such as, for instance, in the rules that invite States to support 'sustainable uses' of water (Article 21) and to promote 'sustainable farming practices' (Article 16).

In this respect, we think that the bundle of rights that the UNDROP sets forth perfectly fits into the category of the so-called biocultural rights.[11] The concept of biocultural rights was developed by Sanjay Kabir Bavikatte in his seminal book Stewarding the Earth[12] while the term biocultural was initially coined by Darrel A. Posey (biocultural diversity).[13] Biocultural rights would characterise our entry into the Anthropocene, a term used for the first time by Crutzen to describe a new geological area,[14] that the Sub-Commission on Quaternary Stratigraphy 2019 defines as that 'interval of geological time in which the collective activities of human beings began to substantially alter many of the Earth's features and processes.'[15]

Biocultural rights are *sui generis* human rights as they are based on a double foundation: on the one hand, they are based on the protection of cultural identity and people's capacity for self-determination. On the other hand, they are built on the protection of the environment and the ecosystem that make such rights correspondent to responsibilities and duties. This attempt at finding integration and harmony between human rights and the ecological issue seems to contain elements of Western culture and cultural ingredients that seemingly belong to non-Western lifestyles, which maintain a non-individualistic approach and revolve around an ideal of harmony with nature that includes the interests of the future generations.

The recognition of such rights draws from environmental legal texts,[16] mainly the Rio Declaration (Principle 22) and Article 8 of the CBD, which encapsulate the idea that some communities' lifestyles are more in harmony with the conservation of the environment and so should be considered guardians of the environment that surrounds them.[17] However, biocultural rights have to reckon with the possibility that such communities do not have sustainable lifestyles at all or no longer wish to have them, and the understanding that such rights, instead, 'are rights that embrace responsibilities – duties – for their recipients – indigenous populations and local communities. A responsibility for being and remaining environmentally sustainable.'[18] It is precisely the double foundation of these rights that sparks off the debate about not levelling these rights to the category of human rights, in order to avoid sacrificing the environmental prerequisite of the biocultural rights when faced with the predomination of different human rights. In this regard, some propose to define such rights as having inherent limitations, so that they will fall within the so-called human rights but, as 'ecological human rights' will still preserve the possibility of being declined both from an anthropocentric perspective, with the interest of the present human beings prevailing over the interest of the environment, and from a non-anthropocentric perspective, with the interest of the environment prevailing over the interest of the present or future human beings.[19]

Having said that, I argue that certain rights laid down by the UNDROP should be included in biocultural rights and in this perspective, they can be seen as 'ecological human rights.' In fact, it would be unthinkable to imagine the UNDROP proposing a reading of biocultural rights that is ready to sacrifice the ecological interest to the interest of mere exploitation, based on an 'obsolete' view of development. Where the UNDROP refers to the right to development without mentioning 'sustainable development,' it underlies that such a right must be accomplished with the aforementioned double foundation in mind (protecting the community and protecting the ecosystem), so that the community's self-government continues to adhere to sustainable practices. This would therefore bring the rights laid down by the UNDROP into the category of biocultural rights, while providing an evolutionary reading that accommodates the objections raised against this category of rights. As Giulia Sajeva puts it:

> Biocultural rights are always accompanied by the idea that it is necessary to take seriously the environmental crisis, the scarcity of resources and the impact of mankind on the earth even when taking into consideration interests that are so important that they rise to the role of human rights. The elevation of the environment to an interest that is protected by the law suggests a different way to consider human rights, a way of thinking that is steeped in the characteristics of the Anthropocene . . . [and that] allows the entrance of a penetrating environmental limit on positions that, framed as indigenous rights, were not subject to this limit and which probably, taking into account the real responsibility for the actual environmental crisis, should not be subject to it.[20]

Such an approach has consequences, as it can pave the way to restrictions on lifestyles and customs and practices that are regarded as traditional in certain contexts and that are not sustainable (e.g., hunting and fishing). In this respect, we deem it essential to retain a level of flexibility to biocultural rights, just because such flexibility allows them to be looked at from an anthropocentric or non-anthropocentric perspective, depending on the context, the country, and the legal environment.[21]

Sustainability and food sovereignty

A practical dilemma stems from the conception of biocultural rights. What production and consumption model or models should be promoted in order to implement such rights? A starting point for the construction of such models could be Article 15 of the UNDROP, in that it affirms the right of peasants and other people working in rural areas to determine their own food and agricultural systems, which hints at the concept of food sovereignty.

The social process from which the concept of food sovereignty emerged is well known. Articulated in 1996 at the International Conference of the Coalition in Tlaxcala (Mexico), it was officially put forward in the United Nations system at the parallel Forum of the FAO World Food Summit in Rome, in November of the same year. Then, the concept of food sovereignty was defined as

> the right of peoples, communities, and countries to define their own agricultural, labour, fishing, food and land policies which are ecologically, socially, economically and culturally appropriate to their unique circumstances. It includes the right to food and to produce food, which means that all people have the right to safe, nutritious and culturally appropriate food, to the resources to produce it and the ability to sustain themselves and their societies.[22]

The notion of food sovereignty was also reaffirmed in the Declaration of Nyéleni, dated 27th February 2007,[23] the main outcome of the Forum for Food Sovereignty. This declaration revolves around different components that are clearly stated in Article 15 of the UNDROP, amongst which are the role of self-determination and participation in decision-making processes in agricultural and food policies; the right to healthy and culturally adequate food; and the right of access to and use of resources in ways that are ecologically safe, sustainable, and in keeping with local cultures. This definition of food sovereignty is a definition that complements those of food democracy,[24] food justice,[25] and procedural environmental rights.[26] It may also be interesting to combine food sovereignty with biocultural rights, because in the context of biocultural rights, it must be verified whether sovereignty is also exercised with respect for environmental interests.[27]

There can be many ways[28] for achieving food sovereignty and fulfilling the rights recognised by the UNDROP in the light of environmental preservation. The use of indicators of sustainable development is one of them. Economists have been struggling with indicators of sustainable development for a long time, and even now the answers are not univocal. The *Club of Rome* had already dealt with this problem in the attempt to find tools and criteria to measure sustainability, and more recently, it took up the issue again.[29] Reflections continue, and in November 2016, the European Commission published a Communication called *Next steps for a sustainable European future, encompassing the economic, social, environmental dimensions of sustainable development, as well as governance, within the EU and globally*[30] that outlines the process for the implementation of Agenda 2030. This communication was followed up by the Eurostat Report on *Sustainable development in the European Union – Monitoring report on progress towards the SDGs in an EU context – 2018 edition*, which, based on 100 indicators, shows statistics of the trends of each indicator in the 17 goals of Agenda 2030.[31] It is clear that finding the right indicators is still a challenge, since the concept of sustainable development is not only multidimensional but also dynamic.[32]

More food for thought could possibly be provided by an analysis of some new models of trademarks, logos, signs, and certification marks related to geographical indications, in particular those that are often based on solutions that embrace ethical-ecological-legal choices related to sustainability that somehow can be regarded as suggesting some pillars of the food sovereignty. In the panorama of trademarks (signs, design, label, logo, or expression, etcetera), which identifies products or services connected to the territory, some are strictly of a private nature. As they are still related to proprietary approaches, we think such models are the least suitable for building sustainable food systems, since they certainly provide no general freedom of access and no participation in the development of the signs, while both factors are essential to developing really sustainable models. Other signs (such as the designations of origin and geographical indications) have a public nature and there is much discussion about the role they can play as promoters of the rights of communities linked to the development of certain products.[33]

In this context, the implementation of the rights laid down by the UNDROP should account for those rule-making models that take inspiration from co-production and co-design processes, which help overcome the classic regulatory frameworks of legal normativity and some dichotomies, such as those between *hard law* and *soft law* or between the public and the private sphere, while questioning the governmental levels, the expressions of democracy, the places, and the ways of building consensus[34] that are public in scope.

Among the signs promoting a more global approach to the theme of sustainability, the expected introduction of a sixth sign in France is noteworthy. This 'sixth sign' would be added to the other five that, already existing in France, endorse, each one in its own sphere, a connection between the quality and the origin of food products. The five existing signs known as SIQO (*Signes Officiels de la Qualité et de l'Origine*)[35] include signs that comply with some European schemes, such as Protected Designation of Origin (PDO), Protected Geographical Indication (PGI), Traditional Specialities Guaranteed (TSG),[36] organic agriculture,[37] and the '*label rouge*,' which applies in France only and can also be used for agricultural non-food products, non-processed farm products,[38] and foods that also bear the PGI or TSG marks but not the PDO.[39] The official acknowledgement of the need to test a sixth quality and origin label for foods can be found in the French *Avis du Conseil économique, social et environnemental* (CESE) of June 2018,[40] developed by the *Section de l'agriculture, de la pêche et de l'alimentation* of CESE.

Recommendation no. 6 in the *Avis Les signes Officiels de la Qualitè et l'Origine* calls for

> testing the creation of a new SIQO sign in France, known as *Agriculture biologique locale et équitable*, that includes the current agronomic characters of organic farming and adds criteria regarding proximity between places of production, processing, and commercialisation, the life cycle analysis, the phrase *Haute Valeur Environnementale* (HVE), corporate social responsibility.

As shown in this case, the so-called sixth sign considers not only the farming methods (organic farming) but also a number of elements that could help implement an agricultural model close to the one envisioned in the UNDROP. These further elements are proximity between the places of production, the awarding of the HVE label (a French certification that is only awarded to those who can prove they meet environmental requirements also in terms of impact of production and consumption of resources[41]), the life cycle analysis of the product (and therefore its impact in terms of footprint), corporate social responsibility and related human rights standards, employment practices, the environment (e.g., biodiversity, climate change, resource efficiency, life cycle analysis, and pollution prevention), and local communities engagement.[42] In addition, the creation of such a sign is part of a project that could be defined as widely participatory, since French national confederations of labour, professional and independent trade unions, members of professional unions, associations of farming families, different 'qualified persons,' student organisations, and the *Mouvements de jeunesse*, which usually sit in the *Section de l'agriculture, de la pêche et de l'alimentation*, were all involved. Other parties, such as members of the *Institut national de l'origine et de la qualité* (INAO), of the European Commission, of the National Federation of Rural Families, the *Centre de Recherche pour l'Étude et l'Observation des Conditions de Vie* (CRÉDOC), were listened too as well, in the drafting of the aforementioned *Avis*.[43]

In the same vein, another example could be the 'Slow Food Presidia' registered brand, which, until recently, only applied to the Presidia[44] but has now been extended to the products from the Presidia as well. In this specific case, the brand mark is only granted to those producers or groups of producers who create a self-certification, which is 'controlled' by consumers who are members of the network, based on the association's guidelines, as in a sort of *'participatory certification.'*[45]

From the perspective of biocultural rights, these models and signs should not only be examined for their public or private nature and levels of participation in the design and management of the signs but also in respect of environmentally friendly practices.

Concluding remarks

The implementation of the UNDROP requires the development of a new model of sustainable production and consumption. In fact, the rights recognised by the Declaration are tightly connected to the claims for sustainable development that can be found in different passages of its text. Starting from this premise, we have tried to highlight the central role of the cultural dimension of sustainable development that should characterise the rights recognised in the UNDROP. These rights can then be fitted into the category of the so-called biocultural rights that, in turn, are linked to the notion of food sovereignty which is referred to in the UNDROP. We feel, therefore, that in order to implement the UNDROP and to pursue food sovereignty, it would be useful to have at our disposal specific indicators[46] that take due account of the role that culture plays in the achievement of sustainable development.

This is in tune with the current developments proposing a re-think of payments for ecosystem services towards the recognition of rights for ecosystem services.[47] The term 'payment for ecosystem services' covers different schemes linked to the payment to persons who contribute to the protection of an ecosystem.[48] This system has been increasingly criticised given that it is essentially based on the distribution of economic benefits and the attribution of an economic value to the environment.[49] The recognition of a bundle of rights for ecosystem services is under discussion. These would be based on the recognition of rights (e.g., collective ownership and access to seed banks), not merely monetary compensation, in exchange for ecosystem services and for acting as guardians of the environment.[50] These rights would become a flexible tool to be used by states and non-state organisations alike.[51]

The journey in search of sustainable models is definitely a journey that deserves to be undertaken for present and future generations, for indigenous and non-indigenous peoples, and for the environment and all living beings.

Notes

1 UN Educational, Scientific and Cultural Organisation, 'UNESCO Universal Declaration on Cultural Diversity' (2 November 2001).
2 UN Educational, Scientific and Cultural Organisation, 'UNESCO Convention on the Protection and Promotion of the Diversity of Cultural Expressions' (20 October 2005), art. 4.
3 See UNESCO, 'Culture and Food: Innovative Strategies for Sustainable Development – Programme' (2019) <https://en.unesco.org/system/files/agenda_forum_parma_provisional_en_final.pdf> accessed 25 May 2021.
4 UN Educational, Scientific and Cultural Organisation World Forum 'Culture and Food: Innovative Strategies for Sustainable Development. Parma Declaration' (13 September 2019).
5 UN Human Rights Council, 'United Nations Declaration on the Rights of Peasants and Other People Working in Rural Areas' (8 October 2018) UN Doc A/HRC/RES/39/12, expressly refers to the following instruments in which these rights have been proclaimed: the Universal Declaration of Human Rights, the International Convention on the Elimination of All Forms of Racial Discrimination, the International

Covenant on Economic, Social and Cultural Rights, the International Covenant on Civil and Political Rights, the Convention on the Elimination of All Forms of Discrimination against Women, the Convention on the Rights of the Child, the International Convention on the Protection of the Rights of All Migrant Workers and Members of Their Families, relevant conventions of the International Labour Organization, the Declaration on the Right to Development, and the United Nations Declaration on the Rights of Indigenous Peoples.

6 Stefania Errico and Priscilla Claeys, 'Human Rights and the Commons: Bridging Gaps and Exploring Complementary Approaches to the Governance of Land and Natural Resources: Paper Presented at the International Association for the Study of the Commons Conference in Utrecht in July 2017' (2017) *XVI Biennial IASC Conference*.

7 Lorenza Paoloni and Simone Vezzani, 'La Dichiarazione ONU sui diritti dei contadini e delle altre persone che lavorano nelle aree rurali: prime riflessioni' (2019) 1 *federalismi.it* <www.federalismi.it/focus/index_focus.cfm?FOCUS_ID=104&focus=humanrights> accessed 31 August 2020, 24.

8 Anna Coote, 'Intergenerational Equity Briefing Review of Social Determinants of Health and the Health Divide in the WHO European Region' (WHO 2015).

9 Marc Edelman, 'What Is a Peasant? What Are Peasantries? A Briefing Paper on Issues of Definition' (2013) *First Session of the Intergovernmental Working Group on a United Nations Declaration on the Rights of Peasants and Other People Working in Rural Areas*.

10 UNDROP, supra note 5, art. 1, par. 2.

11 Giulia Sajeva, 'Rights with Limits: Biocultural Rights – between Self-determination and Conservation of the Environment' (2015) 6(1) *Journal of Human Rights and the Environment* 30ff; Giulia Sajeva, *When Rights Embrace Responsibilities. Biocultural Rights and the Conservation of Environment* (Oxford University Press 2018).

12 Sanjay Kabir Bavikatte, *Stewarding the Earth. Rethinking Property and the Emergence of Biocultural Rights* (Oxford University Press 1999).

13 Addison Darrel Posey (ed), *Cultural and Spiritual Values of Biodiversity* (United Nations Environmental Programme & Intermediate Technology Publications 1999).

14 Paul J. Crutzen and Eugene F. Stoermer, 'The "Anthropocene"' (2000) 41 *IGBP Newsletter*.

15 Subcommission on Quaternary Stratigraphy, 'Working Group on the "Anthropocene"' (2019) <http://quaternary.stratigraphy.org/working-groups/anthropocene/> accessed 25 May 2021.

16 United Nations Conference on Environment and Development, Agenda 21 (14 June 1992) A/CONF.151/26, vol. II; the Convention on Biological Diversity (adopted 5 June 1992, entered into force on 29 December 1993) 1760 UNTS 69 (CBD) (in particular article 8j); the Nagoya Protocol on Access to Genetic Resources and the Fair and Equitable Sharing of Benefits Arising from their Utilization to the Convention on Biological Diversity (adopted 29 October 2010, entered into force 12 October 2014) 30619 UNTS 27; International Treaty on Plant Genetic Resources for Food and Agriculture (adopted 3 November 2001 entered into force on 29 June 2004) 2400 UNTS 303 (ITPGRFA).

17 On this point, see Adriana Bessa and Jérémie Gilbert in this volume.

18 Giulia Sajeva, 'Un passo avanti e un passo indietro nell'Antropocene: rights for ecosystem services, comunità locali e REDD' (2019) 19(1) *Diritto e questioni pubbliche* 277.

19 Ibid., 275–89; Sajeva, *When Rights Embrace Responsibilities. Biocultural Rights and the Conservation of Environment*, supra note 11.

20 Sajeva, 'Un passo avanti e un passo indietro nell'Antropocene: Rights for Ecosystem Services, comunità locali e REDD,' supra note 18, at 280.

21 Sajeva, *When Rights Embrace Responsibilities. Biocultural Rights and the Conservation of Environment*, supra note 11, at cap. 2.

22 NGO/CSO Forum for food sovereignty, *Food Sovereignty: A Right For All. Political Statement of the NGO/CSO Forum for Food Sovereignty. The Failure Since 1996 and the New Official Declaration*, Rome (2002).

23 Declaration of the Forum for Food Sovereignty, Nyéléni (2007).

24 Neva Hassainen, 'Practicing Food Democracy: A Pragmatic Politics of Transformation' (2003) 19(1) *Journal of Rural Studies* 77–86; Tim Lang, 'Food Control or Food Democracy? Re -engaging Nutrition with Society and the Environment' (2005) 8(6a) *Public Health Nutrition* 730–37. See Smita Narula in this volume.

25 Robert Gottlie and Anupama Joshi, *Food justice* (MIT Press 2010) 320.

26 Svitlana Kravchenko, 'Procedural Rights as a Crucial Tool to Combat Climate Change' (2010) 38 *Georgia Journal of International & Comparative Law* 613ff; Benjamin W. Cramer, 'The Human Right to Information, the Environment and Information about the Environment: From the Universal Declaration to the Aarhus Convention' (2009) 14(1) *Communication Law and Policy* 73ff.

27 In our opinion, even the recent European Union Commission's Communication (Communication from the Commission, A Farm to Fork Strategy for a fair, healthy and environmentally-friendly food system COM/2020/381 final) 'A Farm to Fork' seems to support the central role played by the complex notion of food sovereignty as a model to understand and implement a transition to sustainable food contexts that should be inspired by 'policy coherence' and 'mainstream sustainability in all food-related policies and strengthen the resilience of food systems.' Moreover, when the European Commission at 4 refers to 'food environment,' it explains the concept with these words: 'The "food environment" is the physical, economic, political and socio-cultural context in which consumers engage with the food system to make decisions on acquiring, preparing and consuming food' (in doing so the Commission refers to the 2017 report by the High-Level Panel of Experts on Food Security and Nutrition titled Nutrition and food systems).

28 Luc Bodiguel, 'Agriculture durable: la poursuite de la transition juridique' in Benoît Grimonprez (ed), *La réforme du droit foncier rural: demander l'impossible* (LexisNexis 2018) 212; Eloisa Cristiani, 'Modelli di agricoltura "sostenibile" con particolare attenzione al settore vitivinicolo' in Roman Budzinowsk (ed), *Contemporary Challenges of Agricultural Law: Among Globalization, Regionalization and Locality* (Poznan 2018) 133–43; Alessandra Di Lauro, *Le denominazioni geografiche protette e le indicazioni geografiche protette di fronte alla sfida dello sviluppo sostenibile* (2018) 97(3) *Rivista di Diritto Agrario* 382; Matteo Ferrari, 'Profili di auto-regolazione nella tutela del paesaggio vitivinicolo' (2019) 98(3) *Rivista Di Diritto Agrario* 456.

29 On this point, Jorgen Randers et al, *Transformation is Feasible. How to Achieve the Sustainable Development Goals within Planetary Boundaries. A report to the Club of Rome, for its 50 Years Anniversary 17 October 2018* (Stockholm Resilience Centre 2018).

30 Communication from the Commission to the European Parliament, the Council, the European economic and social Committee, and the Committee of the regions, *Next steps for a sustainable European future. European action for sustainability*, COM(2016) 739 final, SWD(2016) 390 final.

31 Eurostat, *Sustainable Development in the European Union: Monitoring Report on Progress towards the SDGS in an EU Context* (Publications Office of the European Union 2017).

32 Also, ASVIS (Italian Agency for Sustainable Development) is working on the indicators of the different Goals of Agenda 2030, <https://asvis.it/public/asvis2/files/Comunicati_stampa/CS_Rapporto_ASviS_2019_FINAL.pdf>.

33 FAO, *Strengthening Sustainable Food Systems through Geographical Indications. An Analysis of Economic Impacts* (FAO 2018); FAO, *Identification of Origin-linked Products and Their Potential for Development. A Methodology for Participatory Inventories* (FAO 2015); Di Lauro, *Le denominazioni geografiche protette e le indicazioni geografiche protette di fronte alla sfida dello sviluppo sostenibile,* supra note 28, at 382.

34 The literature on co-production and co-design is vast. See, in particular, on the forms of government of science: Robert King Merton, 'Science and Democratic Social Structure' in *Social Theory and Social Structure* (The Free Press 1968) 604–15; Sheila Jasanoff, 'The Essential Parallel between Science and Democracy' (2009) *Seed Magazine*; Sheila

Jasanoff, 'Science and Democracy' in Ulrike Felt, Rayvon Fouché, Clark A. Miller and Laurel Smith-Doerr (eds), *The Handbook of Science and Technology Studies* (The MIT Press 2017) 259ff; Mariachiara Tallacchini, 'Scienza e diritto. Prospettive di co-produzione' (2012) 1(2) *Rivista di filosofia del diritto* 313; Mariachiara Tallacchini, 'Dalla «scienza come democrazia» alle «società democratiche della conoscenza»' (2017) 32(126) *Notizie di Politeia* 6; Mariachiara Tallacchi, 'Il governo della scienza dall'autoreferenzialità alle interazioni sistemiche tra scienza, policy e democrazia' (2018) 4 *Rivista di filosofia Neo-Scolastica* 727; Véronique Champeil-Desplats, Jérome Porta and Laurent Thévenot, 'Introduction: une experience de recherche cooperative et transverse entre droit et sciences sociales' (2019) 16 in *La Revue des droits de l'homme*. See also on Legal design: <*www.legaltechdesign.com/*>.

35 Article L. 640–2 du *Code rural et de la pêche maritime* (amended by Ordonnance n. 2015–1246, 7 octobre 2015 – art. 2 e art. 6).

36 Regulation (EU) 1151/2012 of the European Parliament and of the Council of 21 November 2012 on quality schemes for agricultural products and foodstuffs (2012) OJ L 343/1.

37 Regulation (EU) 2018/848 of the European Parliament and of the Council of 30 May 2018 on organic production and labelling of organic products and repealing Regulation (EC) No 834/2007 (2018) OJ L 150/1; Council Regulation (EC) 834/2007 of 28 June 2007 on organic production and labelling of organic products and repealing Regulation (EEC) No 2092/91 (2007) OJ L 189/1.

38 Article L. 641–1 du *Code rural et de la pêche maritime* (modified with Ordonnance n. 2006–1547, 7 décembre 2006, art. 2, JORF 8 décembre 2006).

39 Article L. 641–2 du *Code rural et de la pêche maritime* (modified with n. 2010–459, 6 mai 2010 – art. 4) and *Code rural et de la pêche maritime*, art. R. 641–10.

40 Full text available at: <www.lecese.fr/sites/default/files/pdf/Avis/2018/2018_18_qualite_origine_produits_alimentaires.pdf>.

41 With law no. 2010–788 of 22 July 2010 on the national commitment for the environment ('engagement national pour l'environnement'), France has adopted an environmental certification called Haute Valeur Environnementale (HVE). HVE certification is part of the list of mentions valorisantes that can be affixed to food products to indicate their quality and origin, and is included in Article L. 611–6 of the Code rural et de la pêche maritime; the modalities of its recognition have been specified by decrees n. 2016–2011 of 30 December 2016 and n. 2011–1914 of 20 December 2011, now included in Articles D. 617–1 et seq. of the Code rural et de la pêche maritime. This environmental certification of agricultural enterprises constitutes a voluntary path articulated around four main axes: biodiversity, phytosanitary strategy, fertilization management, and water management. It is a progressive commitment divided into three levels, the third of which is considered the level of excellence called Haute Valeur Environnementale (HVE).

42 Communication from the Commission to the European Parliament, the Council, the European economic and social Committee, and the Committee of the regions A renewed EU strategy 2011–14 for Corporate Social Responsibility, Brussels, 1 COM(2011) 681 final 'According to these principles and guidelines, CSR at least covers human rights, labour and employment practices (such as training, diversity, gender equality and employee health and well-being), environmental issues (such as biodiversity, climate change, resource efficiency, life-cycle assessment and pollution prevention), and combating bribery and corruption. Community involvement and development, the integration of disabled persons, and consumer interests, including privacy, are also part of the CSR agenda. The promotion of social and environmental responsibility through the supply-chain, and the disclosure of non-financial information, are recognised as important cross-cutting issues. The Commission has adopted a communication on EU policies and volunteering in which it acknowledges employee volunteering as an expression of CSR.1,' 5.

43 On the so-called sixth sign and the HVE sign see Alessandra Di Lauro, 'Segni e territorio: quale co-design per quali modelli giuridici' (2020) 1(1) *Rivista di diritto agroalimentare* 51–80.

44 Serena Milano, Raffaella Ponzio, Piero Sardo, *I presìdi Slow Food – Come avviare un Presidio, stabilire relazioni con i produttori, organizzare le attività del progetto* (Slow Food 2018) 5.

45 Ibid., 32; Silvia Rolandi, 'Slow Food e il progetto dei Presìdi, verso l'adozione di sistemi partecipativi di garanzia?' in Alessandra Di Lauro and Giuliana Strambi (eds), *Le funzioni sociali dell'agricoltura* (ETS 2020). On participatory guarantee systems and the definition, see IFOAM, 'Participatory Guarantee Systems: 5 Case studies from Brazil, India, New Zealand, USA, France' (2008) <www.ifoam.bio/our-work/how/standards-certification/participatory-guarantee-systems> accessed 26 May 2021; Paulo Niederle, Allison Loconto, Sylvaine Lemeilleur and Claire Dorville, 'Social Movements and Institutional Change in Organic Food Markets: Evidence from Participatory Guarantee Systems, in Brazil and France' (2020) 78 *Journal of Rural Studies* 282–91.

46 See the text mentioned earlier about indicators.

47 Sadra Díaz, Unai Pascual, Marie Stenseke et al, 'Assessing Nature's Contributions to People: Recognizing Culture, and Diverse Sources of Knowledge, Can Improve Assessments' (2018) 359(6373) *Science* 270ff.

48 Sven Wunder, 'Payments for Environmental Services: Some Nuts and Bolts' (2005) 9 *Center for International Forestry Research*; Kanchan Chapara, Rik Leeman, Kumar Pushpam and Simons Henk, *Ecosystems and Human Well-Being: Policy Responses* (Volume 3, Findings of the Responses Working Group of the Millennium Ecosystem Assessment, Island Press 2005).

49 FAO, *Payments for Ecosystem Services and Food Security* (FAO 2011); Joshua Farley and Robert Costanza, 'Payments for Ecosystem Services: From Local to Global' (2010) 69(11) *Ecological Economics* 2060–8; Sven Wunder, Stefanie Engel and Stefano Pagiola, 'Taking Stock: A Comparative Analysis of Payments for Environmental Services Programs in Developed and Developing Countries' (2008) 65(4) *Ecological economics* 834–52.

50 Díaz, Pascual, Stenseke et al, 'Assessing Nature's Contributions to People: Recognizing Culture, and Diverse Sources of Knowledge, Can Improve Assessments,' supra note 47.

51 Rights for Ecosystem Services (RES): A Framework to Protect the Environment and Sustainable Local Communities in the EU (Giulia Sajeva MSCA - IF 2018), Principal investigator, Elisa Morgera <https://cordis.europa.eu/project/id/841546/it>.

Selected bibliography

Bodiguel L, 'Agriculture durable: la poursuite de la transition juridique' in Benoît Grimonprez (ed), *La réforme du droit foncier rural: demander l'impossible* (LexisNexis 2018).

Champeil-Desplats V, Porta J and Thévenot L, 'Introduction: une experience de recherche cooperative et transverse entre droit et sciences sociales' (2019) 16 *La Revue des droits de l'homme*.

Di Lauro A, 'Segni e territorio: quale co-design per quali modelli giuridici' (2020) 1(1) *Rivista di diritto agroalimentare*.

Sajeva G, 'Rights with Limits: Biocultural Rights – between Self-determination and Conservation of the Environment' (2015) 6(1) *Journal of Human Rights and the Environment*.

Sajeva G, *When Rights Embrace Responsibilities. Biocultural Rights and the Conservation of Environment* (Oxford University Press 2018).

PART IV

The declaration on the ground

14

LAND TENURE

Assessing Kenya's legal approach and the practice towards landlessness amongst rural smallholder farmers including women's rights

Robert Kibugi

Introduction

The Third Medium Term Plan for implementing Vision 2030, Kenya's current national development plan, recognises that land is a key driver of rapid economic transformation with secure access to land, proper land use planning, and equitable access to land being essential ingredients to food security.[1] Recognition of their tenure rights on natural resources (land, fisheries, etc.) is central to effectively fulfil the role that small-scale farmers are given in the national development plan and enable the protections they are accorded under national and international law. Smallholder farmers require land for agricultural and other livelihood activities. In Kenya, the average smallholder farm size is 1.16 acres, with a range of 1.2–12 acres, and accounting for 66% of all the farms nationally.[2]

Therefore, this chapter focuses on the land rights of smallholder farmers and other rural workers in Kenya. Using the provisions of the UNDROP to map out key obligations, the chapter examines how these rights to land and its use are protected under Kenyan law. This gives valuable indications on the relevance and suitability of the UNDROP as a legal tool for protecting the land rights of smallholder farmers and other rural workers in Kenya. Specifically, the chapter pays attention to the problem of landlessness amongst rural people who rely on land as a basic factor of production for subsistence. Consideration is given to special circumstances facing the land rights of women in law and practice, especially the rights of widows during succession of land ownership in rural communities.

Right to land in Kenya's Constitution

Kenya's Constitution (Article 40) guarantees a human right for all persons to own and hold property of any description in any part of the country.[3] While traditionally

DOI: 10.4324/9781003139874-19

this right includes land (and other types of property), its scope was expanded by the High Court in 2018 such that property under Article 40 now includes traditional fishing rights.[4]

This right must, however, be activated by acquisition of land, purchase of land, succession to property of a deceased person, transfer through inheritance, ascertainment of customary rights (through adjudication of rights), and occupation of public land or private land belonging to others (squatters on private or public land).

Once lawfully acquired, land rights are capable of protection under Article 40(3) of the Constitution, from deprivation by the State, or any other person. Compulsory acquisition (eminent domain) is only permitted for a public purpose, or in the public interest, and upon just compensation that is paid promptly.[5] Regarding this aspect, the first issues arise. First, amendments to the Land Act in 2019 have permitted the State to take possession of compulsorily acquired land while giving leeway for compensation to be paid within a period not exceeding one year.[6] The effect is that if this compulsory acquisition involves all or most of a person's, family's, or community's land, it could result in landlessness in the intervening period before the compensation (which includes resettlement) is paid. Second, the compensation, to be just, should exceed the mere value of the land, to cover the whole economic injury caused to the landowner(s) as a consequence of acquisition.[7] This is important to avoid constructive landlessness whereby even with resettlement to other land, it may be unsuitable for their livelihoods (e.g., land too dry for farming) or inadequate (e.g., land too small) as a result of lesser valuation.

There is no constitutional protection of unlawfully acquired land.[8] However, judicial interpretation by the Supreme court has provided guidance on how squatters on public land should be treated. As it will be explained later, the doctrine of adverse possession has been applied by the Court of Appeal over private land, subject to strict legal requirements, which has argued it may resolve some of the landlessness since the (erstwhile illegal) squatters are using the land productively, whereas the lawful owners were not doing so for 12 years. Before addressing this case, the next section gives an account of the historical roots of landlessness.

Landlessness in rural Kenya from history to the present and potential legal solutions

Kenya has a problem with landlessness in rural areas, which has, in certain instances, resulted in individuals and families occupying private land. Landlessness is defined by Kenya's land policy as 'a situation in which a person has no access to land necessary for their sustainable livelihood and has no means of acquiring such access.'[9] Landlessness is the converse or opposite of having land tenure rights that would legally confer entitlements and define who can use what resources for how long, and under what conditions.[10] Land tenure rights are a source of incentive, opportunity, equity, and patterns in land use,[11] consistent with the Constitution which provides that land in the country shall be held, used, and managed in a manner that is equitable, efficient, productive, and sustainable.[12] Resolving landlessness surpasses merely acquiring tenure rights but extends to ensuring security of those rights, which is the certainty that a person's rights to land will be recognised by others and protected in cases of specific challenges.[13] Landlessness has given rise to the problem of squatters in rural areas and informal settlements (also known as slums) in urban areas.

Colonial origins of landlessness and squatters

A squatter is defined by Kenya's land policy as a person who occupies land that legally belongs to another person or institution without the owner s consent.[14] The squatter problem in rural Kenya is traceable to decisions and legal enactments by British Colonial authorities which started with relocation of the radical title (ultimate ownership) to land from indigenous inhabitants of present-day Kenya to the British Crown.[15] This resulted in expropriation of land held by the indigenous population and its allotment to white settlers and other private agencies who otherwise would not have qualified to receive that land under customary law.[16] A law was enacted which, in consequence, meant that only allottees received land under English Property law, and the indigenous population lost claims to all land to the colonial power, becoming in effect, mere tenants at the will of the crown.[17] A similar situation arose along the Kenyan coast where land held by the Sultan of Zanzibar (not privately allocated but in reality taken from indigenous populations) was transferred to the British crown, and the indigenous populations were invited to submit claims.[18] Most people did not do so because they were unaware of the 1908 ordinance that required submission of claims, and failing to do so the land became crownland.[19]

Kenya's mismanaged land resettlement after independence

In the 1960s, after independence, Kenya set up land consolidation and adjudication laws and procedures to undertake first registration for people with ascertainable ancestral land claims over returned land.[20] Settlement schemes were set up to take pressure off for land redistribution exerted by the landless and squatters.[21] As an example, the Squatter Settlement Schemes (renamed Haraka) were carried out as low-cost settlement schemes primarily on mismanaged or abandoned farms in former European areas and were designed principally for subsistence cultivation.[22] Individual families received between 9 and 12 acres including 6 acres of arable land. But a study in one of those Squatter settlement schemes, Ngoliba, by the Mbithi & Barnes study found that the unrealistic size of the plots, the poor soils, and overall ecology made the settlement scheme unviable as the new owners could barely meet their basic food needs.[23]

These settlement schemes soon gave way to organised market-driven redistribution programmes supported by the Agricultural Development Bank and the

Agricultural Development Corporation.[24] Kanyinga argues that the resettlement programme had a class dimension too as those with the ability to pay or those with capital began to accumulate more land at the expense of the poor.[25] It also became corrupt. A 2004 Commission of Inquiry reported that its most glaring finding with regard to settlement schemes was that land was allocated to personalities who were entirely undeserving as they were neither 'landless' nor in possession of any unique skills and facilities to be able to use the land in an agriculturally productive manner to the benefit of the country's economy.[26]

Official policy on land allocation for resettlement was disregarded as 'politically correct' individuals in the government were allocated lands in settlement schemes at the expense of the deserving poor from the respective areas.[27] The goal of the resettlement effort was thus defeated from the outset, translating into a basis for ethnic conflicts.[28] Thus in present times, the landed elites are in control of both the politics and the economy in the highlands and they finance violence to deflect attention from inequalities in land ownership to inequalities in distribution of political power.[29] A 2014 World Bank report estimated that, in Kenya, three powerful political families are estimated to own more than 1 million acres of rural land, while at least 4 million rural Kenyan citizens are landless and at least 11 million own less than 1 hectare.[30]

In 2018, the Kilifi county government reported that the formalisation of squatter settlements in the privately registered land in the ten-mile coastal strip remains a thorny issue.[31] Kwale County, also on Kenya's Coast, reported in 2018 that it has historically faced a land issue, with the majority of the land held by absentee landlords and, as a result, there are high incidences of landlessness, with many families with no access to land having to live as squatters in public or private land.[32] Most households are landless specifically in Matuga and Msambweni sub-counties such that 16.6% of households were located on land where the family were squatters.[33] Taita Taveta County reported that a huge chunk of the county is occupied by private farms and ranches.[34] This, together with the long process of land adjudication and settlement in schemes, has contributed to the increase in and high rate of landlessness.[35] A 2013 Truth, Justice and Reconciliation Commission (TJRC) Report stated that, in Taita Taveta county, one family that owned more than 30,000 acres of land offered this to the government between 1996 and 2003, for purchase at the low rate of only KSh600 per acre to settle landless squatters.[36] However, after purchasing the land, the government began to settle other people, instead of the local community for whom the land was intended.[37] The county government argues that the long process of land adjudication and settlement in schemes has also contributed to the increase in landlessness.[38] Nakuru County, in the Rift Valley, also reported in 2018 that just under 20% of households are considered to be landless, including those evicted from the Mau Forest complex; and people who were internally displaced by Kenya 2007–8 post-election violence.[39] The effect of this failed land resettlement programme is that many smallholders farmers in rural Kenya and the coastal belt continue to experience landlessness and live as squatters.

The role of adverse possession in promoting equitable and productive ownership and use of land for the landless

The problem of rural landlessness has created the secondary challenge of squatters, as people and their families occupy any available land for purposes of settlement, and livelihood activities such as smallholder farming or livestock keeping. Legally, no land rights crystallise in favour of squatters, and without public policy decisions to resettle them, the private owners of the land can trigger evictions. However, in an important judgement in 2015, in *Mtana Lewa v Kahindi Ngala Mwagandi*, on adverse possession and squatters, the Court of Appeal of Kenya has refined how this can support equitable and productive land redistribution. In the case, one Kahindi Ngala had made a claim of adverse possession on 16 acres of land belonging to Mtana Lewa, arguing to have been a squatter on the land for more than 12 years. The judges linked the validity of adverse possession to the constitutional perspective that land should be used productively. The court argued that according to the Constitution, all land belongs to the people of Kenya, whether as a nation, communities, or individuals, and further proclaims that land in Kenya shall be held, used, and managed in a manner that is equitable, efficient, productive, and sustainable.[40]

According to the Court, Kenya's land policy aims, among other things, to secure the rights over land while at the same time guaranteeing all citizens the opportunity to access and beneficially occupy and use land; allocation and use of land that is economically viable, socially equitable and economically sustainable; and efficient and effective utilisation of land and land-based resources.[41] The Court of Appeal ruled that the law on adverse possession gives a land owner a period of 12 years to assert their rights over land. Where such a landowner instead allows this land to be occupied by another person, openly, peacefully, as of right, and with a clear intention of dispossessing him *(animus possidendi), the* landowner cannot subsequently claim that this law is in conflict with the purpose, values, and principles of the Constitution.[42] Other case law in Kenya nonetheless has imposed restrictions on the application of adverse possession in certain instances:

(i) A person cannot claim to have acquired land by adverse possession if they claim that the land they are occupying is their ancestral land on which they were born and brought up and that the registered owner has never been in possession of such land.[43] In this case, the claimant should seek first registration of private land rights through ascertainment of that unregistered customary land through a process of adjudication.[44] For community land (owned collectively), the registration of previously unregistered community land should be preceded by registration of the community 'that is claiming an interest in or right over community land' which is then adjudicated, surveyed, demarcated, and registered.[45]

(ii) A person who seeks to acquire title to land, by the method of adverse possession for the applicable statutory period (12 years), must prove non-permissive or non-consensual actual, open, notorious, exclusive, and adverse use by him or those under whom he claims for the statutory prescribed period without interruption.[46] Possession and occupation with the owner's permission do not give rise to adverse rights.

This application of adverse possession by the Court of Appeal as an equitable remedial tool to fulfil the constitutional promise of equitable land management is therefore innovative and warrants further investigations to ascertain the extent of its assertion by squatters across the nation. In any event, since utilisation of adverse possession requires adversarial judicial petitions that many landless people may not be inclined to file or have access to, the government ought to assess the state of landlessness and develop a suitable programme to address it as promised by Kenya's land policy.[47] The Court of Appeal did, however, mark a jurisprudential milestone, recognizing adverse possession as one of the reasons contemplated in article 24 of the Constitution as permissible bases to limit fundamental human rights and freedoms, including the right to acquire and own land.[48]

Protection of the rights of landless people occupying public land

In January 2021, the Supreme Court of Kenya pronounced itself on rights of persons occupying public land, without title. This was in *Mitu-Bell Welfare Society v Kenya Airports Authority & 2 others* which was a final appeal by members of *Mitu-Bell Welfare Society* representing residents of Mitumba Village, with 3065 households or approximately 15,325 men, women, and children, which was situated on (public) land belonging to Wilson Airport in Nairobi and who after receiving a seven-day notice to vacate the land, sought protection of the courts.[49] The court recognised that each person in Kenya enjoys a right to accessible and adequate housing, but that the problem arises because realisation of the right depends on the availability of land and other material resources, and predicated one's ability to own land.[50] According to the court, this scenario has resulted in informal settlements, which describes a habitation by people considered landless, with some structures on private or public land, with a permanent threat of eviction by the private owners or state agencies.[51] While the illegal occupation of private land cannot create prescriptive rights over that land in favour of the occupants, the same cannot be said of a similar occupation of public land.[52] This, according to the Supreme Court, is because when landless persons occupy public land and establish homes on it, they acquire not a title to the land but a protectable right to housing on the same.[53] The rationale of this holding is because under Kenya's Constitution (article 61(1)), 'all land in Kenya belongs the people of Kenya collectively as a nation, communities and individuals,' and the Constitution 'now creates a specific category of land known as public land. Therefore, every individual as part of the collectivity of the Kenyan nation has an interest, however indescribable, however unrecognizable, or however transient, in public land.'[54]

Thus, according to the Supreme Court, this right to housing on public land crystallises by virtue of a long period of occupation by people who have established homes and raised families on the land and the right derives from the principle of equitable access to land under Article 60 (1) (a) of the Constitution.[55] When faced with an eviction on grounds of public interest, such potential evictees have a right to petition the Court for protection and this need not necessarily be in the form of an order restraining the State agency from evicting the occupants, given the fact that the eviction may be entirely justifiable in the public interest.[56] Indeed, the Supreme Court affirmed that pursuant to article 23 (3) of the Constitution (which empowers courts to grant reliefs in petitions concerning enforcement of fundamental rights), the Court may craft orders aimed at protecting that right, such as compensation, the requirement of adequate notice before eviction, the observance of humane conditions during eviction, the provision of alternative land for settlement, etc.[57] The effect of this judgement, in the context of the UNDROP, is that those subject to unlawful evictions (from public and private land) have protection under the Kenyan Constitution.[58] Further, where such eviction is unavoidable, the State must provide or ensure fair and just compensation for any material or other losses, as framed by Kenya's Supreme Court as being within the scope of Article 23(3) of the Constitution.[59] This provides a limited sword or shield to other landless persons occupying public land in Kenya.

Women's land rights and landlessness resulting from discriminatory succession laws and customary practices

Article 4(1) of the UNDROP requires States to take all appropriate measures to eliminate all forms of discrimination against peasant women and other women working in rural areas and to promote their empowerment to ensure, on the basis of equality between men and women, that they fully and equally enjoy all human rights and fundamental freedoms and that they are able to freely pursue, participate in, and benefit from rural economic, social, political, and cultural development. This aligns with Kenya's Constitution, which prohibits any form of discrimination and affirms gender equality.[60]

A 2013 study reported that a key cause of this landlessness was the sale of family land by the head of the house without due consultation within the family.[61] Thus, many people had become landless because land was sold without consultation within the family.[62] The sale of most rural land, often categorised as agricultural land, must be preceded by obtaining approval of a Land Control Board (LCB), which is located at Sub-County level.[63] When deciding whether to grant or refuse consent for the sale of agricultural land, an LCB is required to have regard to the effect their decision is likely to have on the economic development of the land concerned or on the maintenance or improvement of standards of good husbandry within the land control area.[64]

The LCBs have interpreted their mandate to include protection of the families and dependents of vendors, requiring spouses to be physically present during applications for the sale of land, and satisfy the board on whether they have any objections.[65] In Kilifi County, for instance, there is a written requirement that, where family land is concerned, both the wife and the husband should present themselves before the LCB.[66] This is the same case in Kiambu County.[67] However, a member of the Kilifi LCB told Human Rights Watch in 2020 that, 'women are scared. If I challenge a man, the whole community looks at me as a bad person.'[68] While it appears that LCBs are evolving into tools to protect spouses and families from landlessness resulting from sale by the sole registered owner without consultations, the risk of landlessness from failure to consult prior to selling family land remains real and serious.

The demise of a father and the subsequent taking of property by relatives is also an important cause of landlessness.[69] This is despite the Constitution requiring enactment of legislation to protect the dependants of deceased persons holding interests in any land, including the interests of spouses in actual occupation of land.[70] The case concerning the landlessness that is faced by widows in some parts of rural Kenya is worth examining.

Where the deceased dies intestate and is survived by one spouse and children, the surviving spouse is entitled to a life interest of the property that is not personal (e.g., clothes).[71] The ultimate destination of the property, upon the death of the surviving, is to the children.[72] Importantly, this law provides that where the surviving spouse is a widow, that life interest terminates upon her remarriage to any person.[73] No equivalent restriction is placed for widowers, and even if there were, it would have little effect because due to the patriarchal nature of Kenya's society, land is most likely registered in the man's name. This provision, in law, is discriminatory as it treats men and women differently when it concerns intestate succession. Its effect, in restricting the inheritance of widows to a life interest, is also devastating.

Experts on women's land and property rights say widow inheritance and dispossessions are still widespread in Kenya.[74] The process of applying for and obtaining Grants of Letters of Administration (required by law during intestate succession) over the property depends on who, out of the wife or the deceased husband's relative, applies for a grant first, who has possession of the deceased's birth and death certificates and other identification documents, and is supported by the traditional leader.[75] In some communities, women face massive resistance from traditional elders and when a widow tries to succeed the property after the death of the husband, elders believe they [widow] can't manage it and need 'help' from a brother-in-law.[76] A 2018 study of intestate succession by widows in Makunga Ward in Kakamega County revealed, according to respondents who were widows, they believed they were not trusted with land ownership because of the fear that they would sell out the land left by their deceased husbands and end up remarrying.[77] Some widows are dispossessed of land and sent away to go back and ask for land from their parents.[78]

The widows interviewed for the 2018 study blamed their deceased husbands for failing to divide land or leaving wills behind to guide the process of land division.[79] They also reported being compelled to follow discriminatory cultural practices by agreeing to be inherited by a male member of the family/community upon the demise of the husband. This decision is unlawful and undermines the equality between men and women that is enshrined in the Constitution. In addition, the motive for men wanting to inherit widows is for them to gain an interest in the land that may be up for inheritance, with the hope that after some time, they could obtain ownership of the land in question.[80] If the widow declines to be inherited, she is likely to be ejected and expelled from the homestead all together.[81] Once dispossessed of family land, the widow has to rent a house close to the market centres and engage in small business or casual labour, having lost economic use of the family land.[82]

The Law of Succession Act, in limiting the inheritance rights of a widow when the husband dies intestate, is discriminatory under article 27(3) of the Constitution since no similar provision exists for widowers. Indeed, the Maputo Protocol on the Rights of Women in Africa affirms the rights of widows to remarry persons of their choice, including the right to inherit an equitable share of her husband's property, and to continue living in her matrimonial home including after remarriage if the house belonged to her, or she inherited it.[83] Kenya's Constitution imposes a duty on all state organs and public officers to address the needs of vulnerable groups within society including women.[84] This protection integrates the needs of widows with respect to inheritance and dispossession of matrimonial homes and properties.

The Community Land Act has attempted to legislate for gender equality in its provisions. Every member of the community has the right to equal benefit from community land, and this equality is defined to include full and equal enjoyment of rights of use and access.[85] More specifically, this law provides that women, men, youth, minorities, persons with disabilities, and marginalised groups have the right to equal treatment in all dealings in community land.[86] Every man or woman married to a member of the community gains automatic membership of the community and such membership shall subsist until the spouses legally divorce and the woman remarries or the woman remarries after the death of a spouse.[87] Since the community land is owned collectively in equal undivided shares by all members, this law is silent on what happens to the land rights of the woman who may get a divorce, or is widowed and subsequently remarry. In effect, the silence in this provision creates a legal loophole to landlessness that undermines the economic capabilities of the concerned women. This goes against the spirit of the UNDROP whose preamble stresses that peasant women and other rural women play a significant role in the economic survival of their families and in contributing to the rural and national economy. The legal and practical challenges highlighted in the forgoing discussion affirm that denying women, especially widows, right to land because of their remarriage, or upon unlawful conditions imposed by family members, hurts their families as well. The examples discussed in this section show that implementation of constitutional and UNDROP standards of equality, in securing the land rights of women, in Kenya is yet to be achieved.

Lessons learnt and conclusion

Smallholder farmers, including women, working in rural areas depend on land rights (or fisheries rights for coastal communities) to produce for subsistence and/

or market. For this reason, fulfilment of the relevant property rights (land, fisheries, etc.) is central to effectiveness of protections they are accorded under national law, or the UNDROP. This is because the productivity of peasants depends on having land tenure rights, and in those rights being secure.

Kenya's experience with landlessness resulting from colonial dispossession and poorly managed resettlement programmes after independence shows the divergence between law as written and in practice. Further, despite constitutional guarantees of gender equality, statute law contains provisions discriminatory to women during spousal intestate inheritance. Provisions on community land ought to be further unbundled to address land rights of women, upon divorce, or of widows who chose to remarry. The ability of widows to realise even these rights is constrained by custom and extra-legal conditions, but action by some LCBs has protected families from landlessness by requiring consent from family members before approving sale of agricultural land. But this is not yet a universal practice across Kenya. Kenyan courts have added value to constitutional provisions concerning squatters on public and private land.

Provisions of Kenya's legal framework are, for a significant part, in tune with the UNDROP within the context of this study. The Declaration contains many valuable provisions that enhance protection of tenure rights, and security, for peasants and rural workers. This includes the right to own land and the obligation on the State to respect, protect, and fulfil this right. Other key provisions include protection of the rights of women, including from discriminatory practices. The Declaration does not, however, adequately stress that widow's rights should be protected.

Indeed, the Declaration could aid the enhancing of fulfilling land rights of smallholder farmers and women in Kenya and dealing with landlessness. The development of UNDROP technical guides is a viable option. These should be analytical, in a thematic form, to provide normative guidance to users of UNDROP. The scope could include options for resolving landlessness including resettlement while learning from poorly managed processes; expanding the content of the rights of women to include those of widows; and guidelines to manoeuvre intestate succession. This guide should be usable in different legal jurisdictions including Kenya.

Notes

1 Kenya, *The Third Medium Term Plan 2018–2022* (The National Treasury and Planning 2018) 26.
2 George Rapsomanikis, *The Economic Lives of Smallholder Farmers: An Analysis Based on Household Data from Nine Countries* (FAO 2015) 5. See also Kenya, Agriculture Sector Growth and Transformation Strategy (Ministry of Agriculture 2019) 36.
3 Constitution of Kenya (2010) art. 40(1).
4 *Mohamed Ali Baadi and others v Attorney General & 11 others* (2018) eKLR, para 296.
5 Constitution of Kenya (2010), art. 40(3).
6 Land Value (Amendment) Act No. 15 of 2019, section 2.
7 Five Star Agencies Limited v National Land Commission (2014) eKLR, 8.
8 Constitution of Kenya (2010) art. 40(6).
9 Kenya, Sessional Paper No. 3 of 2009 on National Land Policy (Ministry of Lands 2009) 64.

10 Paul Munro-Faure et al, 'Land Tenure and Rural Development' (2002) 3 *FAO Land Tenure Studies*, Chapter 3, section 3.1 <www.fao.org/3/y4307E/y4307e05.htm#bm05.4> accessed 31 May 2021.
11 Francis Kariuki and Raphael Ng'etich, 'Land Grabbing, Tenure Security and Livelihoods in Kenya' (2016) 9(2) *African Journal of Legal Studies* 79–99, 84.
12 Constitution of Kenya (2010), art. 60.
13 Paul Munro-Faure et al, 'Land Tenure and Rural Development' supra note 10.
14 Kenya, *Sessional Paper No.3 of 2009 on National Land Policy* (Ministry of Lands 2009) 64.
15 Kenya, *Report of the Commission of Inquiry into the Land Law System of Kenya on Principles of a National Land Policy Framework, Constitutional Position of Land and New Institutional Framework for Land Administration* (Government Printer 2002) 24.
16 Ibid.
17 Ibid.
18 Ibid., 22.
19 Ibid.
20 Section 14 of Registered Land Act, Cap 300 Laws of Kenya (now repealed), provided for first registration.
21 Kenya, *Report of the Commission of Inquiry into the Land Law System of Kenya on Principles of a National Land Policy Framework,* supra note 15, 31.
22 P. Mbithi and C. Barnes, 'Land Settlement in Kenya; A review of the Squatter Problem' (1972) 8 *World Bank Studies in Employment and Rural Development* 80.
23 Ibid.
24 Kenya, *Report of the Commission of Inquiry into the Land Law System of Kenya on Principles of a National Land Policy Framework,* supra note 15, at 31.
25 Karuti Kanyinga, 'The Legacy of the White Highlands: Land Rights, Ethnicity and the Post-2007 Election Violence in Kenya' (2009) 27(3) *Journal of Contemporary African Studies* 325–44, 341.
26 Kenya, *Report of the Commission of Inquiry into the Illegal/Irregular Allocation of Public Land* (Government Printer 200) 26.
27 Ibid., 127.
28 Kanyinga, 'The Legacy of the White Highlands: Land Rights, Ethnicity and the Post-2007 Election Violence in Kenya,' supra note 25, at 341.
29 Ibid.
30 Frank Byamugisha, *Agricultural Land Redistribution and Land Administration in Sub-Saharan Africa: Case Studies of Recent Reforms* (World Bank 2014) 2.
31 Kilifi, *County Integrated Development Plan 2018–2022* (Kilifi County Government 2018) 18.
32 Kwale, *County Integrated Development Plan 2018–2022* (Kwale County Government 2018) 18.
33 Ibid.
34 Taita Taveta, *County Integrated Development Plan 2018–2022* (Taita Taveta County Government 2018) 23.
35 Ibid.
36 Kenya, *Final Report – TJRC Report Volume 2B. Truth, Justice and Reconciliation Report* (Truth Justice and Reconciliation Commission 2013), para. 231.
37 Ibid.
38 Taveta, *County Integrated Development Plan 2018–2022*, supra note 34, at 23.
39 Nakuru, *County Integrated Development Plan 2018–2022* (Nakuru County Government 2018) 24.
40 *Mtana Lewa v Kahindi Ngala Mwagandi* (2015) eKLR, 25.
41 Ibid.; Kenya, *Sessional Paper No.3 of 2009 on National Land Policy* (Ministry of Lands 2009) 1.
42 *Mtana Lewa v Kahindi Ngala Mwagandi* (2015) eKLR, p. 26.
43 *Haro Yonda Juaje v Sadaka Dzengo Mbauro & Kenya Commercial Bank* (2014) eKLR
44 Land Adjudication Act, Cap 284 Laws of Kenya. See also *Peter Kimandiu v Land adjudication officer Tigania West District & 4 others* (2016) eKLR. See also Land Adjudication

Act, Cap 284 Laws of Kenya, section 23(2). The outcome of this process of adjudication results in a 'first registration' which the first recording of the formal land rights either as individual land or community land.
45 Community Land Act, No. 27 of 2016, section 7 and 8.
46 *Mbira v Gachuhi* (2002) IEALR, p. 137.
47 Kenya, *Sessional Paper No. 3 of 2009 on National Land Policy* (Ministry of Lands 2009) 19.
48 *Mtana Lewa v Kahindi Ngala Mwagandi* (2015) eKLR, 26.
49 *Mitu-Bell Welfare Society v Kenya Airports Authority & 2 others*, Initiative for Strategic Litigation in Africa (Amicus Curiae) (2021) eKLR, para. 1, 2, 3, 4, and 5.
50 Ibid., para. 149.
51 Ibid., para. 150.
52 Ibid., para 151.
53 Ibid.
54 Ibid.
55 Ibid., para. 152.
56 Ibid.
57 Ibid.
58 UN General Assembly, United Nations Declaration on the Rights of Peasants and Other People Working in Rural Areas' (17 December 2018) UN Doc A/RES/73/165.
59 Ibid.
60 Constitution of Kenya (2010) artt. 27(1) and (3).
61 Bernard Omboi and Lucy Gachaba, 'Factors Leading to Squatter Problem in Rift Valley Province in Kenya' (2013) 6(3) *Journal of Law and Conflict Resolution* 48–55, 51.
62 Ibid.
63 Land Control Act, Cap 302 Laws of Kenya.
64 Ibid., section 9(1)(a).
65 Robert Robert, 'Governing Land Use in Kenya: From Sectoral Fragmentation to Sustainable Integration of Law and Policy' (Doctoral thesis, University of Ottawa 2011) <https://ruor.uottawa.ca/bitstream/10393/20268/3/Kibugi_Robert_Machatha_2011_thesis.pdf> accessed 12 June 2021, 172–3.
66 See online: <https://eregulations.invest.go.ke/procedure/230/149/step/850?l=en> accessed 26 May 2021.
67 Author experience, August 2019.
68 Human Rights Watch, *Once You Get Out You Lose Everything: Women and Matrimonial Property Rights in Kenya* (Human Rights Watch 2020) <www.hrw.org/report/2020/06/25/once-you-get-out-you-lose-everything/women-and-matrimonial-property-rights-kenya> accessed 26 May 2021.
69 Omboi and Gachaba, 'Factors Leading to Squatter Problem in Rift Valley Province in Kenya,' supra note 60, at 52–3.
70 Constitution of Kenya (2010) art. 68(c)(vi).
71 Law of Succession Act, Cap 160 Laws of Kenya, section 35.
72 Ibid., section 35(1).
73 Ibid.
74 Human Rights Watch, *Once You Get Out You Lose Everything: Women and Matrimonial Property Rights in Kenya*, supra note 67.
75 Ibid.
76 Robert Kibugi and Carolyne Ouko, *The Case for a Human Rights Based Land Governance Monitoring Framework: Country Assessment on the Land Rights Experiences of Widows in Makunga Ward, Kakamega County, Kenya* (Think Tank for Sustainability 201) unpublished, 30.
77 Ibid.
78 Ibid.
79 Ibid.
80 Ibid., 31.

81 Ibid.
82 Ibid., 33.
83 African Union, *Maputo Protocol on the Rights of Women in Africa* (African Union 2003), article XIX(c), XX (1).
84 Constitution of Kenya (2010) art. 27(3).
85 Community Land Act, No. 27 Laws of Kenya, section 30(1) and (2).
86 Ibid., section 30(3).
87 Ibid., section 30(4).

Bibliography

Byamugisha F (ed), *Agricultural Land Redistribution and Land Administration in Sub-Saharan Africa: Case Studies of Recent Reforms*, Directions in Development-Agriculture and Rural Development (World Bank Studies 2014).

Kanyinga K, 'The Legacy of the White Highlands: Land Rights, Ethnicity and the Post-2007 Election Violence in Kenya' (2009) 27(3) *Journal of Contemporary African Studies* 325–44.

Mbithi P and Barnes C, 'Land Settlement in Kenya: A Review of the Squatter Problem' (1972) 8 *World Bank Studies in Employment and Rural Development.*

15

REIMAGINING AGROBIODIVERSITY CONSERVATION AND AGRICULTURAL INNOVATION FROM THE GRASSROOTS UP

The case of the Peliti seed network in Greece

Elsa Tsioumani

Agricultural innovation and farmers' rights

Agricultural development throughout history has been based on farmers' innovation. Farmers have developed and maintained a high diversity of plants (and animal breeds), as well as different location-specific bodies of traditional knowledge and farming practices. Farmers' crop varieties tend to be heterogeneous (not genetically uniform) and thus contain high levels of genetic diversity. They are currently recognised as reservoirs of agricultural biodiversity and related knowledge, and their genetic material is used as the basis of 'modern scientific' varieties. They are thus considered a safety valve in the face of pests, diseases, and environmental stresses.[1]

Farmers' varieties have been developed through collective systems of seed sharing, conservation, and use.[2] Seed exchanges have been regulated on a customary basis, largely at the community level, and have included both informal ones and more organised systems, such as seed fairs and community seed banks.[3] Extensive literature shows the importance of such informal seed systems for smallholders' livelihoods, particularly in developing countries, as well as their role in building ecological and socio-economic resilience, promoting biodiversity conservation, supporting community cohesion, and even sustaining peace in countries affected by conflict.[4]

The green revolution transformed agriculture dramatically, including through the emergence of professional plant breeding and the commercial seed sector.[5] It led to increased food production, at least in some regions, but came with a series of environmental and socio-economic impacts.[6] The high-yielding, genetically uniform crop varieties it promoted led to monocultures and the erosion of agricultural biodiversity, making agricultural production vulnerable in the face of threats such as pests and extreme environmental conditions. The professionalisation of plant breeding also led to the marginalisation of farmer breeding practices, varieties, and

DOI: 10.4324/9781003139874-20

seed systems, in favour of 'scientific' varieties and public or corporate-led innovation.[7] Farmers were expected to be mere end-users of seeds rather than innovators; they gradually became alienated from the innovation process, lost knowledge and confidence in their capacities, and were disempowered vis-à-vis other sectors of society.[8] The establishment and application of intellectual property rights (IPRs) on plant varieties, mainly through plant breeder rights and patents, as well as of seed certification and marketing regulations outlined later, has made farmers more dependent on commercial seed providers. This loss of autonomy has in some cases put their livelihoods at risk.[9]

The concept of farmers' rights, enshrined in the International Treaty on Plant Genetic Resources for Food and Agriculture (ITPGRFA), emerged partly as a response to the restrictions in use associated with IPRs. Addressing farmers' rights to seeds, traditional knowledge, participation in decision-making, and fair and equitable benefit sharing, the concept aimed at recognising farmers' contribution to global food security and agrobiodiversity conservation and ensuring that this contribution would not only be allowed to continue but also be encouraged and supported.

In this chapter, I focus on the case study of the Peliti seed network, a network of local groups working on the conservation and exchange of traditional crop varieties in Greece.[10] I first provide a brief overview of the legal and socio-economic context, globally and in Greece. I then present the case of Peliti and analyse it in the context of contemporary agrarian struggles for environmental and social justice. I finally question the role of international law, including obstacles but also hopes, as illustrated by the adoption of the UN Declaration on the Rights of Peasants and other People Working in Rural Areas (UNDROP).

The legal context

The international law of relevance to agriculture includes instruments from different fields. The mainstream, neoliberal model of agricultural development and production, which promotes industrialisation and monocultures for high yields, gives emphasis to trade law, including IPR legislation, arguably favouring large-scale agricultural producers. Access to markets is further governed by seed and variety certification and food marketing regulations, including food safety standards. Models valuing agricultural biodiversity, farmer-based innovation, and seed systems, or those more fitted to smallholders and subsistence agriculture, find support in international environmental and human rights law of relevance to sustainable management of biodiversity and natural resources, labour and farmers' rights, and the right to food.

The international legal and policy context is mainly defined by the Agreement on Trade-related Aspects of Intellectual Property Rights (TRIPS Agreement) and the Agreement on Agriculture of the World Trade Organization (WTO),[11] the International Convention for the Protection of New Varieties of Plants (UPOV Convention),[12] the ITPGRFA,[13] the Convention on Biological Development (CBD)[14] and its Nagoya Protocol on access and benefit sharing,[15] as well as a series of instruments related to human rights, including the UNDROP. This list draws a picture of a contradictory legal and policy framework involving a variety of objectives, from trade liberalisation and protection of IPRs to agrobiodiversity conservation and promotion of farmers' rights.[16] As a result of the power imbalance among the stakeholders involved and their increased enforcement potential, trade-related agreements are often applied to the detriment of human rights and environmental norms. In addition, an ever-growing web of bilateral and regional free-trade and investment agreements reduces policy choices for developing countries through IPR- and land-related standards.[17]

The challenge of reconciling contradictory policy objectives is showcased by the EU seed regulations of relevance to the Peliti case study. In the EU, seed certification and variety registration in official catalogues are part of the mandatory authorisation process for entry into the seed market.[18] While the objective is to control varietal identity and purity, and seed health, this results in marginalising farmers' varieties and hindering the conservation of crop genetic diversity.[19] To address genetic erosion, rules on 'conservation varieties' set different requirements for the recognition and inclusion of such varieties in national catalogues;[20] however, they remain too restrictive,[21] largely failing to protect smallholders' rights. Limited legal space for peasant seeds has also been provided under the European Commission's 2014 'temporary experiment,' which authorised the marketing of heterogeneous populations of wheat, barley, oats, and maize,[22] and the 2018 regulation on organic production, which allows the marketing of heterogeneous plant reproductive material.[23] Following the EU approach, the Greek legislative framework on seed production and commercialisation is highly formalised and controlled by the Ministry of Agriculture.

The socio-economic context

Albeit the realities of agriculture and food production vary greatly among regions and countries, the global picture of industrial agriculture is characterised by expansion of IPRs and increasing concentration of the agro-chemical sector. The breakthrough of modern biotechnologies in the 1990s and the wave of privatisation promoted by trade-related agreements led to an ever-increasing number of patents covering not only transgenic plant varieties but also genes and plant breeding methodologies.[24] At the same time, while the number of patent applications increases, the number of applicants decreases. Dozens of mergers, in combination with the dismantling of public support for agricultural development, have resulted in a dramatic concentration of control in a handful of companies, sparking concerns regarding undue control of global food production. Following the 2015–2016 wave of mergers and acquisitions, three conglomerates control more than 60% of the market for commercial seed and agricultural chemicals and own the majority of patent applications for plants.[25]

The concentration of the control of agricultural inputs and knowledge is coupled by increasing concentration in land ownership.[26] Historically linked to economic inequality, social instability, and environmental degradation, land concentration is currently reinforced by market dynamics and legal norms within the mainstream paradigm of neoliberal development. The latest global wave of land rush is marked by rapidly increasing commercial pressure on agricultural land, with the global capital seeking control of agricultural production through direct investments, domestic or transnational.[27] As an illustration, Bill and Melinda Gates have recently become the top owners of farmland in the USA.[28]

At the same time, smallholder agriculture remains an important part of the socio-economic and ecological landscape all over the globe, and the foundation of food security and nutrition in many countries, particularly in Africa. According to the UN Committee on World Food Security (CFS), smallholder agriculture, if adequately supported by policy and public investments, can contribute effectively to food security, sustainable resource management, rural development, and the reduction of socio-economic inequalities, including improved community resilience and the emancipation of neglected and marginalised groups.[29]

The neoliberal model of agricultural development is not uncontested. Agrarian movements call for an alternative development paradigm based on land redistribution, fair use of natural resources, and public support for smallholders against big agri-businesses. The action of the Brazilian Landless Rural Workers' Movement (MST) and the protests of Indian farmers of 2020/2021 represent notable examples. Known for its massive land occupations, MST has also launched the Popular Agrarian Reform, a model of agrarian reform linking land distribution demands with food sovereignty and agro-ecology considerations.[30] In India, hundreds of thousands of farmers have been protesting against a series of agricultural laws dismantling public support, including guaranteed prices for certain crops.[31] Agrarian movements have also been a driving force behind international law development, reflected most prominently in the involvement of global peasant movement La Via Campesina (LVC) in the negotiation and adoption of the UNDROP.[32]

In Greece, the recent financial crisis, as well as the austerity measures and structural reforms that followed, had a particular impact on smallholder farmers, tipping the balance in favour of large food retailers, and resulting in increasing rural poverty. At the same time, however, a range of grassroots initiatives emerged, based on principles of solidarity economy and aiming to secure the right to food. Embodying different political tendencies and operating in different ways, such initiatives included agricultural cooperatives, 'no intermediaries' markets, communal gardens promoting food self-sufficiency, community-supported agriculture schemes supporting networks of consumers and producers, and eco-communities also covering access to land and natural resources.[33] Albeit pre-existent, Peliti can be situated within this wave of mostly self-organised responses to crisis.

The Peliti seed network

Peliti was established in 1995 as a non-profit organisation under Greek civil law. Its aims include the collection, conservation, and distribution of traditional agricultural varieties; the non-monetary exchange of goods and services; and the creation of a community focusing on conserving traditional varieties, including the preservation of seed saving-related knowledge, as well as local animal breeds. Based in Messochori, in the area of the virgin forest around Paranesti in northern Greece, Peliti started as a personal effort of its founder Panagiotis Sainatoudis. It gradually got bigger, resulting in the creation of 20 local seed groups operating in various locations around Greece, one in Bulgaria and one in the USA.[34]

Peliti unites a diverse group of people, including professional farmers and food producers, amateur gardeners, and citizens of different age groups with an interest in seed collection and conservation of traditional agricultural varieties. Members may join for different reasons: love of nature, quest for healthy food, desire for participation in activities considered socially and environmentally beneficial, or a sense of belonging in a community of like-minded individuals. The 'passion for the seed' was quoted as an important rationale, which points towards situating Peliti within a global movement of grassroots agrobiodiversity conservation initiatives in the context of food sovereignty, explored later.

Peliti's activities include seed collection expeditions in mainland Greece and its islands, free distribution of seeds by post, and yearly festivals for seed exchanges and seed distribution. These festivals are considered to be among the most popular globally, with thousands of visitors coming from Greece and abroad every year. The organisation is also engaged in awareness-raising activities, developing publications and online material, and conducting seminars and international exchanges. International networking intensified as a result of the increasing popularity of seed festivals, as well as the recognition of Panagiotis Sainatoudis among the 'Guardians of Diversity in the Mediterranean' by Bioversity International in 2009.[35] Since 2012, Peliti has also created a community seed bank, in an old school provided by a local municipality. The community considers the seed bank to serve as a safety net but not replace conservation and the use of agricultural varieties on-farm, which it actively supports. It is also increasingly involved in campaigns for the legal recognition of farmers' rights at the national and EU level.

With regard to financial and human resources, the network operates on the basis of donations, the income from the sale of publications, and volunteer work. Peliti is opposed to selling seeds and even rejects the idea of providing seeds in exchange for annual subscriptions – a solution that could potentially bypass the prohibition of sale of seeds of varieties not included in the national and European catalogues. Still, while the non-monetary exchange of goods and services is an important ideological basis of the organisation, it strongly supports farmers' right to save, use, exchange, and sell farm-saved seed if they so wish.

The organisation thus distributes solely traditional seeds in the public domain free of charge. This saves them a host of legal challenges but adds to financial ones. Limited finances pose obvious obstacles to its activities, for instance, regarding the operations of the seed bank. However, the range of its activities appears to be disproportionately large compared to its finances, with approximately 50,000 seed pockets distributed yearly by post, several tens of thousands distributed in planned events, and seeds sent to countries in crisis, including Palestine and Syria.

Ironically, the Greek financial crisis had an asymmetrical effect on the organisation; while its finances plummeted, its seed-sharing activities started attracting an ever-increasing degree of attention. This can be attributed to Peliti's collaboration with many of the bottom-up initiatives that emerged as noted earlier. It can also be linked to a rise in amateur gardening, observed during the crisis as well as the subsequent lockdowns associated with the Covid-19 pandemic, addressed later.

As highlighted by Peliti member Vasso Kanellopoulou, the distribution of seeds and thus utilisation and development of genetic resources are among their biggest accomplishments. Peliti strongly believes in farmer-based innovation, recognition of which is lacking in Greece. This is reflected in farmers' denied access to genetic resources stored in the national genebank and lack of legal support for traditional varieties. Participation in law- and policy-making processes at the national and European level is thus becoming increasingly important, aimed at the increased recognition of farmers as innovators and farmers' rights, including the right to seeds.

At the 2019 general assembly, Peliti members decided to decentralise administration of local groups by creating a network, while maintaining regular collaboration and communication activities. Following intense discussions, local groups became independent with a focus on seed-saving and distribution, while the original Peliti non-profit organisation, still coordinated by Panagiotis Sainatoudis together with an administrative board, would focus on networking with international groups and creation of an eco-community.

Discussion: reimagining agrobiodiversity conservation and agrarian justice

The 'passion for the seed' quoted by Peliti's members as a reason for participation situates the organisation within a global grassroots movement of communities and networks engaged in agrobiodiversity conservation, as well as within the context of broader agrarian struggles for environmental and social justice. The contribution of agrarian movements in the negotiation and adoption of the UNDROP is a moment of success in these struggles. This section will place the case study in the context of relevant literature, before discussing the role of and challenges for international law in the quest for justice.

Literature from different disciplines, including political sociology, anthropology, rural studies, and political ecology, increasingly examines the rise of seed activism around the globe, including in Latin America,[36] North America,[37] and Europe.[38] Karine Peschard and Shalini Randeria see seed activism as a 'diverse and growing constellation of networks and coalitions' aiming to respond to processes of seed enclosures, including threats of legal, economic, and technological nature to peasant seed systems, and to the loss of agricultural biodiversity.[39] An anthropological approach to seed activism further reveals interconnections between seed saving and the history and ecology of a place, relationships between people generated by seeds, and alternatives to the unsustainability of the food system.[40] This is consistent with the rise in amateur gardening, observed during the crisis in Greece, as well as the subsequent lockdowns associated with the Covid-19 pandemic: seeds are not only a source of food, but they also provide ways to manage anxieties and imagine alternatives.

Looking at agrarian movements more broadly, Claudio Brenni suggests a taxonomy along three dimensions: genetic resource appropriation, which concerns ownership of biodiversity and related knowledge; production orientation, which concerns the right to choose between different agricultural production and trade models; and agricultural innovation, which concerns participation in and control of the innovation process.[41]

Elise Demeulenaere and Yvonne Piersante on the other hand differentiate 'new agrarian movements' on the basis of their links to the farming sector. Their analysis of the *Red de semillas* in Spain, *Réseau Semences Paysannes* in France, and *Rete Semi Rurali* in Italy underscores their origin in coalitions of farmers' mobilisations and civil society organisations and their principled opposition to the neoliberal development model. These organisations have linked food quality, farmers' autonomy, rural development, and biodiversity management, and have created important collaborations with other social movements.[42]

In this framework, according to Brendan Coolsaet, agrobiodiversity conservation is more than an environmental or agricultural matter: it becomes an instrument for a broader struggle for justice. Agrarian organisations redefine the conditions for participation in decision-making and foster empowerment by creating structures of self-governance. At the same time, they devise strategies to resist corporate and institutional threats and propose alternative mechanisms for agrobiodiversity conservation. Rebuilding rural identities, they offer transformations of the agricultural production towards sustainability,[43] increasingly framing their struggles in the language of the commons.

While each movement or network is unique, a crucial common element is their broad appeal: as the basis of our food system, the seed inspires and carries symbolic power. These organisations manage to bridge the rural–urban divide and show the capacity to unite individuals from different backgrounds, with diverse political agendas.[44] They also bridge various scales, from the local to the transnational. Enabled by communication technologies, they share experiences and coordinate actions. This results in a re-emergence of peasantry as a political actor from the national to global level.[45] At the same time, such diversity in membership increases the potential for conflict based on class and ideologies, including between members who call for a radical restructuring of the food system and those who contest only specific aspects.

These organisations have been using a variety of strategies in opposing threats and reclaiming seeds, engaging in seed conservation and redistribution activities, grassroots action and civil disobedience, as well as participation in law making at the national and international levels. Engagement with the law is another common characteristic of agrarian movements, including Peliti, and includes efforts to create legal space for peasant seeds at the domestic (and EU) level. These efforts award political significance to seemingly technical criteria on seed quality and certification. They highlight the fact that such criteria marginalise farmers from the innovation process, enable corporate consolidation, and fail to reflect the value of agricultural biodiversity within production models.

Engagement with the law has more recently expanded to participation in international policy processes, including the UNDROP. In the words of Priscilla Claeys and Marc Edelman, the negotiation and adoption of the UNDROP represent 'a rare case of direct involvement of social movements in international law-making.'[46] LVC had a leading role in the negotiations, from alliance building and advocacy to the sharing of information and analysis. The value of the UNDROP thus lies first in the bottom-up process of its development, and second, in its comprehensive, substantive approach regarding existing and new rights of rural populations, their contribution to global food security, and their relationship with land, seed, and natural resources.[47]

The UNDROP clearly states that peasants and other people working in rural areas have the right to save, use, exchange, and sell their farm-saved seed or propagating material, and the right to maintain, control, protect, and develop their own seeds and traditional knowledge.[48] These rights are accompanied by a range of obligations for states, including measures to respect, protect, and fulfil the right to seeds, recognise peasants' rights to rely on their own seeds and to decide on the crops they wish to grow, and support peasant seed systems.[49] States must further take measures to ensure that agricultural research and development integrates peasants' needs, and ensure that seed policies, certification schemes, and intellectual property laws respect and take into account peasants' rights, needs, and realities.[50] It thus constitutes a valuable addition to the international legal basis of farmers' rights and a tool for the reinforcement of policy choices to implement seed rights at the domestic level, in view of the contradictory global policy framework described earlier.

The UNDROP also provides additional legal arguments in support of advocacy actions of seed movements such as Peliti, regarding recognition of seed rights and farmer-based innovation. Despite the lack of legally binding force, the UNDROP may influence power structures and relationships and inspire political mobilisations. It can legitimise and provide political leverage to previously marginalised groups and promote legal change at the domestic level or judicial interpretations favouring seed rights and peasant autonomy. Furthermore, to reflect Martha Nussbaum's analysis of women's rights, like other international instruments, the UNDROP helps people network across borders, develop a common language and common demands, and a sense that progress is being made.[51] Its legal significance, however, will be realised through implementation at the domestic level, through legislatures or courts of law.

As Claeys and Edelman put it, the adoption of the UNDROP was a 'huge success' for agrarian movements, providing 'a glimpse of hope that multilateralism can be centred on human rights, anti-colonialism, justice and solidarity . . . even in times of rising authoritarianism.'[52] For the time being, peasant rights enjoy limited acceptance as norms by states, and domestic implementation is far from certain. From the suffragettes to the civil rights movement however, acts of civil disobedience have historically been critical in reassessing a society's moral principles and legal norms, eventually promoting social change. The ultimate impact of the UNDROP on the agriculture and rural development paradigm remains to be seen.

Notes

1 Julie Bélanger and Dafydd Pilling (eds), *The State of the World's Biodiversity for Food and Agriculture* (FAO Commission on Genetic Resources for Food and Agriculture Assessments 2019).
2 Michael Halewood, Isabel López Noriega, and Sélim Louafi (eds), *Crop Genetic Resources as a Global Commons* (Routledge 2013).
3 Eric Garine et al, 'Which Scale to Understand Seed Fluxes in Small-Scale Farming Societies? Snapshots of Sorghum from Africa' in Fabien Girard and Christine Frison (eds), *The Commons, Plant Breeding and Agricultural Research* (Routledge 2018).
4 Connie J.M. Almekinders, Niels Louwaars and Gerardus Hubertus de Bruijn, 'Local Seed Systems and their Importance for an Improved Seed Supply in Developing Countries' (1994) 78 *Euphytica* 207; Shawn McGuire, and Louise Sperling, 'Seed Systems Smallholder Farmers Use' (2016) 8 *Food Security* 179; FAO, *Farmer Seed Systems and Sustaining Peace* (FAO 2018).
5 Robert E. Evenson, and Douglas Gollin, 'Assessing the Impact of the Green Revolution, 1960 to 2000' (2003) 300(5620) *Science* 758.
6 Raj Patel, 'The Long Green Revolution' (2013) 40(1) *The Journal of Peasant Studies* 1.
7 FAO, *Harvesting Nature's Diversity* (FAO 1993); Kenneth A. Dahlberg, *Beyond the Green Revolution. The Ecology and Politics of Global Agricultural Development* (Plenum Press 1979).
8 Ola Tveitereid Westengen and Tone Winge, 'New Perspectives on Farmer – Breeder Collaboration in Plant Breeding' and Rene Salazar, Gigi Manicad, Anita Dohar and Bert Visser, 'Participatory Plant Breeding: Human Development and Social Reform' in Ola T. Westengen and Tone Winge (eds), *Farmers and Plant Breeding: Current Approaches and Perspectives* (Routledge/Bioversity International 2019).
9 UN General Assembly, 'Special Rapporteur on the Right to Food; Seed Policies and the Right to Food: Enhancing Agrobiodiversity and Encouraging Innovation' (23 July 2009) UN Doc A/64/170; Pat Roy Mooney, *The Parts of Life. Agricultural Biodiversity, Indigenous Knowledge and the Role of the Third System* (Dag Hammarskjöld Foundation 1998).
10 The case study description largely draws from the relevant section in Elsa Tsioumani, *Fair and Equitable Benefit-Sharing in Agriculture: Reinventing Agrarian Justice* (Routledge 2020). It is based on information found on Peliti's website and an interview with Peliti member Vasso Kanellopoulou, whom I thank profoundly for sharing her experiences and valuable insights.
11 The Agreement on Trade Related Aspects of Intellectual Property Rights (adopted 15 April 1994, entered into force on 1 January 1995) 1869 UNTS 299 (TRIPS).
12 1961 UPOV Convention, later amended in 1972, 1978, and 1991.
13 International Treaty on Plant Genetic Resources for Food and Agriculture (adopted 3 November 2001, entered into force on 29 June 2004) 2400 UNTS 303 (ITPGRFA).
14 Convention on Biological Diversity (adopted 5 June 1992, entered into force on 29 December 1993) 1760 UNTS 69 (CBD).
15 Nagoya Protocol on Access to Genetic Resources and the Fair and Equitable Sharing of Benefits Arising from their Utilization to the Convention on Biological Diversity (adopted 29 October 2010, entered into force 12 October 2014) 30619 UNTS 27.
16 For the full analysis, see Tsioumani, *Fair and Equitable Benefit-Sharing in Agriculture: Reinventing Agrarian Justice*, supra note 10.

17 Lorenzo Cotula, 'Do Investment Treaties Unduly Constrain Regulatory Space?' (2014) 9 *Questions of International Law, Zoom-in* 19–31.
18 For the purposes of this analysis, these include Council Directive 2002/53/EC of 13 June 2002 on the Common Catalogue of Varieties of Agricultural Plant Species, and Council Directive 2002/55/EC of 13 June 2002 on the Marketing of Vegetable Seed.
19 Niels Louwaars, *Seeds of Confusion: The Impact of Policies on Seed Systems* (PhD dissertation, Wageningen University 2007).
20 See Commission Directive 2008/62/EC of 20 June 2008 and Commission Directive 2009/145/EC of 26 November 2009.
21 Regine Andersen, 'Norway's Path to Ensuring Farmers' Rights in the European Context' in Regine Andersen and Tone Winge (eds), *Realising Farmers' Rights to Crop Genetic Resources: Success Stories and Best Practices* (Routledge 2013).
22 Commission Implementing Decision 2014/150/EU of 18 March 2014.
23 Regulation (EU) 2018/848 of the European Parliament and of the Council of 30 May 2018.
24 Elsa Tsioumani et al, 'Following the Open Source Trail outside the Digital World: Open Source Applications in Agricultural Research and Development' (2016) 14(1) *triple C: Communication, Capitalism and Critique* 145.
25 The two US corporations DuPont and Dow Chemical merged; ChemChina bought the Swiss company Syngenta; and the German chemical giant Bayer took over the US company Monsanto. See Sylvie Bonny, 'Corporate Concentration and Technological Change in the Global Seed Industry' (2017) 9 *Sustainability* 1632.
26 European Economic and Social Committee, *Land Grabbing: A Warning for Europe and a Threat to Family Farming (Own-Initiative Opinion)* (20–21 January 2015) NAT/632.
27 Katie Sandwell et al, *A View from the Countryside: Contesting and Constructing Human Rights in an Age of Converging Crises* (TNI Issue Brief 2019); Liz Alden Wily, 'The Global Land Grab: The New Enclosures' in David Bollier and Silke Helfrich (eds), *The Wealth of the Commons: A World beyond Market and State* (Levellers Press 2012).
28 Eric O'Keefe, 'Bill Gates: America's Top Farmland Owner' (2021) *The Land Report* <https://landreport.com/2021/01/bill-gates-americas-top-farmland-owner/> accessed 25 April 2021.
29 UN Committee on World Food Security – High Level Panel of Experts, 'Investing in Smallholder Agriculture for Food Security' (2013).
30 Camila Azevedo, 'Gaining Land and Gaining Ground? The Popular Agrarian Reform by the Landless Rural Workers' Movement (MST) in Brazil' (2016) 52 *International Colloquium on Global Governance/Politics, Climate Justice and Agrarian/Social Justice: Linkages and Challenges*.
31 Hannah Ellis-Petersen, 'Farmers' Protests in India: Why Have New Laws Caused Anger?' (2021) *The Guardian* <www.theguardian.com/world/2021/feb/12/farmers-protests-india-why-laws-caused-anger> accessed 25 April 2021.
32 Priscilla Claeys and Marc Edelman, 'The United Nations Declaration on the Rights of Peasants and Other People Working in Rural Areas' (2020) 47 *The Journal of Peasant Studies* 1.
33 Stephan Backes et al, *Democracy Not for Sale: The Struggle for Food Sovereignty in the Age of Austerity in Greece* (Transnational Institute, FIAN International and Agroecopolis 2018).
34 See Peliti's webpage at <https://peliti.gr/>.
35 See Bioversity International, *Geneflow* (Biodiversity International 2009).
36 Laura Gutiérrez Escobar and Germán Vélez, 'The Struggle for Peoples' Free Seeds in Latin America: Experiences from Brazil, Ecuador, Colombia, Honduras, and Guatemala' (2016) (8) *Right to Food and Nutrition Watch: Keeping Seeds in People's Hands* 68–76.
37 Brian C. Campbell and James R. Veteto, 'Free Seeds and Food Sovereignty: Anthropology and Grassroots Agrobiodiversity Conservation Strategies in the US South' (2015) 22(1) *Journal of Political Ecology* 357.

38 Elise Demeulenaere and Yvonne Piersante, 'In or out? Organisational dynamics within European 'peasant seed' movements facing opening-up institutions and policies' (2020) 47(4) *The Journal of Peasant Studies* 767.
39 Karine Peschard and Shalini Randeria, '"Keeping seeds in our hands": the rise of seed activism' (2020) 47(4) *The Journal of Peasant Studies* 613.
40 Campbell and Veteto, 'Free Seeds and Food Sovereignty: Anthropology and Grassroots Agrobiodiversity Conservation Strategies in the US South,' supra note 37.
41 Claudio Brenni, 'Where are the Local Communities? Food Sovereignty Discourse on International Agrobiodiversity Conservation Strategies' in Amy Trauger (ed), *Food Sovereignty in International Context: Discourse, Politics and Practice* (Routledge 2015).
42 Demeulenaere and Piersante, 'In or Out? Organisational Dynamics Within European "Peasant Seed" Movements Facing Opening-up Institutions and Policies,' supra note 38.
43 Brendan Coolsaet, *Farming Justice: Rights-based Approaches to Collective Agrobiodiversity Conservation* (Université Catholique de Louvain 2016).
44 Peschard and Randeria, '"Keeping Seeds in Our Hands": The Rise of Seed Activism,' supra note 39.
45 Paolo Perasso, 'Connecting through the Seed: Alternative Agriculture and Peasant Resistance in the Colombian Coffee Axis' (Leiden University 2014).
46 Claeys and Edelman, 'The United Nations Declaration on the Rights of Peasants and Other People Working in Rural Areas,' supra note 32, at 6.
47 Tsioumani, *Fair and Equitable Benefit-Sharing in Agriculture: Reinventing Agrarian Justice*, supra note 10.
48 UN Human Rights Council, 'United Nations Declaration on the Rights of Peasants and Other People Working in Rural Areas' (8 October 2018) UN Doc A/HRC/RES/39/12, artt. 19.1(d) and 19.2.
49 Ibid., artt. 19.3, 19.5, and 19.6.
50 Ibid., artt. 19.7 and 19.8.
51 Martha C. Nussbaum, 'Women's Progress and Women's Human Rights' (2016) 38 *Human Rights Quarterly* 589.
52 Claeys and Edelman, 'The United Nations Declaration on the Rights of Peasants and Other People Working in Rural Areas,' supra note 32, at 29.

Selected bibliography

Backes S, Gkiougki J, Kay S et al, *Democracy Not For Sale: The Struggle for Food Sovereignty in the Age of Austerity in Greece* (Transnational Institute, FIAN International and Agroecopolis 2018).

Claeys P and Edelman M, 'The United Nations Declaration on the Rights of Peasants and Other People Working in Rural Areas' (2020) 47 *The Journal of Peasant Studies* 1.

Peschard K and Randeria S, '"Keeping Seeds in Our Hands": The Rise of Seed Activism' (2020) 47(4) *The Journal of Peasant Studies* 613.

Tsioumani E, *Fair and Equitable Benefit-Sharing in Agriculture: Reinventing Agrarian Justice* (Routledge 2020).

16

THE RIGHTS OF SMALL-SCALE FISHERWOMEN IN THE EU

The potential impact of the UNDROP

Adriana Bessa and *Miguel Ángel Martín López*[*]

Introduction: El palmar and women's struggle for equal rights

'Un cura y una mujer en un barco traen mala suerte' ('a priest and a woman on a boat bring bad luck'). This belief has long permeated the imagery of Spanish small-scàle fishers and served as an extra justification for discriminating against and excluding women from boat fishing. It is estimated that women represent 5.4% of extractive fishing, 62.9% of all shellfish collectors, and the majority of workers in the fish processing industry in Spain.[1] Yet, gender stereotypes and superstitions persist in the sector significantly reducing job opportunities for fisherwomen and directly and indirectly hindering their access to boat fishing licenses.

Spanish fisherwomen face a number of other challenges. These include the lack of recognition for their work and contribution to the fishing sector, scarce quantitative and qualitative data that take due account of female workers, their needs and interests, the need to improve workplaces and identify occupational diseases that affect women especially, the gender wage gap, the unequal access to governmental subsidies, and the insufficient participation of female fishers in decision-making processes and bodies at local, national, and regional (especially the European Union) levels.

Spanish judicial courts have affirmed women's rights to access local fisheries on an equal footing with men. In the *El Palmar* case, the customary laws of the local fishing community prevented women from inheriting fishing rights, thus denying them any rights to fish in the lake of *La Albufera de Valencia*.[2] Due to its outstanding ecological and cultural value, *La Albufera* is protected as a nature reserve. Artisanal fishing traditions date back to the Middle Ages, as does the exclusion of local women from fishing practices. As it has been explained: 'The Community

* The authors would like to thank Francisco Manuel Escortell Mayor, for suggesting the analysis of the case, and Margherita Brunori and Mitchell Lennan for valuable comments on a previous draft of this chapter.

DOI: 10.4324/9781003139874-21

of Fishermen of *El Palmar* only admitted the male children of fishermen, since the General Board of the body interpreted the privilege of exclusive fishing in the lake that King Jaume I granted to the Community in the 12th century literally.'[3]

This restriction was maintained for over 750 years. It should, however, be noted that the fact that local women were not admitted to the Community of Fishermen did not mean that they did not work doing tasks related to fishing. They were seen as assistants, helping their husbands, and as such, were kept out of decision-making processes and the allocation of fishing rights in the lake. This gender-based division of labour has traditionally followed a very marked pattern in Spanish fishing families. As has been explained, '[m]en usually took the roles of fishing, boat building, management, and public representation. Women also took on or shared these tasks, and performed others like weaving and repairing nets, shellfish gathering, or marketing.'[4]

For many years, local women tried unsuccessfully to be admitted to the Community of Fishermen and have their fishing rights recognised. Finally, in 1998, they lodged a complaint in the local court that affirmed the rights of women to be members of the association and exercise equal rights to access the lake and fisheries. Upon the Community's appeals, the case was submitted to the Spanish Supreme Court that endorsed women's rights. It was not until 2008, however, that the Community of Fishermen started admitting women members and allocating fishing positions to women and men in the community on equal terms.[5]

This was the first step towards gender equality in access to fisheries in *La Albufera*. To date, the work is still male-dominated. Nonetheless, the decision of the Supreme Court has paved the way for substantial changes in the fishing sector in terms of gender equality, not only in the nature reserve but also across the country. In addition to gender equality, pollution and lake water stagnation, as well as lack of official recognition as small-scale fishers in accordance with Spanish laws, have haunted community members and affected women fishers especially. For fishing on continental waters – as opposed to maritime waters – the Community of *El Palmar* does not have the same social guarantees or subsidies as sea fishers, augmenting vulnerability in times of economic crisis.[6]

The following sections of this chapter will depart from the *El Palmar* case and look at the current normative body and some public policies relevant to small-scale fisheries in the European Union (EU), paying attention to the Mediterranean. The activity has long been developed in the region and the adoption of the United Nations Declaration on the Rights of Peasants and Other People Working in Rural Areas (UNDROP) has brought new perspectives to its regulation at regional and national levels. It will draw special attention to a continuing controversial thematic in the sector: gender equality. In this sense, this chapter aims to discuss the extent to which the UNDROP has taken this element into account while inquiring how new standards may contribute to increasing the protection and empowerment of fisherwomen.

Small-scale fisheries in the European Union

The European fishery industry has been subject to a number of laws, policies, and regulations at the EU level, particularly the EU Common Fisheries Policy.[7] The

2013 version of the EU Common Fisheries Policy (CFP) provided specific considerations to small-scale fisheries[8] and prescribed a number of measures to protect the rights and interests of small-scale fishers operating in the EU zone. First, small-scale fishers were exempted from a number of obligations and certain restrictions are not applicable to them. These include, for example, having to request fishing authorisations, the submission of landing declarations, sales notes transmissions, and separate stowage of species.[9] Second, small-scale fishers may be granted preferential access to waters and marine resources in the 12-nautical miles coastal belt.[10] Third, in pursuit of more democratic governance of fishing resources in the region, the participation of representatives of small-scale fisheries is now encouraged in Advisory Councils.[11] Having said that, questions related to gender equality and fisherwomen's equal access to fisheries have remained largely neglected.

In 2012, as part of the discussions on the reform of the CFP, the European Parliament shed important light on fisherwomen and the need to improve economic and social rights recognition in regulations and policies concerning small-scale fisheries across the region. It called for

> Taking steps to promote and achieve greater recognition, both legal and social, for the work of women in the fisheries sector, and to ensure that women who work full- or part-time for family undertakings or assist their spouses, thereby contributing to their own economic sustainability and that of their families, receive legal recognition or social benefits equivalent to those enjoyed by people with self-employed status, in particular by applying Directive 2010/41/EU, and that their social and economic rights are guaranteed, including equal pay, the right to unemployment benefit if they lose their jobs (temporarily or permanently), the right to a pension, work-life balance, access to maternity leave, access to social security and free healthcare, and workplace health and safety, as well as other social and economic rights including insurance covering risks at sea.[12]

The resolution manifests a commitment to rectify the oversight of the CFP to gender and to improving women's visibility, recognition, and well-being in the EU fisheries, especially in the small-scale fisheries sector. Unfortunately, this was not incorporated into the latest reform of the CFP.

There is no general working definition for small-scale fishers agreed at the regional level. Vessels length, near-shore fishing, labour-intensive methods, and rudimentary equipment, the maintenance of small-scale production units with low input, low-profile technology use and low capital investment, local socio-economic conditions, and environmental impacts are a just few elements that may be considered in the development of a possible definition. More attention appears to be drawn, however, to vessels length compared to other socio-economic and cultural aspects. For the purposes of the European Maritime and Fisheries Fund (EMFF), for instance, the Regulation (EU) No 508/2014 considers 'small-scale coastal fishing' all 'fishing carried out by fishing vessels of an overall length of less than 12 meters and not using towed fishing gear.'[13]

It is worth noting that this regulation encourages that EMFF resources are applied in a way to promote gender equality and reduce discrimination against fisherwomen. Among other measures, it includes

Networking and exchange of experiences and best practices between stakeholders, including among organizations promoting equal opportunities between men and women, promoting the role of women in fishing communities and promoting under-represented groups involved in small-scale coastal fishing or in on-foot fishing.[14]

The previous regulation used more assertive language, however. It did not only encourage but also demanded the Member States and the EU Commission

> ensure equality between men and women and the integration of the gender perspective [were] promoted during the various stages of implementation of the EFF [European Fisheries Fund, now European Maritime and Fisheries Fund], including the design, implementation, monitoring and evaluation.[15]

In doing so, Member States should 'ensure that operations to enhance the role of women in the fisheries sector [were] promoted.'[16] Indeed, the 2014 Regulation appears to be more lenient with inequalities between men and women in the fisheries sectors, by adopting flexible language, i.e., 'may' and 'shall.'[17] What is more, the document replicates the wording used in CFP in referring to 'fishermen' and 'organizations of fishermen.' Gender-neutral expressions such as 'fishers' and 'fishers organizations' would better reflect the block's commitment to gender equality in the sector.

More specifically on the Mediterranean region, the General Fisheries Commission for the Mediterranean is a regional fisheries management organisation with the aim 'to ensure the conservation and the sustainable use of living marine resources as well as the sustainable development of aquaculture in the Mediterranean and in the Black Sea.'[18] Established in 1949, pursuant to article XIV of the FAO Constitution,[19] it is composed of 22 member countries and the EU. It has the mandate to adopt binding recommendations for fisheries conservation use and management and aquaculture development and plays an important role in supporting small-scale fisheries in the region.

In 2018, a plan of action for small-scale fisheries in the Mediterranean and the Black sea was adopted.[20] Eighteen ministers of fisheries[21] and the EU Commissioner for Maritime Affairs and Fisheries undertook the commitment to support sustainable small-scale fisheries and aquaculture by promoting research and data collection relevant to and with the involvement of small-scale fishers, designing and implementing management plans that facilitate equitable access to marine resources, improving small-scale fisheries value chains, and promoting more democratic resource governance. The promotion of gender equality is only briefly addressed in the document as countries commit to support projects dedicated to fisherwomen, to secure equal participation in decision-making processes, and to support the development of new technologies to improve women's work.[22]

The impact of UNDROP: the outlook

Small-scale fishers' movements joined the UNDROP negotiation process at a later stage. This explains the reason why, although Article 1 makes clear reference to these groups and includes them in the notion of peasants,[23] the Declaration fails to take full account of their special modes of production, needs, and interests in a number of passages, especially those relating to access to waters and fishing resources, mobility and freedom of movement across borders, gender equality, and social security.

Having said that, the UNDROP brought important advancements that are expected to contribute to the reformulation of laws and public policies related to small-scale fishers. The first and most visible one is the emphasis of the UNDROP in promoting gender equality[24] in rural settings and guaranteeing women's equal participation in local economic activities and access to productive resources. These rights are usually overlooked in national normative bodies and policies as small-scale fishing is 'perceived to be a "male only" domain.'[25] As well noted, '[s]mall-scale fisheries sit within gendered social and cultural systems that perpetuate well-being disparities between men and women and introduce vulnerability within processes of ecological and social change.'[26] Article 4.1 of the UNDROP mandates states 'take all appropriate measures to eliminate all forms of discrimination against peasant women and other women working in rural areas' and ensure 'they fully and equally enjoy all human rights' and 'are able to freely pursue, participate in and benefit from rural economic, social, political and cultural development.' Particularly relevant to the case of *El Palmar*, states must guarantee women peasants (and artisanal fishers) enjoy 'equal access to use of and management of land and natural resources,' 'decent employment, equal remuneration and social protection benefits, and . . . have access to income-generating activities' (Article 4.2).

The UNDROP's strong emphasis on the collective dimension of peasants' activities and rights represents an important push for the reformulation of small-scale fishing-related laws and public policies, thereby officially acknowledging community-based production and commercialisation systems, and collective tenure rights.

The human rights standards established in the UNDROP have the potential to guide the reformulation of labour and social security norms at the national level, crucial for the fishing sector. Article 13.5 on the 'right to work,' for example, encourages states to take due account of the specific characteristics of small-scale fisheries and 'monitor compliance with labour legislation by allocating, where required, appropriate resources to ensure the effective operation of labour inspectorates in rural areas.'

Moreover, the following paragraph (Article 13.6) makes a strong case regarding states' obligation to protect peasants in general and fishers and fish workers in particular from all sorts of abusive economic exploitation. Although the provision draws special attention to serious human rights violations, such as forced labour and human trafficking, it is evident that these include the official recognition of all fish workers (including womenfishers) and the ascertaining of their adequate remuneration and access to social security schemes.

Safety in the workplace is also addressed in the UNDROP, which demands States implement policies and enforce national laws and regulations on occupational safety and health in fisheries (Article 14.3). Finally, the UNDROP recognises the rights of small-scale fishers to all social security guarantees (Article 23.1). For the purpose of the discussion around the *El Palmar*, this would entail the legal recognition of women of the community as fishers on an equal footing with men and equalising marine and continental fishing in terms of social security guarantees.

Along the same lines, the UNDROP stresses the connection between the right to an adequate standard of living and access to productive resources. As it states, peasants have the right to 'facilitated access to the means of production necessary to achieve [an adequate standard of living for themselves and their families], including production tools, technical assistance, credit, insurance and other financial services.' Taking the *El Palmar* case into consideration, this doubtlessly includes the rights of women to engage in traditional boat fishing.

Likewise, Article 17 underlined the linkage between the right to land and other natural resources and the right to an adequate standard of living. Clearly, in expanding the scope of this right so as to encompass the needs and interests of artisanal fishers, the UNDROP acknowledges the rights of peasants and small-scale fishers

> to have access to sustainably use and manage land and the water bodies, coastal seas, fisheries, pastures and forests therein, to achieve an adequate standard of living, to have a place to live in security, peace and dignity and to develop their cultures (Article 17.1).[27]

Directly related to the case of *El Palmar*, Article 17.2 demands states 'take appropriate measures to remove and prohibit all forms of discrimination relating to the right to land' (and water bodies), including discrimination on the basis of gender. Lastly, Article 17 also evokes the social function of land (and water bodies) and prescribes that 'small-scale fishers should be given priority in the allocation of . . . fisheries' (Article 17.6). This clarification is key in light of the frequent tensions between artisanal and industrial fishing, other large-scale activities, and development projects.[28]

Articles 18, 20, and 21 protect the rights of small-scale fishers to a living and work environment free from pollution. Article 18 provides for a general right to 'the conservation and protection of the environment and the productive capacity of their lands' (and water bodies), and of the natural resources peasants use and manage. Article 20, in turn, focuses on the rights to biodiversity, which includes, among others, marine and freshwater biodiversity. For communities reliant on coastal and riverine ecosystems, the preservation, access and use of biological diversity is fundamental to guaranteeing not only economic opportunities, but also food security, social and cultural practices and human well-being.[29] This right extends to traditional knowledge, innovations and practices of small-scale fishers, and artisanal fishing (Article 20.2).

Article 21 on the right to water and sanitation takes an important step forward as compared to classic and restrictive formulations of this right. It not only recognises peasants' rights to water for domestic uses but it also acknowledges for the first time in international human rights law the rights to fishing and other water-related livelihoods as an element of the right to water. Accordingly, it demands states ensure the conservation, restoration, and sustainable use of water courses and bodies (Article 21.1). Moreover, the UNDROP prescribes the protection and restoration of 'water-related ecosystems, including mountains, forests, wetlands, rivers, aquifers and lakes, from overuse and contamination by harmful substances, in particular by industrial effluent and concentrated minerals and chemicals that result in slow and fast poisoning' (Article 21.4). Once more, the fight against all forms of discrimination is emphasised here: 'States shall respect, protect and ensure access to water, including the customary and community-based water management systems, on a non-discriminatory basis' (Article 21.3).

A number of issues have not received enough attention in the UNDROP, however. Mobility, for instance, is one of these. Not much has been said to this regard in the Declaration beyond the general statement in Article 7.3. This provision requires states to 'cooperate with a view to addressing transboundary tenure issues affecting peasants and other people working in rural areas that cross international boundaries.' Considering the constant tension on the demarcation of territorial boundaries and the control of waters in border zones,[30] a more assertive language affirming the rights of subsistence and small-scale fishers to access fisheries would have been welcomed.

Conclusion

Departing from the case of *El Palmar*, this chapter identified some of the challenges around the rights of fisherwomen to access fisheries and to participate in fisheries governance on an equal footing with men. The current normative body at the EU level remains evasive with respect to gender equality in the fishery sector, thus leaving space for discriminatory practices in Member States.

The adoption of the UNDROP created the momentum for the revision of laws and public policies to adjust them to human rights standards in rural contexts. It is about time that the EU takes the lead in this process and fully incorporates the fundamental rights of rural women in general and the rights of fisherwomen in particular in its regional regulations.

Notes

1 According to the Spanish Social Institute of the Navy (*Instituto Social de la Marina*), data of 2017. On this, see Europa Azul, Las mujeres representan el 5,4% de la pesca extractiva y el 62,9 % del total de mariscadoras (10 March 2019), at <https://europa-azul.

es/las-mujeres-representan-el-54-de-la-pesca-extractiva-y-el-629-del-total-de-marisca doras/> accessed 23 June 2021.

2 For a general appraisal of the case, see, for example, Francisca Ramón Fernández, 'La Comunidad de Pescadores de El Palmar y la Gestión del Redolí o Derecho de Pescaen la Albufera de Valencia' (2019) 73 *Revista Iberoamericana de Autogestión y Acción Comunal* 283ff; Ana María Gómez, 'Pesca Artesanal y Desarrollo Local en la Albufera de Valencia: Al Caso de la Comunidad de Pescadores del Palmar' (2017) *ECADIM;* Antonio Bellón Climent, 'Las actividades económicas en El Palmar (Valencia): entre la tradición y la modernidad, Universidad de Alicante' (2014) *Grupo Interdisciplinario de Estudios Críticos y de América Latina (GIECRYAL)*; FEP España, '758 años después. Las mujeres reescriben la Historia de la Pesca en El Palmar' (2008) *FEP España* 18–9; Cecilia Delgado, *La Gestión en el bien común de la pesca artesanal en la Albufera de Valencia: El caso de la comunidad de pescadores del Palmar. ECADIM Estudio de caso a dimensión múltiple* (Fundaciòn Daniel & Nina Carasso / Càtedra Tierra Ciudadana 2017).

3 FEP España, '758 años después. Las mujeres reescriben la Historia de la Pesca en El Palmar' (2008) *FEP España* 18, our translation.

4 José J. Pascual-Fernández, David Florido-del-Corral, Raquel De la Cruz-Modino and Sebastián Villasante, 'Small-Scale Fisheries in Spain: Diversity and Challenges' in José J. Pascual-Fernández et al (eds), *Small-Scale Fisheries in Europe: Status, Resilience and Governance* (MARE Publication Series 23, Springer 2020) 253ff.

5 FEP España, '758 años después. Las mujeres reescriben la Historia de la Pesca en El Palmar' (2008) *FEP España* 18.

6 Cecilia Delgado, 'La Gestión en el bien común de la pesca artesanal en la Albufera de Valencia: El caso de la comunidad de pescadores del Palmar' (2017) *ECADIM* 13.

7 Regulation (EU) No 1380/2013 of the European Parliament and of the Council of 11 December 2013 on the Common Fisheries Policy, amending Council Regulations (EC) No 1954/2003 and (EC) No 1224/2009 and repealing Council Regulations (EC) No 2371/2002 and (EC) No 639/2004 and Council Decision 2004/585/EC (as consolidated on 14/08/2019). For an overview of the European Common Fisheries Policy, see, among others: Mike Holden, *The Common Fisheries Policy: Origin, Evaluation, Future* (Fishing New Books 1994); Sebastiaan Princen, 'Venue Shifts and Policy Change in EU Fisheries Policy' (2010) 34(1) *Marine Policy* 36–41.

8 For a thorough appraisal of the EU Common Fisheries Policy and small-scale fisheries, see Jeremy Percy and Brian O'Riordan, 'The EU Common Fisheries Policy and Small-Scale Fisheries: A Forgotten Fleet Fighting for Recognition' in José J. Pascual-Fernández et al (eds), *Small-Scale Fisheries in Europe: Status, Resilience and Governance* (MARE Publication Series 23, Springer 2020) 23ff.

9 On this point also see, European Commission, 'Small-scale Fisheries' <https://ec.europa.eu/oceans-and-fisheries/fisheries/rules/small-scale-fisheries_en> accessed 21 June 2021.

10 Ibid.

11 Ibid.

12 European Parliament Resolution of 22 November 2012 on small-scale coastal fishing, artisanal fishing and the reform of the common fisheries policy (2011/2292(INI)), para. 42.

13 Regulation (EU) No 508/2014 of the European Parliament and of the Council of 15 May 2014 on the European Maritime and Fisheries Fund and repealing Council Regulations (EC) No 2328/2003, (EC) No 861/2006, (EC) No 1198/2006 and (EC) No 791/2007, and Regulation (EU) No 1255/2011 of the European Parliament and of the Council, Article 3.2 (14). Also see: Council Regulation (EC) No 1198/2006 of 27 July 2006 on the European Fisheries Fund. No longer in force, date of end of validity: 31/12/2013, Article 26.1.

14 Regulation (EU) No 508/2014 of the European Parliament and of the Council of 15 May 2014 on the European Maritime and Fisheries Fund and repealing Council Regulations (EC) No 2328/2003, (EC) No 861/2006, (EC) No 1198/2006 and (EC) No

791/2007, and Regulation (EU) No 1255/2011 of the European Parliament and of the Council, Article 29.1(b).

15 Council Regulation (EC) No 1198/2006 of 27 July 2006 on the European Fisheries Fund. No longer in force, date of end of validity: 31/12/2013, Article 11.

16 Ibid.

17 See, for instance, Articles 29.1(b), 97.1(b), and 113(e).

18 See General Fisheries Commission for the Mediterranean's purposes and policy and activities at <www.fao.org/gfcm/about/es/> accessed 22 June 2021.

19 This provision reads: 'The Conference [of the FAO] may approve arrangements placing other public international organizations dealing with questions relating to food and agriculture under the general authority of the Organization on such terms as may be agreed with the competent authorities of the organization concerned.'

20 Ministerial Declaration on a Regional Plan of Action for Small-Scale Fisheries in the Mediterranean and the Black Sea (Malta, 26 September 2018), full text available at <www.fao.org/gfcm/news/detail/en/c/1154586/> accessed 22 June 2021.

21 These include ministers from Malta, Cyprus, Hellenic Republic, Georgia, Croatia, Montenegro, Italy, Romania, Morocco, Spain, Algeria, Bulgaria, French Republic, Lebanon, Slovenia, Libya, and Turkey.

22 See Ministerial Declaration on a Regional Plan of Action for Small-Scale Fisheries in the Mediterranean and the Black Sea (Malta, 26 September 2018), paras. 47–9.

23 See UN Human Rights Council, 'United Nations Declaration on the Rights of Peasants and Other People Working in Rural Areas' (8 October 2018) UN Doc A/HRC/RES/39/12, art. 1.2.

24 UNDROP, supra note 23, art. 5. On this point, also see Joanna Bourke-Martignoni and Priscilla Clayes' chapter in this volume.

25 Poh Sze Choo, Barbara S. Nowak, Kyoko Kusakabe, and Meryl J. Williams, 'Guest editorial: Gender and Fisheries' (2008) 51 *Development* 176–9.

26 Danika Kleiber, Katia Frangoudes, Hunter T. Snyder, Afrina Choudhury, Steven M. Cole, Kumi Soejima, Cristina Pita, Anna Santos, Cynthia McDougall, Hajnalka Petrics, and Marilyn Porter, 'Promoting Gender Equity and Equality Through the Small-Scale Fisheries Guidelines: Experiences from Multiple Case Studies' in Sevein Jentoft, Ratana Chuenpagdee, María José Barragán-Paladines, and Nicole Franz (eds), *The Small-Scale Fisheries Guidelines: Global Implementation* (MARE Publication, Springer 2017) 739.

27 For a discussion on small-scale fishers' tenure rights, see Jokim Kitolelei, Sive Einarsson, Harry Lord, and Taiki Ogawa, *Global Conference on Tenure and User Rights in Fisheries 2018: Achieving Sustainable Development Goals by 2030* (FAO 2018).

28 The complexities involving the access to and the use of water courses and bodies have been subject to international courts, and decisions tended to overlook the rights of small-scale fishers. In the Pulp Mill case, for example, the International Court of Justice disregarded the rights of small-scale fishers to have access to relevant information on the potential contamination of the Uruguay river and be consulted prior to the installation of the industrial plant (ICJ, Case Concerning Pulp Mills on the River Uruguay (Argentina v. Uruguay), Judgement of 20 April 2010). In Costa Rica versus Nicaragua, the same court found Nicaragua's objections to subsistence and small-scale boat fishing in the San Juan river reasonable (ICJ, Case Concerning the Dispute Regarding Navigational and Related Rights (Costa Rica v. Nicaragua). Judgment of 13 June 2009). The impact of the UNDROP remains to be seen in future decisions of the court.

29 On this point, also see Sarah Lawless, Philippa J. Cohen, Sangeeta Mangubhai, Danika Kleiber and Tiffany H. Morrison, 'Gender Equality is Diluted in Commitments Made to Small-scale Fisheries' (2021) 140(105348) *World Development*.

30 See, for instance, *Case of the Maritime Boundary Delimitation between Eritrea and Yemen (Eritrea v. Republic of Yemen)* (1999) PCA.; *Case Concerning the Dispute Regarding Navigational and Related Rights (Costa Rica v. Nicaragua)* (2005) ICJ.

Selected bibliography

Holden M, *The Common Fisheries Policy: Origin, Evaluation, Future* (Fishing New Books 1994).

Jentoft S, Chuenpagdee R, Barragán-Paladines MJ and Franz N (eds), *The Small-Scale Fisheries Guidelines: Global Implementation* (MARE Publication, Springer 2017).

Kitolelei J, Einarsson S, Lord H and Ogawa T, *Global Conference on Tenure and User Rights in Fisheries 2018: Achieving Sustainable Development Goals by 2030* (FAO 2018).

Lawless S, Cohen PJ, Mangubhai S, Kleiber D and Morrison TH, 'Gender Equality is Diluted in Commitments Made to Small-scale Fisheries' (2021) 140(105348) *World Development.*

Pascual-Fernández JJ et al (eds), *Small-Scale Fisheries in Europe: Status, Resilience and Governance* (MARE Publication Series 23, Springer 2020).

17

THE UNDROP AND THE CASE LAW OF THE INTER-AMERICAN HUMAN RIGHTS SYSTEM

Potential impacts and insights from Hacienda Brasil Verde case

Naiara Posenato

Introduction

According to the Observatory for the Eradication of Slave Labor and Trafficking in Persons, a digital tool developed by the Brazilian Public Labour Prosecution Office (in Portuguese: *Ministério Público do Trabalho*) and the International Labour Organization (ILO) – Brazil, 45,028 people were rescued in conditions of forced labour in the country from 2003 to 2018. Seventy-three per cent of the victims worked in the agricultural sector.[1] Most of them are national or foreign migrants who have left their homes for agricultural expansion regions searching for new opportunities or attracted by false promises. In the Americas, the ILO estimates that more than 1.2 million people were affected by forced labour in 2016.[2] In addition to forced labour, other contemporary forms of slavery involving debt bondage and human trafficking continue to haunt the country and are major human rights issues in rural Latin America.[3]

In this context, on 20 October 2016, the Inter-Americal Court of Human Rights (IACHR) delivered the judgement in the case of the Hacienda Brasil Verde Workers *v*. Brazil (hereinafter Hacienda Brasil Verde), by which it declared the Brazilian State internationally responsible for the breach of the obligation to ensure the prohibition of slavery, forced labour, and human trafficking, established in Article 6.1 of the American Convention on Human Rights (ACHR).[4] In what became a leading case in the field, the IACHR comprehensively addressed the theme and reaffirmed some of the Inter-American System's known peculiarities: a universalistic legal approach, recognising the universal nature of international human rights, the *pro personae* principle on the interpretation of human rights norms, and the recognition of categorical rights to vulnerable groups.

Since the United Nations Declaration on the Rights of Peasants and other people working in rural areas (UNDROP) had not been adopted at the time the

DOI: 10.4324/9781003139874-22

decision was delivered, it was not expressly taken into account in the ruling. Nevertheless, the values and principles that guided the drafting of the UNDROP were evoked in the Court's reasoning and largely informed its decision.

This chapter has a two-fold purpose. First, it discusses the IACHR's ruling and identifies insights that may illuminate the implementation of the UNDROP. Second, it will look at the UNDROP standards on rural workers and point out normative and interpretative tools this new international instrument provides for in cases similar to the Hacienda Brasil Verde, which deal with new forms of slavery.

The Hacienda Brasil Verde case: briefing facts and legal reasons

In the decision of the Hacienda Brasil Verde case of 2016, Brazil was declared unanimously responsible for violating the prohibition of slavery and trafficking in persons established in Article 6.1 of the ACHR. The majority of the judges recognised that the violation occurred in the context of historical structural discrimination, based on the economic status of the 85 rural workers identified and rescued by officers of the Brazilian Ministry of Labour in March 2000.[5] Moreover, the State was also condemned for violating Articles 8.1 and 25 of the ACHR, which, respectively, prescribe the right to a fair trial with due guarantees and within a reasonable time and the right to judicial protection.

The case relates to a practice of forced labour and debt bondage in *Fazenda Brasil Verde* (Green Brazil Farm), a cattle ranching company located in Sapucaia, state of Pará, in Northern Brazil.[6] For many years, rural workers from the state of Piauí, one of the poorest in the country, were recruited with false promises of a job with decent remuneration and good working conditions. The two states are more than one thousand kilometres away from each other, and workers had to travel for several days to arrive at the workplace. Upon their arrival, workers' work permits were retained, and they were obliged to sign blank papers and prohibited from leaving the place under the threat of violence.

Working conditions were extreme: 12 working hours or more, a 30-minute break for lunch, and only one day of rest per week. Activities mainly consisted of clearing the dense Amazonian vegetation, and workers' performed their duties with their feet submerged in water. Workers were continuously under threat of armed surveillance. The accommodation was poor, unsafe, and unhealthy: workers slept in hammocks in wooden huts without electricity, beds, and sanitary facilities. The food and water were insufficient and of low quality. Due to severe working conditions and contaminated food and water, workers were often ill. What is more, food and medicines were deducted from workers' salaries, which they would not receive if high production goals were not met. So, victims have received no payment for their work.

Since 1989, the State had been aware of these practices in general, specifically in the Fazenda Brasil Verde. Authorities had been informed about alleged rights violations, several complaints of slave work had been made and the disappearance of young people in the farm had been reported. Despite all the police investigations and inspections as well as court proceedings, the practice of forced labour persisted.[7] The government did not take effective actions to remediate ongoing violations of workers' rights or prevent new cases. Nor did the country provide victims with judicial protection or punish rights violators.

In 1998, two NGOs, the Brazilian Pastoral Land Commission *(Comissão Pastoral da Terra)* and the regional non-governmental organisation Centre for Justice and International Law (CEJIL) lodged a petition with the Inter-American Commission of Human Rights (IACmHR) and international proceedings within the Organisation of American States (OAS) were initiated. On 3 November 2011, IACmHR delivered its decision according to which Brazil was responsible for a situation of forced labour and servitude for debt analogous to slavery and made some recommendations. Unsatisfied with Brazil's response, the IACmHR referred the case to the IACHR on 4 March 2015.[8]

The 2016 Inter-American Court decision first analyses preliminary objections of alleged lack of jurisdiction *ratione personae*, *ratione materiae*, and *ratione temporis*. It concluded that in light of Brazil's acceptance of the Court's contentious jurisdiction dated December 1998, the Court was entitled to examine the state's acts and omissions occurring during investigations and proceedings related to 128 workers in 1997 and 2000, but only the acts performed after 1998 could be taken into account.[9]

Brasil Verde is the first contentious case before the Inter-American Court related to Article 6.1 of the ACHR. Hence, on the merits, primarily, the Court conducted a lengthy analysis of the notion and scope of the prohibition of slavery in binding international instruments and the practice of international tribunals to give content to the conventional concepts of slavery, servitude, forced labour, and practices similar to slavery from a human rights perspective. In doing so, it reaffirmed the prohibition of slavery as *ius cogens* in international law and the right to be free from slavery or servitude as a non-derogable and non-limited human right in the light of the American Convention.[10]

The Court also discussed the notion of contemporary slavery. The new forms of slavery are no longer limited to a right of ownership over a person as prescribed in the 1926 UN Slavery Convention. The notion of slavery must thus be interpreted in light of the principle of evolutive interpretation of human rights norms so as to encompass all occasions in which there is a *de facto* exercise of the right of ownership over the enslaved person to the point of nullifying his/her will or considerably reducing his/her autonomy. As asserted in the judgment:

> The absolute prohibition of traditional slavery and its interpretation have evolved so that it also includes certain analogous forms of this phenomenon, which are revealed in different ways at the present time, but retain certain essential characteristics that are common to traditional slavery, such as the exercise of control over an individual by physical or psychological coercion in such a way that it entails the loss of the individual's autonomy and his unwilling exploitation.[11]

As regards forms of slavery, the decision stated that servitude *ex* Article 6.1 of the ACHR includes 'the obligation to perform certain services for others . . . that is imposed by the use of coercion,' and 'the obligation for the "serf" to live on another person's property and the impossibility of altering his condition.'[12] Forced or compulsory labour, prohibited by Article 6.2 of the ACHR, had already been the object of analysis in the *Ituango Massacres* v. *Colombia* case, which interpreted the provision in the light of Article 2.1 of ILO Convention No 29.[13] Lastly, according to the Court, the expression 'trafficking in slaves and women' in Article 6.1 of the ACHR must be interpreted broadly to refer to 'trafficking in persons,' in conformity with the evolution of international law and 'based on the interpretation that is most favourable to the individual and the *pro persona* principle,'[14] as explained later.[15]

The decision aligned with the UNDROP. Article 6.2 of those Rules prescribes the rights of peasants and other rural actors, including rural workers 'not be subjected to . . . cruel, inhuman or degrading treatment or punishment, and . . . not be held in slavery or servitude.' In Article 13, which refers to peasants' right to work, the UNDROP categorically affirms that: 'No one shall be required to perform forced, bonded or compulsory labour, be subjected to the risk of becoming a victim of human trafficking or be held in any other form of contemporary slavery.' Fundamentally, the Inter-American Court recognises the exploitative working conditions peasants face in many parts of the world and goes further to detail the working and living conditions that are analogous to forced labour and slavery. In doing so, the Court also endorses the Protocol to Prevent, Suppress, and Punish Trafficking in Persons Especially Women and Children, supplementing the UN Convention against Transnational Organized Crime (the Protocol of Palermo) and its definitions of contemporary slavery and ratifying States' international obligations.[16]

Based on these elements, the Court concluded that the working and living conditions of the 85 rural workers rescued in March 2000 corresponded to new forms of slavery. Likewise, considering the recruitment of workers through fraud and false promises, judges found that the rescued workers were also victims of human trafficking.[17]

Another important aspect highlighted in the decision is that the 85 victims shared specific characteristics of a particular vulnerability that made them more susceptible to recruitment through deception. They were poor, illiterate with little or no schooling. They were mostly afro descendants from the country's poorest regions and had thus lower employment opportunities. As the State did not consider the vulnerability of these workers, under the historical structural discrimination based on race and the economic constrains to which they were subjected,[18] Brazil was deemed to be responsible for the violation of Article 6.1 in conjunction with Article 1.1 of the ACHR.

The Court also concluded that Brazil violated Article 8.1 of the ACHR, for not guaranteeing the first 43 workers rescued by state officers the right to a hearing, with due guarantees and within a reasonable time. Finally, it acknowledged the violation of the right to judicial protection provided for in Article 25 of the ACHR, in conjunction with Articles 1.1 and 2 of the same instrument, to the detriment of the 43 workers rescued in 1997 and the 85 workers rescued in 2000.[19]

Historical structural discrimination, vulnerability, and categorisation

During the first years of its functioning, the Inter-American Human Rights System mainly focused on limiting and sanctioning the public authority's repressive activity, often responsible for gross, massive civil and political rights violations. However, it has gradually extended its attention to severe inequality and social exclusion matters that characterise Latin America.[20]

In line with this general direction, since the beginning of 2000, the IACHR analyses the individual right's violation in light of the particular condition or situation of fragility, poverty, and marginalisation of the victims of human rights abuses. Formally based on the principle of equality (Article 11 of the American Declaration on Human Rights and Articles 1 and 24 of the ACHR), and on the related prohibition of discrimination, categories of subjects in a (structural or even only transitory) situation of *vulnerability* have been progressively identified as women and children, migrants, prisoners, people with disabilities, human rights defenders, indigenous peoples, and tribal groups, as well as ethnic minorities.[21]

The practical importance of this praetorian creation is to extend the scope of the treaty provisions beyond the aim of preventing states from adopting laws or measures of a discriminatory nature, in conformity with Article 1.1 of the ACHR,[22] but to impose reinforced positive obligations to guarantee adequate respect for the right to equality and the protection of rights. The substantial dimension of the principle of equality requires states to take positive action to avoid or correct degrading situations at the origin of vulnerability and provide adequate protection to vulnerable groups.

In concrete terms, the first rulings on the matter revealed the recognition, on behalf of vulnerable groups, of the innovative right to live a dignified life (*vida digna*):[23] according to the reading of the IACHR, the right to life must be understood, especially in case of vulnerable groups, not only as the right of each individual not to be deprived of his/her life but also as the right to have access to the minimum conditions for a dignified life and personal development (nutrition, health, education, work), *i.e.*, also called the right to life *lato sensu*.[24] As the system evolved, it prescribed states adopt positive measures to ensure the protection of various vulnerable groups, based on the principle of non-discrimination and concerning different types of rights.[25]

An area of particular interest in Inter-American jurisprudence in this sense is that relating to the protection of indigenous peoples and tribal communities. By virtue of the historical conditions of marginalisation and discrimination suffered by indigenous peoples, the Court expressly recognised them as particularly vulnerable, arguing that states must guarantee more intense protection of their rights by adopting special measures. On this basis, mostly since the case Mayagna (Sumo) Awas Tingni v. Nicaragua of 2001,[26] the case jurisprudence of the IACHR has continuously developed, recognising indigenous communities the right to live a dignified life and to collective property and providing further guarantees in exercising their right to judicial protection, legal personality, freedom of expression, protection of family life, and political rights.[27]

In the Hacienda Brasil Verde case, by five votes to one,[28] the Court considered Brazil responsible for the violation of Article 6.1 in conjunction with Article 1.1 of the ACHR, by shedding special light on the context of historical structural discrimination, based on the economic status of the 85 workers.[29] For the first time, 'poverty' has been considered part of the discrimination on the grounds of the 'economic status,' a criterion expressly listed in Article 1.1 and a factor of a group's vulnerability that intensifies the impact of the human rights violations. The Court also recalled that when dealing with persons in a situation of vulnerability, it is essential that the state adopts affirmative measures. It stated that Brazil failed to address structural discrimination and adopt specific protective actions concerning individuals recruited to work in *Fazenda Brasil Verde*.[30]

This chapter has deliberately overlooked the difficulties of defining vulnerability and categories needing special protection, or the more general critics to such an approach's sustainability.[31] In our opinion, however, it is important to highlight that this argument bears an element of openness to recognizing the need to protect better the category of peasants and other people working in rural areas.[32]

In this perspective, this IACHR decision and the UNDROP mutually strengthen each other's legitimacy, as the preamble of the UNDROP recognises that peasants and rural workers 'suffer disproportionately from poverty, hunger and malnutrition' and 'are often denied tenure and ownership of land, equal access to land, productive resources, financial services, information, employment or social protection.' Also, as 'several factors make it difficult for peasants . . . to make their voices heard [and] to defend their human rights' or to 'gain access to courts, police officers, prosecutors and lawyers,' UN Declaration asserts that States shall adopt appropriate measures to eliminate conditions that cause or help to perpetuate discrimination to safeguard the peasants' and rural workers' rights.

Universalistic approach and interpretation according to *pro personae* principle

A further element of relevance to the discussion on the rights of rural workers and the implementation of the UNDROP is the *pro homine* or *pro personae* principle for the interpretation of human rights norms.[33] This principle corresponds concretely to the prohibition of restrictive interpretation imposed by Article 29(b) of the ACHR, according to which:

> [N]o provision of this Convention shall be interpreted as: . . . b. restricting the enjoyment or exercise of any right or freedom recognized by virtue of

the laws of any State Party or by virtue of another convention to which one of the said states is a party.

In accordance with the Inter-American System case law, this provision establishes that the conventional rules for norms interpretation must be in line with the canons of national and international law and, precisely, must offer the broadest protection for human beings. Furthermore, according to its practical application by the IACHR, the norm authorises an extensive interpretation of the rights protected by the ACHR, prioritising the interpretation that is more favourable to the individual, and harnessing elements contained in national and international laws. So, the *pro personae* principle has a double feature: it is an interpretative principle that establishes a meaning primacy and a normative conflict-resolution technique that establishes a priority of norms.[34] According to this second corollary, the *pro personae* principle indicates that amidst different existing human rights norms, the one bringing more benefits to the person concerned should prevail.

This perspective is similarly embraced when implementing the evolutionary interpretation of the norms entrenched in the ACHR, based on the recognition of its nature as a living instrument.[35] Consequently, the lack of a regional consensus regarding a specific practice has a marginal impact, not substantially limiting the interpretation scope.[36]

Indeed, the Inter-American Court generally adopts a universalistic approach, according to which the unity of human nature and the universal nature of guaranteed rights and freedoms underpin every international protection system.[37] This individual-centred focus legitimises selecting the most advantageous norm from the universal human rights *corpus juris*, for protecting human rights and supplements its interpretation of the American Convention provisions.

For this interpretation which is open to sources external to the regional system, the Court uses both binding international law and soft law and the jurisprudence of other universal or regional judicial bodies, the doctrine of committees, and good practices in particular.[38]

The *pro personae* principle was applied in the Hacienda Brasil Verde case. The significance of 'slave trade and traffic of women' as provided for in Article 6.1 of the ACHR has been broadly interpreted so as to encompass 'trafficking in persons' and augment the protective scope of the provision. Global and regional treaties, the jurisprudence of the European Court of Human Rights, and the reports of the UN special rapporteurs served as a basis for the court's reasoning.[39]

Therefore, the frequent use of sources external to the American regional system, regardless of their hierarchy, represents a potential further opening of this system to the UN Declaration on Peasant's rights. Also, in accordance with the *pro persona* principle, the UNDROP could be taken into consideration by the Court as representing the most advanced interpretation of peasants and rural workers' human rights already affirmed by international and regional conventions.

Broad interpretations of protective measures in tune with the *pro personae* principle have informed a more recent decision of the IACHR. In the case of the

Indigenous Communities of the Lhaka Honhat Association (Our Land) v. Argentina, the court made express reference to the UNDROP, whose new standards on the rights of peasants and other rural actors informed the findings of the court.[40] In this decision, Argentina was declared responsible for the violation of a number of human rights of 132 indigenous communities from the province of Salta, including the right to communal property over land *criollo* families (families of mixed European and indigenous heritage) also inhabited. Drawing upon the definition of peasants offered in the UNDROP, the *criollo* population was considered as 'vulnerable rural settlers.' The Court clarifies that, although they were not included as victims in the proceedings, the State has obligations towards them and must take positive steps to ensure their rights.[41]

Conclusion

Considering the relevance of the violence in rural areas in Latin America, the Inter-American human rights system is expected to pay greater attention to peasants and other rural actors, including rural workers, and provide legal remedies to victims.

Thanks to some particularities of this regional system already tested in the Court's jurisprudence, as in the Hacienda Brasil Verde case, such as the categorisation of vulnerable groups and the application of the *pro personae* principle, the UN Declaration on Peasant's rights can constitute a crucial reference for the interpretation of the ACHR and other regional instruments, as well as an important source of principles for the protection of the rights of rural populations in the Americas. Conversely, UNDROP utilisation by the Court may legitimise and encourage its use in the courts of other jurisdictions.

The new human rights standards entrenched in the UNDROP constitute an essential reference for the interpretation of the ACHR and other regional instruments, as well as an important source of principles for the protection of the rights of rural populations in the Americas. Conversely, the utilisation of the UNDROP by the Court may legitimise and encourage its use in the courts of other jurisdictions.

Notes

1 See, Iniciativa SmartLab: Promoção do Trabalho Decente Guiada por Dados, at <http://smartlabbr.org>.
2 ILO and Walk Free Foundation, *Global Estimates of Modern Slavery: Forced Labour and Forced Marriage* (ILO 2017) 19.
3 For an analysis of the cases already adjudicated by the Inter-American Court of Human rights concerning human rights violations committed in rural areas, focusing on Colombia and Peru events, see Luis van Isschot, 'Assessing the Record of the Inter-American Court of Human Rights in Latin America's Rural Conflict Zones (1979–2016)' (2018) 22 *The International Journal of Human Rights* 1144ff.
4 *Case of the Hacienda Brasil Verde Workers v. Brazil. Preliminary Objections, Merits, Reparations and Costs,* Inter-American Court of Human Rights Series C No. 318 (20 October 2016).
5 For commentaries on the decision, see Tatiana Gos, 'Hacienda Brasil Verde Workers v. Brazil: Slavery and Human Trafficking in the Inter-American Court of Human

Rights' (2017) *Oxford Human Rights Hub* <https://ohrh.law.ox.ac.uk/hacienda-brasil-verde-workers-v-brazil-slavery-and-human-trafficking-in-the-inter-american-court-of-human-rights/> accessed 6 May 2021; Patricia Grazziotin Noschang, 'Aplicação do controle de convencionalidade para erradicação da escravidão no Brasil: o caso Trabalhadores da Fazenda Brasil Verde c. Brazil' in Paulo Pinto de Albuquerque and João Otávio de Noronha (eds), *Comentários à Convenção sobre Direitos Humanos* (Editoria Forense 2020) 491ff; Valentina Milano, 'Human Trafficking by Regional Human Rights Courts: An Analysis in Light of Hacienda Brasil Verde, the First Inter-American Court's Ruling in this Area' (2018) *Revista Electrónica de Estudios Internacionales*, <www.reei.org/index.php/revista/num36/notas/human-trafficking-by-regional-human-rights-courts-an-analysis-in-light-of-hacienda-brasil-verde-the-first-inter-american-courts-ruling-in-this-area> accessed 6 May 2021; Jeffrey S. Vogt, 'The Legacy of Slavery in Brazil: Interpreting Article 6 of the American Convention on Human Rights' (2017) 3 *International Labor Rights Case Law* 393ff.; Irit Weiser, 'Case Comment: Inter-American Court Issues its First Decision on Modern Day Slavery: Case of Hacienda Brasil Verde' (2018) *Philippe Kirsch Institute* <www.kirschinstitute.ca/hacienda-brasil-verde/> accessed 6 May 2021.

6 The facts detailed here are taken from *Case of the Hacienda Brasil Verde Workers*, supra note 4.

7 *Case of the Hacienda Brasil Verde Workers*, supra note 4, para. 129ff.

8 Ibid., paras. 1–4.

9 Ibid., paras. 16–98. On this point, also see Saulo José Casali Bahia, 'O caso Fazenda Verde (CADH), a proteção contra o trabalho escravo e o cumprimento do julgado pelo Brasil: competência, contraditório, ampla defesa, presunção de inocência e imprescritibilidade' in Paulo Pinto de Albuquerque and João Otávio de Noronha (eds), *Comentários à Convenção sobre Direitos Humanos* (São Paulo 507ff, Tirant lo Blanch 2020).

10 This acknowledgement was important from a practical standpoint because crimes involving *ius cogens* human rights norms violations cannot be subject to prescription. See ibid., para. 454.

11 *Case of the Hacienda Brasil Verde Workers*, supra note 4, para. 276.

12 Ibid., para. 280.

13 *Case of the Ituango Massacres v. Colombia*, Inter-American Court of Human Rights Series C No. 148 (1 July 2006), paras. 155–60.

14 *Case of the Hacienda Brasil Verde Workers*, supra note 4, para. 289.

15 See infra, section 3 in this chapter.

16 See Human Rights Council Open-ended intergovernmental working group on the rights of peasants and other people working in rural areas Fourth session 15–19 May 2017, 37.

17 *Case of the Hacienda Brasil Verde Workers*, supra note 4, para. 305.

18 For some reflection on the topic, see infra, section 2 in this chapter.

19 *Case of the Hacienda Brasil Verde Workers*, supra note 4, paras. 360–420.

20 The list of rights expressly guaranteed by the ACHR does not contain the guarantees of fundamental social rights. The 1998 approval of the Additional Protocol to the American Convention on Human Rights in the Area of Economic, Social and Cultural Rights (Protocol of San Salvador), entered into force 16 November 1999 (OAS Treaty Series No. 69), is considered a progress towards the recognition of the fundamental economic, social, and cultural rights in the Region. At the moment, 17 States have ratified the instrument. However, according to Article 19(6) of the Protocol of San Salvador, only potential violations of the Article 8 (Trade Union Rights) and Article 13 (Right to Education) can legitimize Inter-American Court jurisdiction.

21 The literature on this topic is quite vast. See *inter alia*, Bernard Duhaime, 'Vers une Amérique plus égalitaire? L'interdiction de la discrimination et le système interaméricain de protection des droits de l'homme' in Ludovic Hennebel and Hélène Tigroudja (eds), *Le particularisme interaméricain des droits de l'homme* (Pedone 2009) 151ff; Rosmerlin Estupiñan Silva, 'La vulnerabilidad en la jurisprudencia de la Corte Interamericana de

Derechos Humanos: esbozo de una tipología' (2014) *Derechos Humanos de los Grupos Vulnerables, Red de Derechos Humanos y Educación Superior* 193ff; Mariela Morales Antoniazzi, 'La vulnerabilidad en la jurisprudencia de la Corte Interamericana a la luz del *ius constitutionale commune* de la democracia' in Juan Ignacio Ugartemendia Eceizabarrena, Alejandro Saiz Arnaiz and Mariela Morales Antoniazzi (eds), *La garantía jurisdiccional de los Derechos Humanos. Un estudio comparado de los sistemas regionales de tutela: europeo, interamericano y africano* (Onati 2015) 316–7.

22 'Article 1. Obligation to Respect Rights – 1. The States Parties to this Convention undertake to respect the rights and freedoms recognized herein and to ensure to all persons subject to their jurisdiction the free and full exercise of those rights and freedoms, without any discrimination for reasons of race, color, sex, language, religion, political or other opinion, national or social origin, economic status, birth, or any other social condition.'

23 See, on this concept, Steven R. Keener, Javier Vasquez, 'A Life Worth Living: Enforcement of the Right to Health through the Right to Life in the Inter-American Court of Human Rights' (2009) 40 *Columbia Human Rights Law Review* 595ff; Jo M. Pasqualucci, 'The Right to a Dignified Life (*Vida Digna*): The Integration of Economic and Social Rights with Civil and Political Rights in the Inter-American Human Rights System' (2008) 31 *Hastings International and Comparative Law Review* 1ff; Mary Beloff, Laura Clérico, 'The Right to Dignified Living Conditions and the Position of Vulnerability in the Jurisprudence of the Inter-American Court' (2014) *SELA (Seminario en Latinoamérica de Teoría Constitucional y Política) Papers 146* 1ff.

24 The first decision where the right to a dignified life was applied is *Villagrán-Morales et al. Case (The 'Street Children' Case) v. Guatemala, Merits*, Inter-American Court of Human Rights Series C No. 63 (19 November 1999).

25 Among the IACHR capital decisions, see *Juridical Condition and Human Rights of the Child*. Advisory Opinion OC-17/02, Inter-American Court of Human Rights Series A No. 17 (28 August 2002); *Juridical Condition and Rights of the Undocumented Migrants*, Advisory Opinion OC-18/03, Inter-American Court of Human Rights Series A No. 18 (17 September 2003); *Case of the 'Juvenile Re-education Institute' v. Paraguay. Preliminary Objections, Merits, Reparations and Costs*, Inter-American Court of Human Rights Series C No. 112 (2 September 2004); *Case of the Miguel Castro Castro Prison v. Peru. Merits, Reparations and Costs,* Inter-American Court of Human Rights Series C No. 160 (25 November 2006); *Case of Ximenes Lopes v. Brazil. Merits, Reparations and Costs,* Inter-American Court of Human Rights Series C No. 149 (4 July 2006); *Case of Valle Jaramillo et al. v. Colombia. Merits, Reparations and Costs,* Inter-American Court of Human Rights Series C No. 192 (27 November 2008); *Case of González et al. ('Cotton Field') v. Mexico. Preliminary Objection, Merits, Reparations and Costs*, Inter-American Court of Human Rights Series C No. 205 (16 November 2009); *Case of Chinchilla Sandoval et al. v. Guatemala. Preliminary Objection, Merits, Reparations and Costs*, Inter-American Court of Human Rights Series C No. 312 (29 February 2016); *Case of Gonzales Lluy et al. v. Ecuador. Preliminary Objections, Merits, Reparations and Costs*, Inter-American Court of Human Rights Series C No. 298 (1 September 2015); *Gender Identity, and Equality and Non-discrimination with Regard to Same-sex Couples,* Advisory Opinion OC-24/17, Inter-American Court of Human Rights Series A No. 24 (24 November 2017); *Case of Women Victims of Sexual Torture in Atenco v. Mexico. Preliminary Objection, Merits, Reparations and Costs*, Inter-American Court of Human Rights Series C No. 371 (28 November 2018).

26 *Case of Yatama v. Nicaragua. Preliminary Objections, Merits, Reparations and Costs*, Inter-American Court of Human Rights Series C No. 127 (23 June 2005).

27 For a general appraisal of the IACHR case law on indigenous and tribal people's rights see Naiara Posenato, 'La giurisprudenza della Corte interamericana in materia di diritti alla vita e alla proprietà dei popoli indigeni e tribali' (2018) 34(1) *DPCE Online* <www.dpceonline.it/index.php/dpceonline/article/view/489> accessed 6 May 2021, 139ff.

28 Judge Eduardo Ferrer Mac-Gregor Poisot expressed a separate opinion expanding argumentation on Historical Structural Discrimination, to which Judge Elizabeth Odio

Benito had adhered. Judge Eduardo Vio Grossi issued a dissenting opinion on the topic. Despite having voted in favour, Judge Humberto Antonio Sierra Porto also wrote a partially dissenting opinion.

29 'In this case, the Court notes some characteristics of specific victimization shared by the 85 workers rescued on March 15, 2000 . . . This placed them in a situation that made them more susceptible to recruitment by means of false promises and deception. This situation of imminent risk for a specific group of people with identical characteristics, from the same regions of the country, had historical roots.' *Case of Hacienda Brasil Verde Workers*, supra note 4, para. 339.

30 *Case of the Hacienda Brasil Verde Workers*, supra note 4, para. 341.

31 Jean-Yves Carlier, 'Des droits de l'homme vulnérable à la vulnérabilité des droits de l'homme, la fragilité des équilibres' (2017) 79 *Revue interdisciplinaire d'études juridiques* 175ff.

32 In the past, in the decision of the *Case of the Ituango Massacres v. Colombia*, supra note 13, para. 125, 69, the Inter-American Court recognized that the fact that most of the displaced people came from rural areas reinforced their vulnerability.

33 There is a copious body of literature on the theme. See, *inter alia*, Laurence Burgorgue-Larsen, 'El contexto, las técnicas y las consecuencias de la interpretación de la Convención Americana de los Derechos Humanos' (2014) *Estudios Constitucionales* 105ff; Lucas Lixinski, 'Treaty Interpretation by the Inter-American Court of Human Rights: Expansionism at the Service of the Unity of International Law' (2010) 21 *European Journal of International Law* 585ff; Valerio de Oliveira Mazzuoli and Dilton Ribeiro, 'The *Pro Homine* Principle as an Enshrined Feature of International Human Rights Law' (2016) 3 *Indonesian Journal of International & Comparative Law* 77ff; Yota Negishi, 'The Pro Homine Principle's Role in Regulating the Relationship between Conventionality Control and Constitutionality Control' (2017) 28 *European Journal of International Law* 457ff.

34 See Humberto Henderson, 'Los tratados internacionales de derechos humanos en el orden interno: la importancia del principio *pro homine*' (2004) 39 *Revista Instituto Interamericano de Derechos Humanos* 87ff; Hayde Rodarte Berbera, 'The *Pro Personae* Principle and its Application by Mexican Courts' (2018) 4 *Human Rights Law Review (Queen Mary)* 9ff.

35 See, for instance, *Case of the "Street Children" (Villagrán Morales et al.) v. Guatemala. Merits*, Inter-American Court of Human Rights Series C No. 63 (19 November 1999) para. 193; *Case of the Gómez Paquiyauri Brothers v. Peru. Merits, Reparations and Costs*, Inter-American Court of Human Rights Series C No. 110 (8 July 2004) para. 165.

36 See Gerald L. Neuman, 'Import, Export, and Regional Consent in the Inter-American Court of Human Rights' (2011) 19 *European Journal of International Law* 101–7; Gonzalo Candia, 'Interpretation of Regional Human Rights Conventions and Originalism: Different Context, Same Myths' (2017) 31(3) *DPCE Online* <www.dpceonline.it/index.php/dpceonline/article/view/431> accessed 6 May 2021, 600–1.

37 See, for instance, *'Other Treaties' Subject to the Consultative Jurisdiction of the Court (Art. 64 American Convention on Human Rights)*, Advisory Opinion OC-1/82, Inter-American Court of Human Rights Series A No. 1 (24 September 1982) para. 41.

38 Laurence Burgorgue-Larsen, supra note 33, at 120–1.

39 *Case of Hacienda Brasil Verde Workers*, supra note 4, paras. 281–90.

40 *Case of the Indigenous Communities of the Lhaka Honhat Association (Our Land) v. Argentina. Merits, Reparations and Costs*, Inter-American Court of Human Rights, (2020), Series C No. 400.

41 Ibid., paras. 136–7.

18

LAND ACQUISITION AND PUBLIC INTEREST LITIGATION IN INDIA*

Margherita Brunori and Pier Filippo Giuggioli

Introduction

This chapter uses a recent case study, the Mumbai–Ahmedabad High-Speed Train Project developed in the states of Gujarat and Maharashtra, India, in order to critically address two intertwining issues characterising peasants' struggles – land dispossession and access to justice. The study portrays the surfacing tensions surrounding large-scale land acquisitions for public infrastructure in India and the subsequent public interest litigation (PIL) stemming from the affected farmers' protest. It discusses the involved legal instruments in light of the rights affirmed in the United Nations Declaration on the Rights of Peasants and Other People Working in Rural Areas (UNDROP).

This case study was chosen due to the interest regarding the legal instruments deployed in the case. The first is the Right to Fair Compensation and Transparency in Land Acquisition, Rehabilitation and Resettlement Act of 2013 (RFCTLARR 2013 Act), a Union law modernising Indian land expropriation law and introducing progressive procedural safeguards vis-à-vis large scale investments. The second is PIL, a group action of jurisprudential creation, which has enabled the strengthening of social rights in India but is now accompanied by increasing criticism. Both instruments intersect the content of the UNDROP; the first by introducing procedural and substantive standards on land acquisition for a public purpose and the second by constituting a collective tool for strengthening access to justice for the most vulnerable.

The chapter means to unveil, on the one hand, the contradictions that can result from the enactment of liberalist policies aiming to attract investment in a fast-developing country, notwithstanding the attempts to strengthen procedural rights

* Both authors contributed equally to this chapter. Pier Filippo Giuggioli developed sections 2 and 3 and Margherita Brunori sections 1, 2.2, 3.2, 4 and 5. The authors are grateful to Giedre Jokubuskaite and Jordan Treakle for their insightful comments. A special thanks to the Self-Employed Women Association.

DOI: 10.4324/9781003139874-23

for the rural population. On the other, it discusses the potential role of collective legal action in enforcing and progressively interpreting existing laws, as well as its need to be brought in line with the right to access to justice as enshrined in the Declaration. The study contributes to underscoring the challenges of the implementation of UNDROP principles and rights. It also aims to put the UNDROP into the context of a wider legal system and to emphasise its potential contribution to strengthening the legal framework by providing a tool in the light of which assessing the relevant laws.

The bumpy road to strengthening safeguards against land acquisition: the RFCTLARR 2013 Act

The Indian law regulating land acquisition was, until 2013, the Land Acquisition Act of 1894: a law reflecting the colonial period that would only foresee monetary compensation for the holders of land ownership titles for land taken for public purposes.[1] The Land Acquisition Act, based on the doctrine of 'eminent domain' and enacted to allow the acquisition of land for public goals (i.e., railways, roads, bridges, etc.), empowered the government to acquire 'the land in any locality likely to be needed for any public purpose' without the owner's consent.[2]

The Act provided a list of what could be deemed to be included in the expression 'public purpose,'[3] but did not define it, leaving it open to interpretation by central and state governments. The application of the 'public purpose' clause was not highly controversial until 2005; however, land acquisition became critical when land began to be allocated to private enterprises for the development of Special Economic Zones.[4] To remove the allegedly arbitrary land acquisition process provided by the Land Acquisition Act of 1894, the RFCTLARR 2013 Act was adopted. The Act basically intervened on the definition of 'public purpose,' on the conditions for forcible land acquisition, and on the compensation for those affected. In order to widen the possibility for the state to acquire land and to remove the interpretation obstacles of the past, the Act provided a broad and flexible definition of 'public purpose' which allowed states to use it according to their wishes without serious court and/or civil society obstructions.[5]

In order to make the process smooth and resistance-free, the new law prioritised people's consent, limiting the coercive power of the state for certain acquisitions. It introduced several relevant innovations: it increased the number of potential beneficiaries of the new provisions, to include also those land users with secondary tenure rights or those persons whose land constitutes the primary source of livelihoods; it established that the purpose of the land taken had to be public or for a public–private partnership, but in the second case, there are stricter conditions; it fixed high compensation rates (four times the market price for rural areas and twice for urban areas); it provided for the design and enactment of a rehabilitation and resettlement plan including rehousing, job creation, and allocation of alternative land, in some cases; it inserted a food security clause, according to which no irrigated multi-cropped land could be acquired under the Act; it introduced a compulsory social impact assessment; and it established the requirement of prior consent from 70% of the affected population or 80% in case of public–private partnerships. Another important new element of the Act was the broadened concept of 'interested persons,' so that also those land users with secondary tenure rights or those persons 'whose primary source of livelihood is likely to be adversely affected' could benefit from the procedural safeguards.[6] However, the Act excepted the application of these safeguards to thirteen activities, among which railways construction, to which only compensation and resettlement and rehabilitation apply.

The 2013 Act tried to find a compromise between the two conflicting visions of land: a source of food and housing and a necessity for new industries, mines, large-scale agricultural projects, and infrastructure initiatives. On the one hand, it conceded serious and progressive solutions in terms of compensation and rehabilitation to consider the needs of those who have been displaced or have been affected by the acquisition and, on the other hand, it allowed Indian states to deeply restructure their political economy in favour of a more neoliberal and capitalistic approach.[7]

The Act, however, was met with significant resistance by investors (and those with links to investors), and the year after its adoption, the new government announced its intention to repeal it. In December 2014, a temporary Ordinance was controversially promulgated to amend the 2013 Act,[8] mitigating the states' duties in acquiring land.[9] For instance, the Ordinance eliminated the requirement for consent for various initiatives (e.g., industrial corridors, defence projects, rural infrastructures, affordable housing, and infrastructure projects) and for the same initiatives, the Ordinance cancelled the mandatory social impact assessment requested by the 2013 Act. In 2015, when the Ordinance expired, an Amendment Bill was introduced to replace it.[10] Social protests and a judgement of the Supreme Court induced the government to withdraw the Amendment Bill, but several states, including Gujarat and Maharashtra, nevertheless promulgated new laws and rules substantially incorporating most of the changes introduced by the Ordinance and the Amendment Bill,[11] seeking to ease the process of land acquisition in favour of governments and investors.[12] Some of the main guarantees for local population facing land acquisitions are therefore excepted by national amendments to the 2013 Act.[13] From the resulting national amendments, prior consent, social impact assessment, with some variations between States, restrictions for agricultural land use are not applicable with regard to projects attaining to national security or defence, rural infrastructure, social housing, industrial corridors, infrastructure, and social infrastructure, including public–private partnership projects.[14]

The practice of states creating exemptions to the 2013 Act was addressed by the Supreme Court of India in June 2021. The Court, however, dismissed the case and affirmed that a State Act amending the RFCTLARR 2013 Act is legitimate and constitutional.[15] In addition to this debatable flexibility of the Act implementation, the fact that for several projects – including the construction of railways – the majority of procedural rights do not apply remains a controversial issue of the RFCTLARR Act 2013, as it exposes those affected by such projects to unaddressed adverse impacts.

Reading the RFCTLARR Act in the light of the UNDROP

India has endorsed the UNDROP, and its vote in favour should be read as expressing an implicit consent to abide by its contents.[16] Consequently, the UNDROP provides a reference against which it is possible to assess the RFCTLARR Act as well as the subsequent states' amending acts. Firstly, it is useful in interpreting the scope of the 'public purpose.' Even if the 2013 Act introduces a broad definition of the term, the Declaration suggests that any policy and project falling into the 'public purpose' definition should nevertheless contribute to respecting, protecting, and fulfilling the rights of peasants and other people working in rural areas (Article 2). Secondly, thanks to the stake given by Article 5 on the close relationship of rural population to natural resources, and by Article 17 on land, the Declaration stresses the need to extend the application of the standards of social and environmental impact assessment to all cases of land acquisition for a public purpose, and ensuring a sound rehabilitation and resettlement, providing alternative land, with a particular attention to communal land which is equally essential for ensuring communities' livelihoods. It also highlights the need to expand and strengthen the enforcement of benefit sharing, in addition to consultation, in any public or private project affecting rural populations (Article 5.2).

As the UNDROP should be used to orientate the interpretation of Indian legal framework, including the 2013 RFCTLARR Act, likewise it could be used to denounce the States' amendments watering down the Act. Indeed, any departure from the application of procedural standards aimed at fostering the participation of the affected communities, mitigating the impact of large scale projects, and protecting the continued access to natural resources of peasants and other people working in rural areas, clearly departs from the provisions and the aim of the Declaration, and should be condemned and avoided.

The two-edged sword of Public Interest Litigation

The 20th century marks the progressive recognition that socio-economic conditions affect the effective enjoyment of individual freedoms, and that private law needs to be socially oriented in order to contribute to the transition from an individualistic economic society to a structured welfare society.

Overcoming the formal equality of subjects and the emergence of the necessary collective approach to strengthening rights has involved, over the past century, all sectors where changes in economic relations cause the emergence of conflicts that are not adequately portrayed by the traditional, strictly individual approach to relationships. Judicial protection can no longer be invoked only for individual subjective situations but must also be able to be used in events which involve the interests of a multitude of subjects. In fact,

> more and more frequently the complexity of modern societies generates situations in which a single human action can be beneficial or prejudicial to large numbers of people, thus making entirely inadequate the traditional scheme of litigation as merely a two-party affair.[17]

This new kind of litigation, defined by Abram Chayes as 'public interest litigation,' typical of the modern welfare state, usually involves large public institutions.[18] In this context, the courts inevitably assume a new role. While in the traditional model of lawsuits involving two parties, the court is constricted by the selfish interests of the two parties and is focused on the canonical decision of the case in favour of one of them; in the PIL, the court is more active and orientates its attention to solve public problems that other institutions could not, preferring general remedies involving the collective interest and requiring long-term restructuring and monitoring of underperforming public institutions.

India represents a leading state in the expansion of PIL in the world. The Supreme Court of India introduced PIL by modifying the rules of the traditional individual trial. Such a transformation implied radical modification to pre-trial standing and filing rules, evidentiary and burden of proof rules, and post-trial remedies and enforcement mechanisms. The implementation of PIL in India passed through three phases: protection of fundamental rights, protection of the commons, and ensuring good governance. In the third phase, the constitutional court 'broadened the scope of public interest litigation and also entertained petitions to ensure that in governance of the state, there is transparency and no extraneous considerations are taken into consideration except the public interest.'[19] From the 1980s, this modern judicial attitude was combined with an inclusive interpretation regarding fundamental rights.[20]

The Indian Constitution contains both a list of fundamental rights and several socio-economic guarantees and principles. In particular, Part III of the Indian Constitution, entitled 'Fundamental Rights,' includes the rights to equality, freedom, and life, all rights that courts could enforce in cases that directly or indirectly relate to the enjoyment of these rights. Instead, Part IV of the Constitution, entitled 'Directive Principles of State Policy,' provides a detailed list of guidelines for the government to improve the socio-economic conditions of the people, including the 'right to an adequate means of livelihood.'[21] These guidelines are explicitly non-binding and non-justiciable because they are mere aspirational long-term state goals, not suited for judicial review. In this context, the Supreme Court began interpreting the justiciable right to life contained in Article 21 ('No person shall be deprived of his life or personal liberty except according to procedure established by law') through the involvement and absorption of a few non-justiciable socio-economic rights, such as the right to food, workplace protection, health, education, housing, etc.[22] The *People's Union of Civil Liberties (PUCL) v. Union of India and others is an* emblematic case with respect to this judicial approach, based on the favourable constitutional human rights provision and the availability of the PIL mechanism.[23] The case demonstrated the legal dimension of the right to food and paved the way for its implementation in the country.[24]

Despite the merits of the PIL, its late use has raised increasing criticisms. With the extensive use of this instrument, 'courts have emerged as institutions of governance through the device of [public interest litigation],' therefore raising doubts as for the deriving power imbalances between institutions,[25] and the tendency to excessively relax judicial procedures and remedies.[26] Some commentators argue that, since the 1990s, the instrument served the neoliberal turn of the country, to the point of being detrimental to the people which originally it was meant to support.[27] Others argue that what is problematic is the very nature of PIL, that leaves judges with excessive freedom to conduct the process.[28]

Access to justice in the UNDROP

The UNDROP recognises:

> Several factors make it difficult for peasants and other people working in rural areas, including small-scale fishers and fish workers, pastoralists, foresters and other local communities, to make their voices heard, to defend their human rights and tenure rights, and to secure the sustainable use of the natural resources on which they depend. (Preamble)

Collective mechanisms to access justice constitute a valuable tool for those who lack the individual capability to claim their rights to be heard. PIL has played and can continue to play a crucial role in broadening the entitlements of citizens and the progressive realisation of economic, social, and cultural rights in India. Nevertheless, collective litigation should work as an additional instrument for accessing justice and should not substitute fair trials addressing the individual situations of the affected persons, especially the most marginalised.

The UNDROP reiterates that access to justice is crucial for giving effect to human rights and peasants' rights. It reaffirms the right to effective and non-discriminatory access to justice and the right to legal assistance, and calls upon states to set up 'impartial and competent judicial and administrative bodies' providing 'effective and prompt remedies' (Article 12.2). It claims that states 'shall consider measures to strengthen relevant national institutions for the promotion and protection of all human rights' (Article 12.4) and provide peasants and other people working in rural areas with access to 'effective mechanisms for the prevention of and redress for any action' affecting their human rights, dispossessing them of their land or 'depriving them of their means of subsistence and integrity' (Article 12.5). The contents of Article 12 of the UNDROP could constitute the litmus paper for assessing what role PIL plays with regard to the rights of peasants and other people working in rural areas. It may also represent a push for reforming Indian PIL, so as to bring it in line with the purpose it was meant to serve and guarantee the right to access to justice as internationally recognised and reaffirmed in the Declaration.

The Mumbai–Ahmedabad High-Speed Bullet Train Project

The Mumbai–Ahmedabad High-Speed Bullet Train Project entailed the realisation of a high-speed train connecting the two capitals of Maharashtra and Gujarat, Mumbai and Ahmedabad.[29] It is financed by the Japan International Cooperation

Agency (JICA) through an Official Development Assistance loan, signed in agreement with the Government of India, and executed by the National High-Speed Rail Corporation Limited.[30]

According to the JICA ex-ante evaluation report, '[t]he Project involves the land acquisition of 1,022.37 ha and relocation of 1,887 households, and they will be proceeded according to the Resettlement Action Plan based on the domestic laws in India and JICA guidelines.'[31] Although not provided for by the applicable domestic law, for the reasons explained earlier, the ex-ante evaluation report asserts that an impact assessment was conducted according to the JICA guidelines,[32] that affected citizens were informed and 'no dissenting voice on the project has been confirmed in particular,' and that an Indigenous Peoples Plan was implemented and consensus reached.[33] However, the project spurred social protest among affected villages. The Gujarat Khedut Samaj, the farmers' committee, opposing the project, denounced that the calculation for compensation was based on the 2011 land value and that the activities of rehabilitation and resettlement were converted into cash compensation, that no social impact assessment was conducted and the environmental impact assessment as well as consultation was inadequate, as there was no participatory process nor adequate notification. In addition to farmers, it has been claimed that indigenous people's procedural rights were also violated in Maharashtra, and no impact assessment was conducted for fisherfolks.[34]

Members of the Self-Employed Women Association, a well-established women's trade union, reported that the government compensated farmers for the evicted land, but the compensation was deemed not enough to recover revenue losses, and no alternative land was offered.[35] Furthermore, farmers had obtained only a part of the declared compensation rate. Lack of clarity was perceived on how the government estimated the compensation price, so different farmers were receiving different sums for land of equal quality. The situation was worsened by the fact that, for realising the project, the building company needed sand, and acquired it from nearby villages. The extraction of sand made a village pond deeper to the point that animals could not access it for drinking water. Members reported that, as a consequence of the railway construction, many of the villagers were accusing damages to their houses and land because the construction prevented drainage of water and caused floods.[36]

A PIL was initiated in the local court. The petitioners, owners of land in various parts of South Gujarat, contested the forced acquisition of their land for the implementation of the project and challenged the constitutionality of the land acquisition law enacted in 2016 by the Gujarat State. The Gujarat 2016 law provided exemptions to the application of RFCTLARR 2013 Act. On this occasion, the local court found the state legislation was not unconstitutional and the forced acquisition was legitimate, and dismissed the case.[37]

The Mumbai–Ahmedabad High-Speed Train Project case showed that even a powerful collective instrument for accessing justice alone is not sufficient to ensure peasants' rights, especially if the law in force creates exceptions to the standards that itself provides. The tension therefore remains between the respect of the individual and collective rights of peasants and the urge to pursue capitalistic development strategies. The case sheds light on the political nature of land rights and acquisition. Even in the face of strong legal safeguards, investments' interests tend to prevail when the authorities pursue certain (excluding and discriminatory) visions of development.

In the context of development projects such as the one just described, or legal reforms, the Declaration could constitute a leverage for civil society movements to denounce the contradictions and gaps of the framework governing land acquisitions and development agendas. By affirming that '[p]easants and other people working in rural areas have the right to determine and develop priorities and strategies to exercise their right to development' (Article 3.2) and that states shall ensure that investment policies and programmes 'contribute effectively to protecting and strengthening livelihood options' (Article 16.4), the UNDROP helps in reframing the procedural standards put in place for large-scale projects. These standards must be in tune with the human rights of affected populations, the rights to non-discrimination and human dignity especially.

Conclusion

India has voted in favour of the adoption of the UNDROP, adopted progressive laws for the realisation of the right to food and fair investment procedures, and it has strong legal instruments, such as the PIL, to ensure that the most vulnerable are heard. However, the race for modernisation and investment attraction is conflicting with this strengthening of procedural rights, while threats to rural population's livelihoods[38] fuel social conflicts across the country.

The chapter discussed how the UNDROP could be used in the context of land acquisition and access to justice in India. The Declaration may serve several objectives and help untangling tensions in rural settings. Indian's favourable vote may be understood as the country undertaking all human rights obligations detailed in this instrument. Accordingly, national laws should be interpreted so as to accord with the contents of the Declaration, and the UNDROP's contents related to access to justice could be used to evaluate the impact and role of the PIL in respecting, protecting, and fulfilling peasants' rights.

Finally, the Declaration supports solidarity and collective action and gives legitimacy to peasants' struggles. As stated in the preamble:

> the individual, having duties to other individuals and to the community to which he or she belongs, is under the responsibility to strive for the promotion and observance of the rights recognised in the present Declaration and in national law.[39]

In this sense, the Declaration can serve the strengthening the legitimacy of collective action for peasants struggles and ensuring that existing national laws and judicial instruments serve the cause of the most vulnerable and concur in the promotion of social justice.

Notes

1 For a detailed history of the protection of the right to property in India, see Choudhry, Sujit, Madhav Khosla, and Pratap Bhanu Mehta (eds) *The Oxford Handbook of the Indian Constitution* (Oxford University Press 2016) 943–66.

2 Naoyuki Yoshino, Saumik Paul, Vengadeshvaran Sarma, Saloni Lakhia, 'Land Acquisition and Infrastructure Development through Land Trust Laws: A Policy Framework for Asia' (2018) 854 *ADBI Working Paper* 1 <www.adb.org/publications/land-acquisition-and-infrastructure-development-through-land-trust-laws>.

3 Land Acquisition Act, 1894, art 1.3(f).

4 Special Economic Zones Act 2005. See Sazzad Parwez, Vinod Sen, 'Special Economic Zone, Land Acquisition, and Impact on Rural India' (2016) 2(2) *Emerging Economy Studies* 223–39; Meenakshi Gogoi, 'The Nexus between Sovereignty and "Eminent Domain" under the Land Acquisition Act, 1894, and the Land Act, 2013' (2018) 48(2) *Social Change* 179; Kenneth Bo Nielsen and Alf Gunvald Nilsen, 'Law Struggles and Hegemonic Processes in Neoliberal India: Gramscian Reflections on Land Acquisition Legislation' (2015) 12(2) *Globalizations* 203, 204; Kenneth Bo Nielsen and Alf Gunvald Nilsen, 'Law-struggles, Law-making and the Politics of Hegemony in Neoliberal India: Towards a Critical Perspective on the 2013 Land Acquisition Act' in Anthony P D'Costa and Achin Chakraborty (eds) *The Land Question in India: State, Dispossession, and Capitalist Transition* (Oxford University Press 2017) 129–50; Heather Plumridge Bedi, 'Judicial Justice for Special Economic Zone Land Resistance' (2015) 45(4) *Journal of Contemporary Asia* 596–617.

5 RFCTLARR 2013 Act, Section 2.1 'Application of the Act.' The section provides a long, non-exhaustive list of cases that fall into the 'public purpose' definition, including 'strategic purposes,' 'infrastructure projects,' 'projects for projects affected families,' housing, 'planned development,' 'residential purposes,' and also to public–private partnership and 'for private companies for public purpose.'

6 RFCTLARR 2013 Act, Section 2.3(x). See also Anwarul Hoda, 'Land Use and Land Acquisition Laws in India' (2018) 361 *Indian Council for Research on International Economic Relations Working Paper* 1.

7 Abhijit Guha, 'Sociological Context of the New Land Acquisition Law in India: A Case Study' (2018) 15 *Journal of Resources, Energy and Development* 39, 50; Sukhpal Singh, 'Land Acquisition in India: An Examination of the 2013 Act and Options' (2016) 4(1) *Journal of Land and Rural Studies* 66.

8 The Right to Fair Compensation and Transparency in Land Acquisition, Rehabilitation and Resettlement (Amendment) Ordinance, 2014 (n. 9 of 2014). For a detailed report of the dynamics accompanying the adoption of the Ordinance, see G Raghuram, and Simi Sunny, 'Right to Fair Compensation and Transparency in Land Acquisition, Rehabilitation and Resettlement Ordinance 2014: A Process Perspective' (2015) *Indian Institute of Management Ahmedabad Working Paper* No. 2015–07–03.

9 Sanjoy Chakravorty, 'Land Acquisition in India: The Political-Economy of Changing the Law' (2016) 1 *Area Development and Policy* 48.

10 The Right to Fair Compensation and Transparency in Land Acquisition, Rehabilitation and Resettlement (Second Amendment) Bill, 2015.

11 RFCTLARR 2013 Act, art 109.

12 A list of the State deviations from the Union Act is provided by Kanchi Kohli and Debayan Gupta, *Mapping Dilutions in a Central Law* (Centre for Policy Research 2017).

13 For a wider analysis of property rights in India, see Namita Wahi, 'The Fundamental Right to Property in the Indian Constitution' (2015) *SSRN Electronic Journal* <www.ssrn.com/abstract=2661212> accessed 28 February 2019; Namita Wahi, 'The Tension between Property Rights and Social and Economic Rights: A Case Study of India' in Helena Alviar et al (eds) *Social and Economic Rights in Theory and Practice* (Routledge 2014).

14 Hoda, 'Land Use and Land Acquisition Laws in India', supra note 6, 16.

15 *G. Mohan Rao vs. State of Tamil Nadu* (WPC 1411 of 2020) LL 2021 SC 275 <www.livelaw.in/top-stories/supreme-court-dismisses-plea-challenging-tamil-nadu-land-acquisition-laws-176507> accessed 5 July 2021. The Case addressed the Tamil Nadu Land Acquisition Laws (Revival of Operation, Amendment and Validation) Act, 2019.

16 UN General Assembly, 55th plenary meeting, Monday, 17 December 2018, 10 am, New York (Official Records) UN Doc A/73/PV.55, 20–5. Available at <https://digitallibrary.un.org/record/3795602> accessed 6 July 2021. On the legal significance of the Declaration, see the Chapter by Francesco Francioni in this volume.

17 Mauro Cappelletti, 'Vindicating the Public Interest through the Courts: A Comparativist's Contribution' (1976) 25 *Buff L Rev* 643, 647.

18 Abram Chayes, 'The Role of Judge in Public Law Litigation' (1976) 89 *Harv L Rev* 1281. For Indian references, see Upedra Baxi, 'Taking Suffering Seriously: Social Action Litigation in the Supreme Court of India' (1985) 4(6) *Third World Legal Studies* 107, defining PIL as a 'social action litigation.'

19 *State of Uttaranchal v. Balwant Singh Chaufal* (2010) INSC 54.

20 See, for discussion of positive as well as critical aspects of public interest litigation in India: Surya Deva, 'Public Interest Litigation in India: A Critical Review' (2009) 28 *Civil Justice Quarterly* 19; Zachary Holladay, 'Public Interest Litigation in India as a Paradigm for Developing Nations' (2012) 19(2) *Indiana Journal of Global Legal Studies* 555; Francis Xavier Rathinam and A. V. Raja, 'Courts as Regulators: Public Interest Litigation in India' (2011) 16(2) *Environment and Development Economics* 199; Hajime Soto, 'The Universality, Peculiarity, and Sustainability of Indian Public Interest Litigation Reconsidered' (2017) 100 *World Development* 59; Manvendra Singh Jadon and Sourabh Roy, 'The Role of Public Interest Litigation in Shaping Up the Public Policy Regime in India: Over-Reaching or Justified and the Way Ahead' (2019) 3(1) *NLUAL* 1 <https://ssrn.com/abstract=3484794> accessed 9 June 2021.

21 Constitution of the Republic of India, 1949, art 39(a).

22 Deva, 'Public Interest Litigation in India,' supra note 20.

23 Writ Petition n. 196 (2001) <https://main.sci.gov.in/jonew/bosir/orderpdfold/33378.pdf> accessed 9 June 2021.

24 For a description of the case and a general assessment of public interest litigation, see Rehan Abeyratne, 'Enforcing Socioeconomic Rights in Neoliberal India' (2020) 29 *Minn. J. Int. Law* 1–64.

25 Parmanad Singh, 'Promises and Perils of Public Interest Litigation in India' (2010) 52(2) *Journal of the Indian Law Institute* 172, 187.

26 Aparna Chandra, 'Book Review: Anuj Bhuwania, Courting the People: Public Interest Litigation in Post-Emergency India' (2018) 16(2) *International Journal of Constitutional Law* 710–18; Singh, 'Promises and Perils of Public Interest Litigation in India,' supra note 25.

27 Mayur Suresh and Siddharth Narrain, *The Shifting Scales of Justice: The Supreme Court in Neo-liberal India* (Orient Blackswan 2014); Namita Wahi 'The Tension between Property Rights and Social and Economic Rights: A Case Study of India,' supra note 13.

28 Anuj Bhuwania, 'Public Interest Litigation as A Slum Demolition Machine' (2016) 12 *MIT Journal of Planning Projections* 67.

29 This case study was informed by a fieldwork conducted by Margherita Brunori in the ambit of a consultancy financed by the Self-Employed Women Association and FAO, in October 2018 in Gujarat, India.

30 JICA India Office Press Release, 'JICA Supports Project for the Mumbai – Ahmedabad High-Speed Rail by Providing an ODA loan of INR 5,500 Crore as Tranche 1' (September 28, 2018) <www.jica.go.jp/india/english/office/topics/press180928_01.html> accessed 31 May 2021.

31 JICA, *Ex-Ante Evaluation (for Japanese ODA Loan): Project for Construction of Mumbai – Ahmedabad High Speed Rail (I) Loan Agreement*. September 28, 2018 <www2.jica.go.jp/en/evaluation/pdf/2018_ID-P277_1_f.pdf> accessed 31 May 2021.

32 Japan International Cooperation Agency (JICA) Guidelines for Environmental and Social Considerations <www.jica.go.jp/english/our_work/social_environmental/guideline/pdf/guideline100326.pdf> accessed 31 May 2021.
33 JICA, *Ex-Ante Evaluation*, supra note 31.
34 DNA India, 'Legal Cases Mount against New Project' (2 December 2018) <www.dnaindia.com/india/report-legal-cases-mount-against-new-project-2691456> accessed 31 May 2021; DNA India, 'Sacrificing our agriculture for development' (2 December 2018) <www.dnaindia.com/india/report-sacrificing-our-agriculture-for-development-2691448> accessed 31 May 2021.
35 On the Self Employed Women's Association, see Rekha Datta, 'From Development to Empowerment: The Self-Employed Women's Association in India' (2003) 16(3) *International Journal of Politics, Culture and Society* 351.
36 The evidence was collected by Margherita Brunori during a focus group with Self-Employed Women Association Members, in Anand (4 October 2018).
37 *J.A. Patel v. State of Gujarat*, C/CSA/9864/2018 (19 September 2019).
38 Avishek G Dastidar, Kavitha Iyer, Neha Kulkarni, Vishwas Waghmode, Avinash Nair, 'Bullet Train to Start in 2022: A Look at its Progress, Hurdles,' *The Indian Express*, Ahmedabad, Mumbai, New Delhi (11 July 2018) <https://indianexpress.com/article/explained/bullet-train-project-land-acquisition-narendra-modi-mumbai-ahmedabad-bullet-train-5254273/> accessed 31 May 2021.
39 This is particularly relevant in consideration of the farmers protests taking place across India in opposition of the agricultural market reforms. See on this point: Mohammad Amir Anwar and Adnan Shakeel, 'Taking the Bull by its Horns: The Political Economic Logics of New Farm Laws and Agrarian Dissent in India' (2021) *Contemporary South Asia* 1.

Selected bibliography

Singh P, 'Promises and Perils of Public Interest Litigation in India' (2010) 52(2) *Journal of the Indian Law Institute* 172

Sonak I and Kumar Singh S, *Dilution by Design: Status of Land Acquisition Legislation in India* (Centre for Science and Environment, 2018).

Soto H, 'The Universality, Peculiarity, and Sustainability of Indian Public Interest Litigation Reconsidered' (2017) 100 *World Development* 59.

Wahi N, 'The Tension between Property Rights and Social and Economic Rights: A Case Study of India,' in Helena Alviar et al (eds) *Social and Economic Rights in Theory and Practice* (Routledge 2014).

CONCLUSIONS

Mariagrazia Alabrese, Adriana Bessa, Margherita Brunori, and Pier Filippo Giuggioli

The adoption of the United Nations Declaration on the Rights of Peasants and Other People Working in Rural Areas (UNDROP) opens up a new chapter of the international law and policy making concerning food and agriculture under a human rights perspective and triggers several reactions, reflections, and questions. This book aimed to unfold some of these issues through the analysis of a selection of themes, making up the four parts of the book. Accordingly, this volume first gave an account of the subjective scope of the Declaration, by clarifying how the diversity of actors working in rural areas is recognised and addressed, and considered intersectionality while configurating a collective subject, recipient of specific rights. Second, the book offered a closer look at the rights, of a simultaneously individual and collective nature, that compose the distinctiveness and innovative cipher of the UNDROP and collocate natural resources in a sphere of a common interest that conditions their governance, ownership, and use towards the achievement of overarching societal goals: social justice and sustainability. Third, it turned its attention to the broader domain of the governance of global food and agricultural systems, where the UNDROP represents the first articulated international human rights-based document. It explores the crosscutting themes revolving around the need to promote equity and specific knowledges as well as to ensure the long-term ecosystems' ability to regenerate. Finally, it puts the Declaration 'into context,' by exploring several surfacing and potential impacts of the instrument both in global, regional, and national legal systems and in legitimising grassroots claims.

Through a cross-cutting analysis of several phenomena characterising international law and global governance of natural resources, the volume offers a contribution to the interpretation of the UNDROP. The first transversal theme emerging from the contributors' analyses is the need for adequate identification of minority social categories and the recognition of specific rights. As it is argued here, this necessarily involves curbing the fragmentation of international law. By acknowledging

DOI: 10.4324/9781003139874-24

the interconnectedness of social, cultural, economic, and natural phenomena, the UNDROP emerges as an instrument able to remediate the fragmentation of international law regarding the governance of natural resources and food systems. It crosses quite diverse branches of international regulation, while at the same time offering a comprehensive and principled framework to guide states with respect to their human rights obligations.

The rights included in the UNDROP reaffirm and elaborate upon many existing international legal and interpretative instruments. In doing so, the UNDROP cuts across different areas of law: human rights, trade and investment, intellectual property, environmental protection and climate change, food and agriculture, and cultural heritage. The contributing authors of this edited book have unfolded the interconnections of the UNDROP with the multiple legal regimes it intersects and explored how the text incorporates well-established or emerging principles and standards in international law. The rights to food and food sovereignty, for instance, have been identified as key in the UNDROP. Golay argues that they need to be implemented for overcoming structural discrimination against peasants. The book's contributions also stress where UNDROP lacks sound engagement. This more critical aspect is evidenced mainly by Martignoni and Clayes' chapter, which argues that women's rights were partially 'sacrificed' in order to obtain consensus and ensure the adoption of the Declaration. Similarly, Francioni evidences a lack of definition of the jurisdictional scope of states' obligations, which would have brought the Declaration more in line with emerging recognition of the importance of the extraterritorial dimension of state responsibility for the realisation of economic, social, and cultural rights. The limitations of this instrument considered, several chapters allow for the appreciation of the usefulness of the combined reading of the UNDROP with other binding instruments as well as international guidance documents. Indeed, Martignoni and Claeys posit that a reading of the text alongside the General Comment n. 34 of the Committee on the Elimination of Discrimination against Women could ensure an adequate contextualisation of women peasants' rights in light of the most recent authentic interpretation of the contents of the Convention on the Elimination of All Forms of Discrimination against Women. Bessa and Gilbert engage with a comparison of the UNDROP and the UN Declaration on the Rights of Indigenous Peoples, thereby evidencing the different, yet complementary function of the two instruments in recognising collective rights of indigenous peoples and traditional local communities. Morgera and Nakamura explain how the complementary reading of the UNDROP, the FAO Voluntary Guidelines for Securing Sustainable Small-Scale Fisheries, and international environmental law creates a complete regulatory framework for safeguarding small-scale fishers, but they also stress that this could also favour a reconceptualisation of existing standards in light of building a 'more constructive engagement of duty-bearers with the rightsholders.' Le Teno, Frison, and Cogolati offer a combined reading of the farmers' rights as enshrined in the International Treaty on Plant Genetic Resources for Food and Agriculture and the UNDROP and conclude that the Declaration 'can create positive synergies between farmers' rights, . . . and standard [intellectual property rights] legislation, while maintaining predictability for breeders and life science companies, and fill a gap in international law,' provided that peasants organisations are effectively included in a multilateral forum as active subjects instead of as mere recipients of policies. In this sense, the contributing authors highlight how the possible tensions between rightsholders' protection can be overcome or avoided both by looking at the ratio as well as the combined interpretation of different international law instruments.

The contributing authors have also concluded that the fragmentation of international law is among the root causes of inadequate governance of natural resources, and the consequent regulatory gaps have led to relevant negative impacts on local populations and the environment. They position the UNDROP in a broader trend whereby the interaction of international law branches as well as jurisprudential interpretation of treaties allow those existing gaps to be filled. In this context, Cotula affirms that 'building on this tapestry of human rights configurations, the UNDROP's affirmation of peasants' right to land represents a turning point in longstanding efforts to connect land to human rights.' By putting the UNDROP in the context of climate change law, Alabrese and Savaresi affirm that 'the UNDROP falls within the context of a growing body of international human rights practice requesting states to enable traditional knowledge holders' participation in decision-making and science creation.' Similarly, Tignino, Restifo, and Kebebew argue that the UNDROP contributes to broadening the notion and the contents of the right to water and States' obligations in that regard. Notwithstanding the limitations of the UNDROP in addressing the rights of fisherfolk, Bessa and Martín López explore the ways in which the UNDROP should guide a better translation of the rights of women fishers as well as equitable conditions for small-scale fisheries in the context of the EU Common Fisheries Policy.

The second theme that crosses the entire book relates to challenging the politics of law and power. Every legal and policy document is embedded in implicit political visions. The Declaration is rooted in open criticism of the underlying politics of the current international economic and legal systems. Viewed in light of the political economy of food systems, Narula argues that the UNDROP reaffirms the centrality of the States in the agricultural governance in contrast to the neoliberal reforms operated in the last decades in the sector. She also claims that the 'effective implementation of the UNDROP has the potential to significantly alter human rights and environmental struggles, shift the narrative and vision for food systems, and decentralize power therein.'

The keyword of the document – *peasant* – is itself an expression charged with strong historical and political meaning. The use of a term, which has a negative connotation in the English language, is an act of provocation, redemption, and empowerment. In this context, the book addresses questions related to the reasons and consequences of this political use of language in international human rights standard making and to whether the active participation of grassroots movements in drafting and negotiation is a temporary trend or a structural change in international law-making. Furthermore, it is noticeable that the Declaration breaks with established paradigms and welcomes different worldviews, concerning, for instance, land and natural resources and the notion of culture and cultural identity. In this perspective, Edelman explains how the discussion around the expression 'peasants' and its definition unveiled that, contrary to the assumption in the literature that saw them as opposing actors, peasants and farmers are on a 'continuum of technological mixes and economic aspirations, from mainly self-provisioning to exclusively entrepreneurial and commercial enterprises,' and that the pejorative connotation of the word *peasants* points out exactly the subordination and discrimination of the category. Cotula similarly evidences how the recognition of the right to land 'more explicitly connects rights to economic relations and control over the means of production,' works as a reminder that rights are a social construct, and 'highlights the power of agency in shaping human rights.' Discussing the concept of sustainable agriculture, Cristiani and Strambi unveil how the UNDROP takes a stand in the tension characterising the narrative of the sustainable transition of agricultural systems. By recognising the fact that there is no single road to sustainability, the authors argue that the UNDROP makes a strong case that non-sustainable models of agriculture need to be restricted in order to protect and fulfil the rights of peasants and other people working in rural areas. Engaging in the local context, Tsioumani posits that the 'UNDROP helps people network across borders, develop a common language and common demands, and a sense that progress is being made.' Brunori and Giuggioli, in turn, highlight how the UNDROP could be a tool for 'ensuring that existing laws and judicial instruments serve the cause of the most vulnerable and concurs in the promotion of social justice.' Claiming the need to dislodge the neoliberal assumptions of international law, Le Teno, Cogolati, and Frison propose the theory of the commons as a possible theoretical frame for effectively translating the contents of UNDROP on the ground, and Martignoni and Claeys argue that only through the adoption of a feminist perspective can the UNDROP deliver its emancipatory potential for rural women.

The third cross-cutting theme regards the global governance of natural resources and the role of human rights as a transforming agent of food systems. In the light of the crisis of multilateralism in international law, the human rights machinery has never been so active. Considering the persisting lack of binding power and means of implementation, the book aimed to contribute to the debate on whether social challenges in general, and agrarian struggles in particular, should be addressed by statements of new rights and principles or rather by more technical international agreements. Several authors have offered their perspective on how and by which means the UNDROP can deliver its functions. Francioni illustrates the standing of the UNDROP in the international law framework and suggests the possible ways through which the Declaration can be used to unfold states' obligations. Di Lauro argues that the notion of biocultural rights must be fully integrated with food law and suggests the development of indicators 'that take due account of the role that culture plays in the achievement of sustainable development.' Along the same lines, Cristiani and Strambi call for the active engagement of United Nations agencies, and the FAO in particular, in the integration of the UNDROP into all policy strategies for a sustainable transition of food systems. In his analysis of the Kenyan legal framework governing land tenure and the persisting problem of landlessness, Kibugi suggests the possible usefulness of technical guidelines helping states to put into practice the rights enshrined in the Declaration. Posenato, in turn, discusses how regional human rights mechanisms may make use of the Declaration to better enucleate and address rural workers' rights. Morgera and Nakamura find global goals useful for fostering a harmonised interpretation and implementation of emerging standards. Tsioumani sheds light on the importance of promoting the national implementation of the new rights and standards, with specific reference to the rights to seeds. Brunori and Giuggioli unveil the political face of the coin of any legal reform and stress the double role of the Declaration as a reference to assess the legal framework in place, as well as a political instrument to support grassroots claims.

From each attempt to answer a question, another arises, and a close look at one domain unveils the importance of considering other aspects. Only time will tell the role and limitations of the UNDROP and help researchers, lawyers, and activists in understanding the more fruitful paths for bringing equity and justice to the global governance of natural resources and food systems. This book has offered a first attempt to comprehensively unfold the multiple contributions of the Declaration with the aim of triggering further proposals and inspiring new research projects so as to feed the much-needed debate on how to affirm the rights of peasants and other people working in rural areas.

TABLE OF INSTRUMENTS

International instruments

Additional Protocol to the American Convention on Human Rights in the Area of Economic, Social and Cultural Rights (Protocol of San Salvador) (entered into force 16 November 1999) OAS Treaty Series No. 69.

Convention on Biological Diversity (adopted 5 June 1992, entered into force on 29 December 1993) 1760 UNTS 69 (CBD).

Convention on the Elimination of All Forms of Discrimination against Women (adopted 18 December 1979, entered into force 3 September 1981) 1249 UNTS 13 (ICEDAW).

Convention on the Rights of the Child (adopted 20 November 1989, entered into force 2 September 1990) 1577 UNTS 3 (CRC).

International Covenant on Civil and Political Rights (adopted 16 December 1966, entered into force 23 March 1976) 999 UNTS 171 (ICCPR).

International Covenant on Economic, Social and Cultural Rights (adopted 16 December 1966, entered into force 3 January 1976) 93 UNTS 171 (ICESCR).

International Treaty on Plant Genetic Resources for Food and Agriculture (adopted 3 November 2001, entered into force on 29 June 2004) 2400 UNTS 303 (ITPGRFA).

Kyoto Protocol to the United Nations Framework Convention on Climate Change (adopted 11 December 1997, entered into force 16 February 2005) 2303 UNTS 14.

Nagoya Protocol on Access to Genetic Resources and the Fair and Equitable Sharing of Benefits Arising from their Utilization to the Convention on Biological Diversity (adopted 29 October 2010, entered into force 12 October 2014) 30619 UNTS 27.

The Agreement on Trade Related Aspects of Intellectual Property Rights (adopted 15 April 1994, entered into force on 1 January 1995) 1869 UNTS 299 (TRIPS).

United Nations Convention to Combat Desertification in Those Countries Experiencing Serious Drought and/or Desertification, Particularly in Africa (adopted 17 June 1994, entered into force on December 1996) (UNCCD).

United Nations Declaration on the Rights of Indigenous Peoples, UN General Assembly (2 October 2007) UN Doc A/RES/61/295.

United Nations Declaration on the Rights of Peasants and Other People Working in Rural Areas,' UN General Assembly (17 December 2018) UN Doc A/RES/73/165.

United Nations Framework Convention on Climate Change (adopted 9 May 1992, entered into force 21 March 1994) 1771 UNTS 107 (UNFCCC).

Voluntary Guidelines on the Responsible Governance of Tenure of Land, Fisheries and Forests in the Context of National Food Security (VGGT), FAO CFS (11 May 2012) FAO CL 144/9.

General comments

UN Committee on Economic, Social and Cultural Rights (CESCR), 'General Comment No. 3: The Nature of States Parties Obligations (Art. 2 of the Covenant)' (14 December 1990) UN Doc E/1991/23.

UN Committee on Economic, Social and Cultural Rights (CESCR), 'General Comment No. 15: The Right to Water (Arts. 11 and 12 of the Covenant)' (20 January 2003) UN Doc E/C.12/2002/11.

UN Committee on the Elimination of Discrimination against Women, 'General Recommendation No. 34 on the Rights of Rural Women' (7 March 2016) UN Doc CEDAW/C/GC/34.

Other documents

UN Human Rights Council, 'Final study of the Human Rights Council Advisory Committee on the advancement of the rights of peasants and other people working in rural areas' (24 February 2012) UN Doc A/HRC/19/75.

UN Human Rights Council, 'Promotion and protection of the human rights of peasants and other people working in rural areas' (11 October 2012) UN Doc A/HRC/RES/21/19.

UN Human Rights Council, 'Draft Declaration on the rights of peasants and other people working in rural areas' (20 June 2013) UN Doc A/HRC/WG.15/1/2.

UN Human Rights Council, 'Draft declaration on the rights of peasants and other people working in rural areas' (8 March 2016) UN Doc A/HRC/WG.15/3/2.

UN Human Rights Council, 'Draft declaration on the rights of peasants and other people working in rural areas' (6 March 2017) UN Doc A/HRC/WG.15/4/2.

UN Human Rights Council, 'Study on the Normative Sources and Rationale Underlying the Draft Declaration on the Rights of Peasants and Other People Working in Rural Areas' (4 May 2017) UN Doc A/HRC/WG.15/4/3.

UN Human Rights Council, 'Revised draft United Nation declaration on the rights of peasants and other people working in rural areas' (12 February 2018) UN Doc A/HRC/WG.15/5/2.

UN Human Rights Council, 'United Nations Declaration on the Rights of Peasants and Other People Working in Rural Areas' (8 October 2018) UN Doc A/HRC/RES/39/12.

INDEX

For Product Safety Concerns and Information please contact our EU representative GPSR@taylorandfrancis.com
Taylor & Francis Verlag GmbH, Kaufingerstraße 24, 80331 München, Germany

www.ingramcontent.com/pod-product-compliance
Ingram Content Group UK Ltd.
Pitfield, Milton Keynes, MK11 3LW, UK
UKHW020150200926
475702UK00016B/2646
9780367689773